THE RAIN FORESTS

OF

GOLFO DULCE

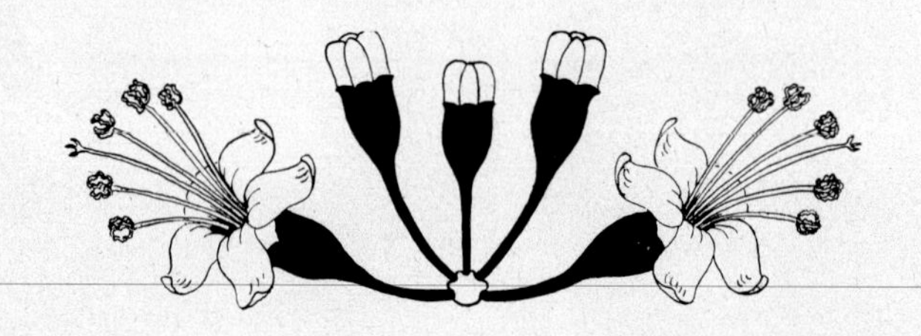

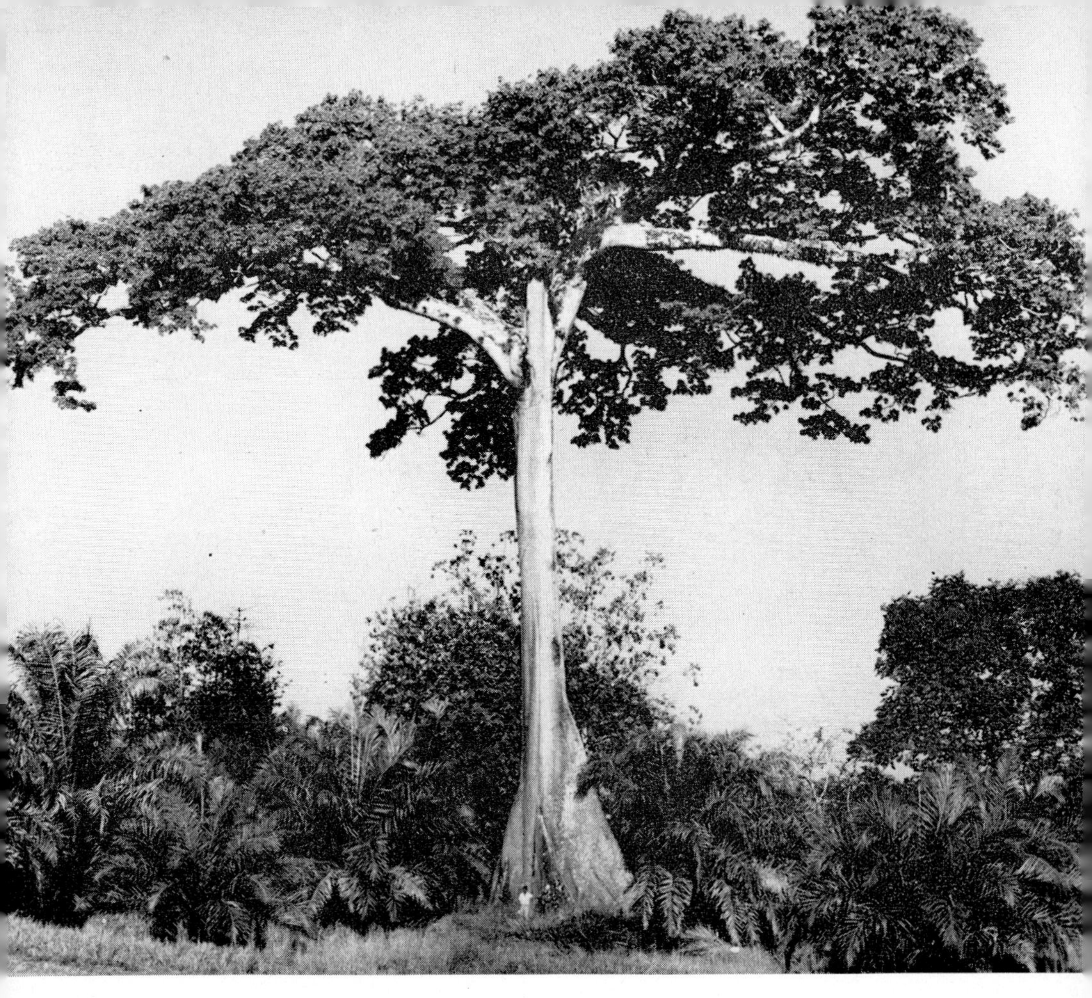

PLATE I

Ceiba pentandra, near Palmar. (See man at base for scale.)

"There are, furthermore in the firme lande, trees of such biggenesse that I dare not speak thereof but in place where I haue so many wytnesses which haue seen the same as wel as I."

(Gonzalo Fernandez de Oviedo y Valdes. *The Natural History of the West Indies.* Richard Eden translation. Emprynted at London in Paules Churchyard at the Signe of the Byble. Anno Domini M.D.LV.)

THE RAIN FORESTS

OF

GOLFO DULCE

by

PAUL H. ALLEN

With drawings by

DOROTHY O. ALLEN

1956

Gainesville

UNIVERSITY OF FLORIDA PRESS

A UNIVERSITY OF FLORIDA PRESS BOOK

Library of Congress Catalogue Card Number: 56-9482

PRINTED BY THE RECORD PRESS, INC., ST. AUGUSTINE, FLORIDA
BOUND BY UNIVERSAL-DIXIE BINDERY, JACKSONVILLE, FLORIDA

PREFACE

THE ASSOCIATIONS OF PLANTS making up the great tropical rain forests remain, until our time, the most difficult and poorly understood of the earth's floristic zones. Unbelievably complex in composition, they have tended to defy detailed analysis. Not without reason did Gonzalo Fernández de Oviedo compare those of Central America to a great and dark sea.

It has long been apparent that such stands can be adequately studied only by a permanent resident in the area under consideration, with opportunities for intimate, almost daily observation. The accumulation of the basic facts needed for a logical approach to tropical forest tree identification is a process which cannot be hurried. The cosmos does not render up its secrets so easily. Before Central American forestry can become a science in the best and fullest sense of the word, the individual species must be studied intensively in the field, both in regard to their physical characteristics and their place in the general ecological pattern.

A little acquaintance with tropical rain forests will demonstrate that some species flower and fruit with predictable regularity, while others may bloom only in exceptionally favorable seasons, or may last but a day or two at each flowering, circumstances which must immeasurably complicate the task of one attempting to base identification on flowering material. The fact of the matter is, of course, that no local woodsman does this in actual practice. The purpose of the present modest work is to attempt to reduce the use of field characters, such as trunk, sap, leaf, and bark differences to some sort of organized system. In the case of previously unexplored areas, or new, rare, or poorly known species, initial identifications must usually rest on floral analysis, but once this has been accomplished there is no logical reason why the method must be continued indefinitely, since the more conspicuous characters can be made to serve the same end.

ACKNOWLEDGMENTS

PUBLICATION OF THE PRESENT WORK has been made possible by a grant from the National Science Foundation, to which I especially wish to express my appreciation.

Most of the field surveys were done under the auspices of the United Fruit Company, as part of their policy of social responsibility in Central America. My heartfelt thanks go to both Dr. V. C. Dunlap, former Director of Tropical Research, and Mr. N. E. Sanderson, Manager of the Golfito Division, who have long had an active interest in the intelligent utilization of the natural resources of Costa Rica.

It is also particularly desired to acknowledge the very active cooperation of Dr. Louis O. Williams and Dr. Paul C. Standley, of the Escuela Agrícola Panamericana in Honduras, who have identified the bulk of our collections and have described most of the new material. They and Dr. and Mrs. Wilson Popenoe have repeatedly provided friendly encouragement and hospitality during my frequent stays at Zamorano while working on keys and text. A great deal of help has also been provided by Dr. C. V. Morton, Dr. L. B. Smith, and Dr. F. C. Leonard, of the Smithsonian Institution. Many of our palm specimens have been examined and named by Dr. H. E. Moore, of the Bailey Hortorium of Cornell University. Dr. Jorge Leon, of the Instituto Interamericano de Ciencias Agricolas at Turrialba, Costa Rica, has undertaken monographic studies of the Ingas, and it is to him that I am indebted for the present names in that genus. Grateful acknowledgments are also due to Dr. Robert E. Woodson, of the Missouri Botanical Garden; Dr. Julian A. Steyermark, of the Chicago Natural History Museum; Dr. Jason Swallen and Dr. A. C. Smith, of the United States National Herbarium; Dr. Reed Rollins, of the Gray Herbarium of Harvard University; Dr. C. F. Kobuski, of the Arnold Arboretum; and Dr. Edwin Navarro, of the Museo Nacional in San José, Costa Rica, for the facilities so kindly provided in the examination of specimens and records in the herbaria under their care.

Particular thanks are also due to Dr. William A. Dayton, of the United States Department of Agriculture Forest Service, and to Dr. Graham Fairchild, of the Gorgas Memorial Laboratory, for critical readings of the manuscript and suggestions.

P.H.A.

TABLE OF CONTENTS

Introduction 1
Physiography 4
Ecology 7
  climatic formations 10
    Evergreen lowland forest 10
    Littoral woodland 22
      Sandy beaches 22
      Rocky seashores 23
    Lower montane rain forest 24
    Upper montane rain forest 60
  edaphic formations 62
    Swamp forest 62
    Palm swamp 66
    Mangrove woodland 66
    Herbaceous swamp 69
    Gallery woodland 69
  transitional formations 70
    Savanna 70
    Pastures 70
    Fence rows and trailside thickets 77
    Second growth 82
Conspicuous or distinctive species 86
  conspicuous flowers 86
  conspicuous vegetative characters 91
  species which harbor stinging ants 98
  species which occur in concentrated stands 98
Common names 101
Scientific names 103
Utilization lists 105
  economic species 105
  medicinal species 115
  poisonous species 115
Alphabetical index to families, genera, species, common names 116
Glossary 367
Plates 375
Appendix 409
References 411
Index 415

LIST OF ILLUSTRATIONS

PLATES *(in numerical order, following page 374)*

	PLATE
Acacia farnesiana	19
Albizzia filicina	21
Anacardium occidentale	14
Andira inermis	21
Apeiba aspera	31
Asterogyne Martiana	4
Astrocaryum alatum	8
Astrocaryum Standleyanum	6
Bactris balanoidea	8
Bactris militaris	10
Batocarpus costaricense	24
Belotia macrantha	23
Bravaisia integerrima	22
Brosimum utile	3 & 7
Byrsonima crassifolia	23
Calycophyllum candidissimum	31
Carapa Slateri	14
Casearia arguta	25
Casearia banquitana var. laevis	31
Cassia grandis	19
Cassia reticulata	13
Cassia spectabilis	19
Cecropia peltata	11
Cecropia Sandersoniana	11
Cedrela mexicana	21
Ceiba pentandra	*(frontispiece)*
Chrysophyllum panamense	24
Cochlospermum vitifolium	12
Cocos nucifera	9
Compsoneura Sprucei	31
Conocarpus erecta	20
Cordia alliodora	23
Corozo oleifera	8
Croton panamensis	18
Cryosophila guagara	6
Dicraspidia Donnell-Smithii	34
Didymopanax Morototoni	12
Diospyros ebenaster	24
Diphysa robinioides	17
Dipterodendron costaricense	18
Dracaena americana	22
Duguetia panamensis	14
Dussia macrophyllata	15
Dussia mexicana	15
Enallagma latifolia	16
Eschweilera calyculata	14
Ficus lapathifolia	5

	PLATE
Geonoma congesta	4
Gliricidia sepium	19
Goethalsia meiantha	24
Guatteria aeruginosa	25
Guatteria chiriquensis	25
Guatteria lucens	25
Guazuma ulmifolia	23
Hampea Allenii	18
Heliocarpus appendiculatus	29
Hibiscus tiliaceus	16
Hieronyma tectissima	29
Hillside rain forest	3 & 4
Hippomane mancinella	7 & 16
Homalium eurypetalum	29
Huberodendron Allenii	5
Iriartea gigantea	9
Ladenbergia Brenesii	30
Laguncularia racemosa	20
Licaria Cufodontisii	27
Macrocnemum glabrescens	32
Meliosma Allenii	33
Miconia argentea	29
Mora oleifera	16
Mortoniodendron anisophyllum	33
Muntingia calabura	28
Neea elegans	28
Ochroma lagopus	12
Ocotea Williamsii	27
Oenocarpus panamanus	6
Pachira aquatica	13
Palmar Norte	2
Palmar Sur	2
Pelliciera rhizophorae	5 & 20
Perebea trophophylla	32
Persea americana	27
Phyllocarpus septentrionalis	13
Pithecolobium austrinum	13
Pouteria heterodoxa	30
Pouteria neglecta	30
Pouteria triplarifolia	30
Pterocarpus Hayesii	15
Pterocarpus officinalis	15
Quararibea guatemalteca	28
Raphia taedigera	8
Rhizophora mangle	20
Rollinia Jimenezii	28
Sapindus saponaria	34

TABLE OF CONTENTS

PLATES (continued)

	PLATE
Scheelea rostrata	6
Simarouba glauca	17
Sloanea laurifolia	32
Socratea durissima	9
Spondias mombin	17
Spondias purpurea	17
Stemmadenia Donnell-Smithii	26
Sterculia apetala	10
Sterculia mexicana	10
Swartzia panamensis	12
Symphonia globulifera	33
Tabebuia chrysantha	11
Tachigalia versicolor	21
Terminalia bucidioides	33
Tetragastris panamensis	18
Tetrathylacium costaricense	26
Theobroma angustifolium	26
Townsite of Palmar	2
Vitex Cooperi	11
Warscewiczia coccinea	26
Welfia Georgii	10
Xylopia sericophylla	34
Xylosma excelsum	34

FIGURES

FIGURE		PAGE
1	Map of Golfo Dulce area	5
2	Anacardium excelsum	123
3	Aspidosperma megalocarpon	129
4	Basiloxylon excelsum	135
5	Brosimum utile	143
6	Carapa Slateri	152
7	Caryocar costaricense	155
8	Ceiba pentandra	166
9	Chimarrhis latifolia	171
10	Chlorophora tinctoria	173
11	Couratari panamensis	185
12	Enterolobium cyclocarpum	199
13	Huberodendron Allenii	227
14	Hura crepitans	229
15	Inga marginata	232
16	Luehea Seemannii	253
17	Peltogyne purpurea	289
18	Pentaplaris Doroteae	290
19	Prioria copaifera	303
20	Vantanea Barbourii	352
21	Virola surinamensis	356
22	Vochysia Allenii	359

EL MARE MAGNO E OCULTO

DIGO QUE EN GENERAL los arboles que en estas Indias hay es cosa para no se poder explicar, por su moltitud; y la tierra esta tan cubierta dellos en muchas partes, é con tantas diferencias y desemejanca los unos de los otros, assi en la grandeça como en el tronco é las ramas é cortecas y en la hoja y aspecto y en la fructa y en la flor, que ni los indios naturales los conosçen, ni saben dar nombres a la mayor parte dellos, ni los chripstianos mucho menos, por serles cosa tan nueva é no conoscida ni vista por ellos antes. Y en muchas partes no se puede ver el çielo desde debaxo destas arboledas (por ser tan altas y tan espessas é llenas de rama) y en muchas partes no se puede andar entre ellas; porque demas de su espessura, hay otras plantas é verduras tan texidas y revueltas é de tantos espinas é bexucos é otras ramas mezcladas, que con mucho trabaxo e a fuerça de puñales y hachas es menester abrir el camino. Y lo que en esto se podria decir es un MARE MAGNO E OCULTO; *porque aunque se ve, lo mas dello se ynora, porque no se saben, como he dicho, los nombres . . . ni sus propiedades. Hay algunos dellos de muy buen olor é lindeça en sus flores, é olerosa la madera o corteças; otros de innumerables é diversas formas de fructas salvajes, que solemente los gatillos monos las entienden é saben las que son a su propossito. Otros arboles hay tan espinosas é armados que no se dexan tocar con mano desnuda; otros de mala vista é salvajes: otros cargados de yedras é bexucos é cosas semejantes: otros llenos de arriba abaxo de cierta manera de hilos, que paresce cubiertos de lana hilada, sin serlo. Los unos tienen fructa é otros flor, é otros comiencan a brotar; é assi como son de diversos generos, assi goçan del tiempo en diferente manera, é se ve todo junto en un saçon é en qualquier parte del año.*

(*Prohemio al Libro Noveno, del Primera Parte de la Natural y General Historia de las Indias, Islas é Tierra-Firme del Mar Oceano*, por Gonzalo Fernández de Oviedo y Valdés. 1526.)

THE GREAT AND DARK SEA

I SAY, IN GENERAL, that the trees of these Indies are a thing that cannot be explained, for their multitude; and the earth is so covered with them in many parts, and with so many differences and dissimilarities between them, both in their great size as well as in the trunk and branches and bark and in the aspect of their fruits and flowers, that not even the native Indians know them nor know how to give names to the majority, and the Christians much less, since they are so new, and not known or seen by them before. And in many parts one cannot see the sky from below these woodlands (for their being so tall, and thick, and full of branches) and in many places one cannot walk between them; because, besides their thickness, there are other plants and herbs so interwoven and twined about them and mixed with so many thorns and vines and other branches that it is only with much effort and by force of knives and axes that it is possible to open a path. And in this respect one could say that this is a GREAT AND DARK SEA; because though part is seen, much more is not, since their names and properties are unknown, as I have said. There are some of them of good odor and beauty of flower and with fragrant wood or bark; others of innumerable and diverse forms of wild fruits that only the little catlike monkeys understand and know which are to their purpose. Other trees are so spiny and armed that they cannot be touched with the naked hand; others are of evil and savage appearance: others loaded with ivy and vines and similar things; others completely filled from top to bottom with a sort of thread, so that they appear covered with woolen yarn, without its being so. Some have fruit and others flowers and others burst into growth and each kind enjoys the season in its own way and one may see every stage of development at a given time and during any part of the year.

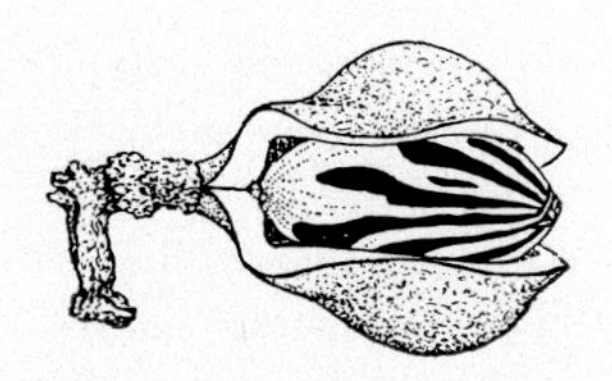

INTRODUCTION

FORESTS COVER about three-quarters of the 19,200 square miles of Costa Rica, and reach their maximum development in areas of high rainfall, particularly along the Atlantic coast, and in the Golfo Dulce region. Here a tremendous arborescent tapestry blankets the land, sweeping almost unbroken from the coastal swamps of Coto and Sierpe to the crests of the highest hills. It is difficult to believe that anyone could view these woodlands without emotion. In many ways they are dark and inscrutable as fate; yet they are also alive, and timeless and magnificent. They were essentially as we see them now long before the first Indian tribes fashioned the mysterious stone spheres found scattered in their shade, and who can say with certainty that they may not remain, when we and all our works are gone. In a certain special sense, they represent the ultimate reality.

No living man can give a completely accurate, detailed account of what these forests may contain, for it is a sobering fact that our knowledge of them is still very incomplete, even after more than four centuries of European contact. When work was begun on the present project, for example, it was expected that the species would prove to be those already known from adjacent mixed stands typical of Panama and other parts of Costa Rica, since both countries have the reputation of being among the more thoroughly explored of the Central American Republics. Surprisingly enough, this has been only partially true, and an astonishing number of plants have been found to be either new to science, or to represent tremendous and often entirely unexpected extensions of range. The affinities of our local flora would seem to be with the Atlantic coast rain forests, and hence South American, even in some cases Amazonian, rather than with the plant populations of closely adjacent Chiriquí or Guanacaste. This is particularly notable in many of the palms, especially in the presence of great tracts of *Raphia taedigera* and in the occurrence of genera of trees like *Huberodendron, Basiloxylon, Peltogyne, Vantanea, Sacoglottis,* and *Couratari.* Although the vast storehouse of our forests obviously must still hold many things unknown to us, it is believed that

we now have a fairly comprehensive understanding of the common, conspicuous, and commercially important species on the floodplains and lower mountain slopes, up to about 2,000 ft.

Selection of material will be found to have been somewhat arbitrary. We have listed most of the conspicuous palms, but have rejected some species, particularly of essentially nonwoody genera such as *Solanum* and *Piper* which are often shrubby, and in any case seldom exceed 12 or 15 ft. in height. The common Mango, and a few other exotics may persist near old house sites, and along trails, and are here given honorary citizenship as members of the local flora, contingent upon good behavior, and as a matter of convenience. In general, those smaller species which seemed to me, from personal experience, to be common, conspicuous, or important for other reasons were included, though it must be admitted that this is a highly subjective method!

It is hoped that it may be possible at some future time to prepare a comprehensive flora, with keys and brief descriptions of all the plants of the area, of which we now have records of 134 families embracing some 1,315 species in 662 genera. Of this total, 72 families, 267 genera, and 433 species are included here. More than 40 species and one genus of trees have been described as new from our collections and are not thus far known from adjacent territory.

These forest stands represent a renewable natural resource of inestimable value, and it is hoped that some of them may be preserved long enough so that some sort of intelligent program can be worked out for their exploitation. The preparation of a utilization manual, such as the present one, is basic research, which, while it does not produce any immediate dollars and cents return, is fundamental to any further progress in the field. It is, in effect, an inventory of available assets, a preliminary phase that might be compared to tooling up in heavy industry. Maximum utilization of these tremendous reserves in Latin America will depend on an intelligent program of continued exploration, together with tests of the more promising materials for strength, durability, toughness, transverse shear, resistance to termites, ease with which they may be worked or processed chemically, and other factors. The business world can be expected to convert such findings rapidly into useful and valuable products. There is, in fact, every indication that the tropical hardwoods will enter world markets at an increasingly accelerated rate as they become better known, and as demand increases with the depletion of softwood stands in the United States and northern Central America. New equipment and new technical skills are now providing solutions to many

of the old problems. The establishment of plywood plants in the tropics may be one way of effectively utilizing the mixed stands, for example, since there are dozens of species suitable for corestock, and more than adequate supplies of the rarer cabinet woods for finish. Simple chemical treatment of railroad ties and construction timbers may be another solution to on-the-spot use of local products, or the complete reduction of some types of hardwoods by chemical means to provide paper pulp, or the basic material for a great range of synthetic products such as plastics, wood alcohol, detergents, paints, acids, insecticides, sugars, and even foodstuffs. Timbers now in little demand will compare favorably with lumber from the United States when properly kiln-dried. There is every indication that wood and wood products will remain in short supply for many years to come, and that tropical forests may provide the bulk of the world's industrial raw material of the future. *Properly managed* forests, which are protected against indiscriminate felling and burning, can be exploited indefinitely. Unfortunately, in many parts of Costa Rica, particularly in the adjacent General Valley, the forest stands are being destroyed just as rapidly as people can get to them. Many invaluable stands are felled for catch crops of rice or corn worth only a fraction of the market price of the lumber destroyed, and other even more extensive tracts are deliberately killed with fire to provide pasture. This is living on capital, which no modern population can afford to do.

The material upon which the present work has been based has been assembled during nearly five years of residence in Palmar, the principal collections having been made near Palmar Norte, in the Esquinas Forest preserve, and in the hills near Golfito. We have also been very fortunate in being able to include several earlier collections, the most notable being those of Henri Pittier and Adolfo Tonduz, who made frequent visits to Buenos Aires, Boruca, and Santo Domingo de Osa between about 1890 and 1901. Dr. C. W. Dodge, of the Missouri Botanical Garden, has spent some time near Puerto Jiménez, and Dr. Alexander Skutch collected at Esquinas for about three months in 1948. More recently, Dr. H. E. Moore, of the Bailey Hortorium, has spent about ten days, devoted to intensive collecting of palms near Palmar and at Esquinas. These and occasional other specimens that have come to our attention in the various herbaria are cited, following the descriptions of the individual species. Duplicates of most of these collections may be found at the Escuela Agrícola Panamericana in Honduras, at the National Herbarium in Washington, and at the Chicago Natural History Museum.

PHYSIOGRAPHY

THE AREA COVERED BY OUR SURVEY is shown on the accompanying map (Fig. 1), on which the forest stands have been roughly indicated. In spite of its relatively limited extent, the territory may be divided into three major topographic sections, each of which has a distinctive floristic cover. These divisions are (1) the floodplains and swamps of the Río Terraba, Río Esquinas, Río Coto, and the northeastern plain of the Peninsula de Osa, (2) the rugged mountain slopes, (3) the broad savannas of San Andrés, Potrero Grande, and Buenos Aires. Many of the high, steep ranges, such as the Cerros de Retinto near Palmar Norte, have the superficial aspect of fault scarps, and the gorge of the Río Terraba between Palmar and Potrero Grande looks suspiciously like a fault trough, or *Graben.* This is separated from the Sierpe swamps and the valley of the Río Esquinas by a broken, precipitous range, which reaches its maximum heights of over 5,000 ft. roughly between Piedras Blancas and Cañas Gordas. These mountains block the warm, moisture-laden winds from Golfo Dulce, and have, on their windward slopes, an exceedingly high precipitation, often exceeding 200″ a year, with only a limited, poorly defined dry season, a condition duplicated elsewhere on the Pacific slope of Central America only near Tilarán, in Guanacaste.

In dramatic contrast to these high, wet forests are the savannas of San Andrés and Potrero Grande, which appear to be in the rain shadow of this somber massif, and where the wide, wind-swept grasslands and open stands of *Curatella americana* proclaim, better than any words, the profound difference in rainfall. Here, in the inner valley, the dry season is severe and long, and the plant population almost as alien to that of the dark, dripping, aroid- and hepatic-laden rain forest as the flora of another world.

PHYSIOGRAPHY

The hills near Boruca are rugged in the extreme, with high, usually bare ridges alternating with valleys filled with forest or second-growth scrub, those near the village usually dominated by thickets of *Zexmenia frutescens,* whose golden yellow flowers are so conspicuous during December and January. Here brawl clear little streams on their stony and precipitous course, often interrupted by sheer drops, down which they cascade, a thin, silvery ribbon against the mountain wall. During the rainy season the landscape is one of soft pastels, varying shades of greens, grays, and browns. Winding trails ascend the long ridges to tiny grass-thatched houses set in small patches of sward, ringed about with occasional coconut palms, plantains, plumed heads of flowering

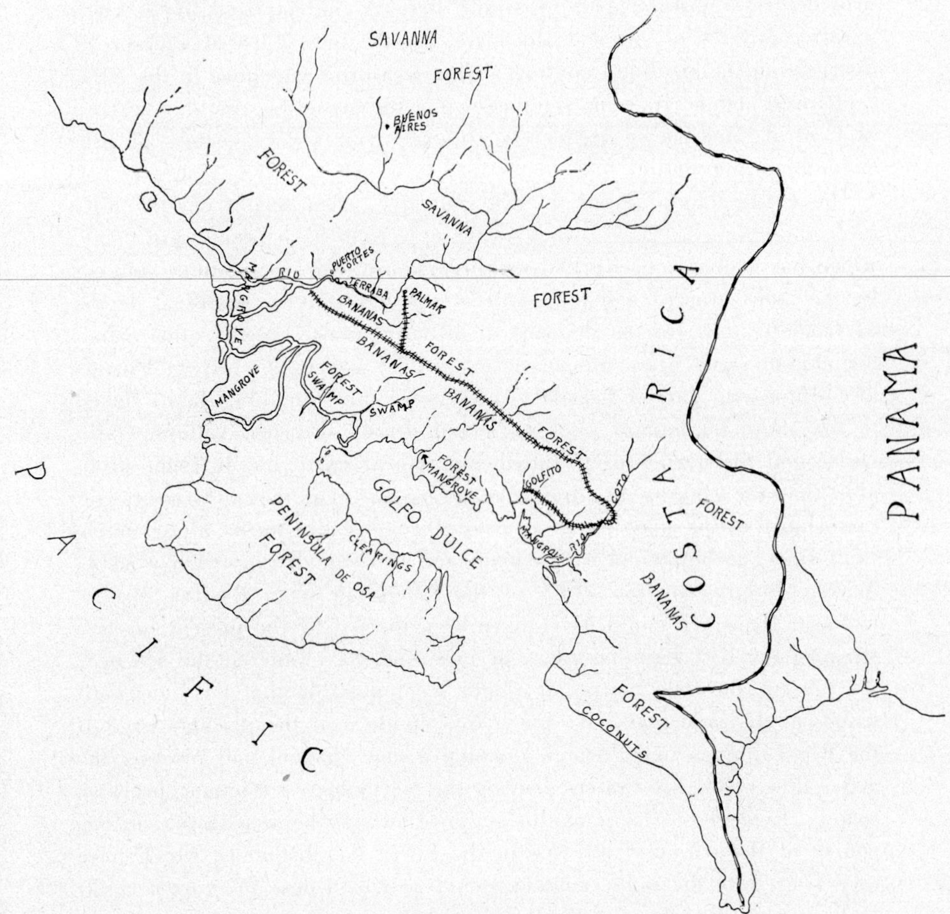

Fig. 1. Map of the Golfo Dulce area.

sugar cane, or the occasional pink flush of the new leaves of a mango. When increasing drouth in March and April has seared the landscape almost beyond recognition, there comes the universal shrill stridulation of the cicada, and the heady fragrance of "Tuete" *(Vernonia patens)*, so familiar to every traveler in Central America. Bare, grassy ridges are covered with thousands of narrow terraces, made by the cattle in search of the bits of sustenance left among the rocks.

The most extensive grasslands are, however, to be found between San Andrés and Terraba, and near Buenos Aires. Their vegetation is, in large measure, determined by the fire resistance of the individual species, since the grass is burned at the end of each dry season to eradicate ticks and provide fresh pasture. This grassland flora is characterized by its comparative poverty of species, which are, for the most part, of widespread distribution, in profound contrast to the wealth of endemics in the adjacent forest stands. It seems significant that the greatest areas of grassland in Central America so frequently coincide with known centers of pre-Columbian population.

The entire floodplain of the Río Terraba, when seen from the hills or from a plane, occupies what would appear to have been a great bay, which has probably been filled within comparatively recent time, judging by the many stumps and logs, still in sound condition, found at depths of from 6–20 ft. in the deposits of silt or ballast. On the other hand, the classic signs of a submergent coast are everywhere, from Quepos to Golfito, with narrow beaches, rocky headlands, tombolo islands linked to the shore by narrow causeways and deeply drowned valleys, Golfo Dulce and the harbor of Golfito being typical examples. It seems probable that the sinking has not been a gradual, continuous process, but that sharp drops may have occurred through adjustment along active fault lines, accompanied by frequent and occasionally destructive earthquakes. Judging by the width of the floodplain from Piedras Blancas to Farm 18, which could never have been formed by the present inactive streams, the Río Esquinas must, at one time, have entered the sea near the present Boca de Sierpe. It would seem possible that the downward tilting of the land may have put a low divide near the present gorge of the Río Esquinas under water, opening a new channel and pirating the active flow of the headwaters, leaving the Sierpe area a stagnant, brackish swamp. Evidence in favor of this supposition may be seen in the curious course of the tributary streams in the lower Río Esquinas, which have their sources to the south, southeast, and west, and flow *away from* Golfo Dulce for their entire length, before entering the main stream.

ECOLOGY

THE HETEROGENEOUS NATURE of our terrain is reflected in the vegetation, which is composed of a number of very distinctive associations that are limited in their distribution by rainfall, slope, drainage, soil type, and other factors. Species found in one climax or transitional association will often be completely lacking in another, even though the distance separating them may not be more than a few yards. Many are, in effect, microfloras, and might quite logically be treated as such, since they tend to form self-contained units. The great majority of our tree species are confined to one, or at most two, of the *major* ecological associations and some are limited to a single, highly specialized faciation of one of these.

Although the first reaction to the mixed stands is one of bewilderment, it gradually becomes apparent that while some species are almost universal in their distribution, others are restricted to difficult situations, such as gravel bars, poorly drained clay, steep ridges, or the acid summits of the highest hills, and are seldom to be found elsewhere. These species have apparently become adapted to conditions too poor to support anything else and will usually be encountered again and again as small, local colonies over great extents of territory.

This basic fact explains statements by observers like Richard Spruce (1908) describing forest stands on the Amazon, and more recently by Dr. H. A. Gleason (1926), who have said that an individual hectare of tropical forest may contain hundreds of species of trees, none of which can be found on an adjoining hectare. This condition is particularly notable in parts of northern Honduras and northeastern

Nicaragua, where tracts of broadleaf woodland are often seen in close proximity to stands of pine and oak. Unfortunately, the implication is sometimes drawn that this situation continues throughout a given area, in the sense that each hectare would tend to contain unique elements not to be expected elsewhere. It has been my experience, however, that the great tropical forests, particularly of Costa Rica, Panama, and Colombia, are actually reasonably homogeneous when viewed as a whole. Collections of specimens and studies with field glasses over a period of years bring to light the genera and species with which one has become familiar, which appear in flower, fruit, and change of leaf, each in their season. The specialized associations will also be found to be repeated over many square miles of terrain, and even the local rarities can be expected when conditions are right for them.

Our formations, as I understand them, are listed in the following outline. *Climatic* formations are those whose principal limiting factor is rainfall, and its distribution. *Edaphic* formations are determined by conditions of soil or topography. *Transitional* formations are those which have been modified by human intervention. The divisions are, to a considerable degree arbitrary, since most of the savannas, for example, are probably the result of clearing and subsequent annual burning over long periods of time, although their maintenance is also due in part to the drier climate of that area. Much of the present savanna soil is nearly sterile, but it was not always so.

Artificial keys have been provided for each of the several ecological formations. They have been written, insofar as possible, so that they may be of use to the field worker. While not in any sense phylogenetic, they are based on the rather obvious assumption that the presence or absence of buttresses, stilt roots, spines, or distinctively colored sap, as well as leaf type and flower color will be of considerably greater aid to the average individual for identification, than the technical position of the species in the *Humiriaceae* or *Nyctaginaceae*, for example.

Plant identification by floral analysis is unquestionably infallible in the hands of an experienced technician, and with these we have no quarrel, but it would seem (at least in restricted areas) that this may not always be either the easiest or most practical method. Relatively few of those interested in tropical trees have had the basic training needed to use a key to the plant families, which is a standard feature and initial point of departure in any orthodox flora. Even to those so trained, obvious practical obstacles are constantly presented. Species of exceptional economic interest may not flower annually, and about

all that the observer is assured of seeing are the conspicuous gross characters of the lower trunks of the larger forest trees, and with somewhat greater difficulty, their leaves and branches.

Many times I have accompanied Doctors of Philosophy in Botanical Science in the field and seen them speculate endlessly and futilely on the identity of some common, conspicuous tree, because it did not happen to be in bloom. Yet a little acquaintance with the tropical forests soon shows that the key to the problem is relatively simple, particularly for the field worker, since he has to do with definite, and often restricted communities of trees, occupying a more or less specific situation. A very simple, and extreme example may be cited in the Mangrove association, in which the elements of the limited heterogeneous population have little in common with one another excepting their tolerance of salt. In the Golfo Dulce area, only about ten arborescent species are represented. Of these, two are palms, and the remainder so strongly dissimilar as to be immediately separable by the most obvious of gross characters. The same, in increasing degrees of complexity, is true of sea beaches, fresh-water swamps, second-growth scrub, and fence rows, and on to the most difficult of all, the rich and varied associations found on the mountainsides above the river plains.

Even here, great, botanically unrelated groups can be arbitrarily segregated because of spiny trunks, blood-red sap, scaling bark, and other equally obvious characters, and separated from one another with comparative facility, even by an untrained person.

A manual, at best, is but a means to an end, a necessary evil whose primary purpose is to enable the user to identify the material at hand as easily as possible. Some manuals bog down in technical jargon, and others, unintentionally become vehicles for the display of the author's personal erudition. In our age of specialization workers in the various natural sciences are becoming increasingly incomprehensible to one another and to the intelligent layman. This is extremely unfortunate, since all nature forms an interrelated whole, whose intimate facets can be completely understood only in their relation to the total environment. Facility of recognition, by whatever means, should be our objective.

In addition to the keys to the ecological formations, artificial keys have also been provided to the various families and genera, which will be found in alphabetical order in the text.

CLIMATIC FORMATIONS

(1) — EVERGREEN LOWLAND FOREST

In this category are to be found the stands on reasonably well-drained river floodplains, of the type considered suitable for the planting of bananas. Precipitation varies from an annual average of about 140″ at Palmar, with a fairly well-defined dry season lasting from mid-December until early May, to an average of nearly 200″ at Esquinas, where a week seldom passes, even at the height of the *verano,* without heavy rains. In all the area, about 40 species reach the canopy, which varies from about 100-140 ft. in height, the dominant element being *Anacardium excelsum,* which may account for as much as 50% of the total stand. Some of these giant species are briefly deciduous, but seldom for any extended period.

Although the woodlands vary somewhat from place to place, they are surprisingly open, when viewed from the air, particularly in the Coto Valley, where many of the crowns of the larger trees do not touch, and where one can look directly into the depths of the forest. Here are many palms, and a total of about 85 arborescent species, all of which are evergreen. Broad-leaved and succulent herbaceous plants are extremely common, particularly *Dieffenbachias,* and various species of *Heliconia* and *Calathea,* so that a considerable amount of chopping is often necessary to get from place to place.

Large trees, which reach the canopy layer—100–140 ft.

Anacardium excelsum—Clovelike fragrance when in bloom. Kidney-shaped seeds.
Brosimum alicastrum—Milky latex.
Brosimum sapiifolium—Milky latex. Very narrow leaves.
Brosimum terrabanum—Milky latex. Undersurfaces of leaves brown when dried.
Bursera simaruba—Thin, red, peeling bark.
Carapa Slateri—Conspicuous terminal flushes of bright-red new leaves.
Cedrela mexicana—Bark and wood bitter.
Ceiba pentandra—Tremendous plank buttresses in old specimens.
Chimarrhis latifolia—Shining, opposite leaves. Wood orange.
Chlorophora tinctoria—Often with spiny branches. Wood yellow.
Chrysophyllum mexicanum—Milky latex. Undersides of leaves brown.
Cynometra hemitomophylla—Leaflets 2.
Dacryodes epiphytica—Red, peeling bark. Epiphytic, in tall trees.
Dipterodendron costaricense—Fernlike foliage.
Enterolobium cyclocarpum—Dark-brown, coiled, earlike pods.
Ficus Bullenei—Milky latex. Leaves velvety brown.
Ficus Tonduzii—Milky latex. Leaves leathery.

Ficus Werckleana—Milky latex. Buttressed trunks.
Hernandia didymantha
Hieronyma tectissima—Black sap. Crowns with scattered blood-red leaves.
Homalium eurypetalum
Hura crepitans—Trunks densely covered with short, sharp spines.
Luehea Seemannii—Leaves reddish brown on the lower surface.
Pithecolobium austrinum—Fernlike foliage. Red, pendulous pods.
Pouteria chiricana—Milky latex.
Pouteria neglecta—Milky latex. Trunks deeply furrowed in old specimens.
Prioria copaifera—Black gum.
Pterocarpus officinalis—Blood-red sap. Trunks strongly buttressed.
Simarouba glauca—Bitter bark and wood.
Spondias mombin—Deeply corrugated, gray bark. Fruits yellow, plumlike.
Sterculia apetala—Very prominently developed buttresses in old specimens. Leaves palmately lobed.
Swartzia panamensis—Flowers bright yellow, on long, pendulous stems.
Tabebuia chrysantha—Leaves digitately compound. Flowers yellow.
Tabebuia pentaphylla—Leaves digitately compound. Flowers pink.
Terminalia lucida—Guava-like bark.
Virola guatemalensis—Red sap. Nutmeg-like fruits.
Virola Koschnyi—Red sap. Nutmeg-like fruits.
Virola nobilis—Red sap. Nutmeg-like fruits.
Virola sebifera—Red sap. Nutmeg-like fruits.
Virola surinamensis—Red sap. Nutmeg-like fruits.
Vitex Cooperi—Brownish-green foliage. Flowers blue.

Understory species—15–90 ft. in height

Andira inermis—Purple, pealike flowers, followed by globose, 1-seeded fruits.
Anisomeris Recordii
Astrocaryum alatum—Palm, with compact clusters of spiny fruits.
Astrocaryum Standleyanum—Tall palm, with very spiny trunks. Fruits in pendulous clusters, bright orange.
Bactris balanoidea—Spiny cluster palms, with pinnate leaves.
Bactris militaris—Spiny cluster palms, with undivided leaves.
Bravaisia integerrima—Trees with multiple trunks and stilt roots.
Capparis Sinclairii—Relatively large white flowers with masses of showy white stamens.
Carica pennata—A wild papaya, with small, orange fruits.
Casearia banquitana var. *laevis*—Tall, slender trees, with short, flat-topped crowns.
Castilla fallax—Large, 2-ranked, pendulous leaves. Milky latex.
Cecropia obtusifolia—Leaves peltate. Stems hollow, with stinging ants.
Cecropia peltata—Leaves peltate. Stems hollow, with stinging ants.
Cecropia Sandersoniana—Leaves peltate. Stems hollow, with stinging ants.
Chamaedorea sp.—*Allen 6262*—Small palm.
Corozo oleifera—Dwarf palms, with massive, often creeping trunks.
Crataeva tapia—3 leaflets.

Croton glabellus
Cryosophila guagara—Palms with fan-shaped leaves and spiny trunks.
Erythrina costaricensis—Leaflets 3. Flowers red.
Ficus Bullenei—Milky latex. Leaves velvety brown.
Ficus costaricana—Milky latex.
Ficus lapathifolia—Milky latex.
Ficus Oerstediana—Milky latex.
Ficus Tonduzii—Milky latex.
Ficus Werckleana—Milky latex.
Gloeospermum diversipetalum
Goethalsia meiantha—Often grows in pure stands. Seeds winged, in 3's.
Grias Fendleri—Leaves very large, clustered at the ends of the branches. Flowers and fruits produced directly from the trunk.
Guarea Hoffmanniana
Guatteria chiriquensis—Leaves pendulous, 2-ranked. Flowers green.
Gustavia angustifolia—Leaves very large, clustered at the ends of the branches. Flowers terminal.
Hamelia magnifolia—Leaf veins and petioles red.
Hampea platanifolia
Heisteria concinna—Drupes white, subtended by a bright-red circular calyx.
Heisteria longipes—Drupes black, subtended by a bright-red, circular calyx.
Herrania purpurea—Leaves digitately compound. Fruits cacao-like.
Inga multijuga
Inga quaternata
Inga Ruiziana
Inga sapindoides
Ixora nicaraguensis
Lonchocarpus guatemalensis
Lonchocarpus minimiflorus
Lonchocarpus sericeus var. *glabrescens*—Sap pink, drying red.
Malvaviscus arboreus—Flowers dark pink, Hibiscus-like.
Neea elegans
Neea Popenoei
Neonicholsonia Watsonii—Dwarf palm.
Ochroma lagopus—Large, tubular, white flowers. Wood very light.
Ocotea Ira
Ocotea rivularis
Ouratea Valerii
Pachira aquatica—Large, showy, Bombax-like flowers.
Picramnia latifolia—Bark bitter.
Pithecolobium glanduligerum—Flowers in showy white, globose heads at the ends of long, slender, pendulous stems.
Pouteria subrotata
Pseudima costaricensis
Quararibea guatemalteca—Pyramidal crown, with horizontal or somewhat pendulous branches.
Randia armata
Randia grandifolia

Rheedia edulis—Yellow, sticky sap.
Rinorea pubipes
Rollinia Jimenezii—Edible, Annona-like fruits.
Sapium jamaicense—Milky latex. Leaves with 2 glands at the base.
Sapium thelocarpum
Saurauia yasicae
Scheelea rostrata—Tall, unarmed palms with very large fronds.
Sickingia Maxonii—Very large, opposite, obovate leaves.
Socratea durissima—Palms with conspicuous spiny stilt roots.
Spondias mombin—Frequently a dominant, but often smaller. Yellow, plumlike fruits.
Stemmadenia Donnell-Smithii—Milky latex. Conspicuous paired fleshy fruits.
Stemmadenia grandiflora—Milky latex.
Swartzia panamensis—Often reaches the canopy, but flowers when much smaller. Flowers bright yellow, on long, pendulous spikes.
Symphonia globulifera—Yellow sticky latex. Flowers red.
Synechanthus angustifolius—Slender, green-trunked palm.
Theobroma angustifolium—Flowers rich orange, produced from the branches. Fruits cacao-like.
Tovomitopsis glauca
Trichilia montana
Triplaris melaenodendron—Showy dark-pink or red fruits.
Trophis chorizantha
Trophis racemosa
Vitex Cooperi—Mature specimens very tall, but often flower when much smaller. Leaves digitately compound. Flowers blue.
Welfia Georgii—Tall, unarmed palms.
Ximenia americana—Woolly petals.
Xylosma excelsum—Trunks with dense clusters of needle-like spines.
Zanthoxylum procerum—Trunk with short, broad-based thorns. Leaves pinnate.

MASTER KEY

Palms .... KEY A, p. 14
Not Palms.
  Trees with thorny or spiny trunks or branches .... KEY B, p. 14
  Trunks or branches not spiny.
    Flowers and fruits produced directly from the trunk or larger branches, as in cacao .... KEY C, p. 15
    Flowers and fruits not produced as above.
      Trees with thin red, tan, or white peeling bark .... KEY D, p. 15
      Bark not as above.
        Sap milky or variously colored.
          Trees with milky white latex .... KEY E, p. 15
          Trees with yellow or orange latex .... KEY F, p. 16
          Trees with red or pink sap .... KEY G, p. 17
          Trees with black sap or gum .... KEY H, p. 17
        Sap not milky or colored.

Leaves compound, with two or more distinct leaflets.
Leaves digitately compound .... KEY I, p. 17
Leaves pinnately compound .... KEY J, p. 18
Leaves simple.
Leaves simple, opposite .... KEY K, p. 19
Leaves simple, alternate .... KEY L, p. 20

KEY A

PALMS

1. Leaves fan-shaped .... *Cryosophila guagara*
1. Leaves not fan-shaped.
  2. Trunks spiny.
    3. Plants with multiple trunks.
      4. Leaves divided into many segments .... *Bactris balanoidea*
      4. Leaves undivided, the apex forked .... *Bactris militaris*
    3. Trunks solitary.
      4. Fruits smooth, orange, in pendulous clusters .... *Astrocaryum Standleyanum*
      4. Fruits spiny, in short, compact clusters .... *Astrocaryum alatum*
  2. Trunks not spiny.
    3. Trunks short and massive, often creeping .... *Corozo oleifera*
    3. Trunks not short or massive, never creeping.
      4. Trees with very prominently developed prickly stilt roots ... *Socratea durissima*
      4. Roots not developed into conspicuous stilts.
        5. Trees tall, usually more than 40 ft. in height.
          6. Flowers produced in deep pits on a fleshy, octagonal rachis. Fruits deep purple .... *Welfia Georgii*
          6. Flowers not produced in pits, the rachis not octagonal in cross section. Fruits dark yellow .... *Scheelea rostrata*
        5. Trees relatively small, usually less than 25 ft. in height.
          6. Plants dwarf, the trunks usually less than 1 ft. high. Inflorescence an unbranched spike .... *Neonicholsonia Watsonii*
          6. Trunks more than 3 ft. in height. Inflorescences branched.
            7. Staminate and pistillate flowers produced on separate inflorescences .... *Chamaedorea* sp.—*Allen 6262*
            7. Staminate and pistillate flowers produced on the same inflorescence .... *Synechanthus angustifolius*

KEY B

TREES WITH THORNY OR SPINY TRUNKS OR BRANCHES

1. Thorns long and needle-like, in dense, branching clusters at intervals along the lower trunk .... *Xylosma excelsum*
1. Thorns short, usually with a broad base, never branching or needle-like.
  2. Leaves compound.
    Leaves digitately compound. Leaflets 5–7 .... *Ceiba pentandra* (Young specimens)

3. Leaves pinnate.
    4. Leaflets 3. Flowers red..*Erythrina costaricensis*
    4. Leaflets more than 3. Flowers white........................*Zanthoxylum procerum*
2. Leaves simple.
    3. Trunks spiny. Branches unarmed.
        4. Trunks intensely spiny. Leaves broadly ovate. Sap watery.........*Hura crepitans*
        4. Trunks with a few scattered spines. Leaves elliptic-oblong. Sap milky white..*Sapium thelocarpum*
    3. Branches spiny. Trunks unarmed.
        4. Leaves alternate. Inner surface of the petals conspicuously woolly..*Ximenia americana*
        4. Leaves opposite. Inner surface of the petals not woolly...............*Randia armata*

KEY C

Flowers and fruits produced directly from the trunk or larger branches, as in cacao

1. Leaves digitately compound..*Herrania purpurea*
1. Leaves simple.
    2. Leaves deeply and conspicuously lobed..*Carica pennata*
    2. Leaves not lobed..*Grias Fendleri*

KEY D

Trees with thin red, tan, or white peeling bark

1. Leaves pinnate. Bark usually red.
    2. Plants epiphytic, growing in the tops of tall trees.................*Dacryodes epiphytica*
    2. Plants never epiphytic..*Bursera simaruba*
1. Leaves simple. Bark tan, guava-like.
    2. Branches hollow. Leaves usually more than 6″ long. Fruits completely enveloped in a showy, red, 3-lobed calyx........................*Triplaris melaenodendron*
    2. Branches not hollow. Leaves usually less than 6″ long. Fruits tan, not enveloped in a 3-lobed calyx..*Terminalia lucida*

KEY E

Trees with milky white latex

1. Leaves opposite. Flowers yellow.
    2. Trees 40–60 ft. in height. Common................................*Stemmadenia Donnell-Smithii*
    2. Shrubs or small trees, 10–15 ft. in height. Rather infrequent..*Stemmadenia grandiflora*
1. Leaves alternate. Flowers not yellow.
    2. Leaves with 2 small glands at the base of the blade.................*Sapium jamaicense*
    2. Leaves without glands at the base of the blade.

3. Leaves in 2 ranks along the branches, pendulous, rough to the touch........ *Castilla fallax*
3. Leaves not as above.
  4. Flowers with definite petals. Fruits with sapote-like seeds.
    5. Leaves conspicuously reddish brown on the lower surface........ *Chrysophyllum mexicanum*
    5. Leaves not reddish brown on the lower surface.
      6. Leaves conspicuously clustered at the ends of the branches. Fruits conspicuously woolly........ *Pouteria neglecta*
      6. Leaves not clustered at the ends of the branches. Fruits not woolly.
        7. Leaf petioles about 1″ long. Leaf blades rounded or subacute at the apex........ *Pouteria subrotata*
        7. Leaf petioles about ½″ long. Leaf blades acute or acuminate at the apex........ *Pouteria chiricana*
  4. Flowers minute, without definite petals.
    5. Flowers borne on the outside of a spherical, fleshy receptacle. Fruits with 1 or 2 relatively large seeds.
      6. Leaves narrowly linear-lanceolate, the tip long-acuminate........ *Brosimum sapiifolium*
      6. Leaves oblong, or elliptic-lanceolate, the tip merely acute or short acuminate........ *Brosimum alicastrum*
    5. Flowers borne on the inside of a hollow receptacle. Fruits with many small seeds.
      6. Leaves very large, usually more than 10″ long. Fruits solitary, never in pairs.
        7. Leaf blades with about 10 pairs of lateral nerves........ *Ficus Tonduzii*
        7. Leaf blades with 20–25 pairs of lateral nerves........ *Ficus Werckleana*
      6. Leaves usually less than 10″ long. Fruits in pairs.
        7. Leaf petioles and veins of the lower surface reddish brown, velvety........ *Ficus Bullenei*
        7. Leaf petioles and veins not reddish brown or velvety.
          8. Leaf blades mostly broadest above the middle.
            9. Fruits sessile, about ⅜″ in diameter........ *Ficus costaricana*
            9. Fruits with a distinct peduncle; about ¼″ in diameter........ *Ficus Oerstediana*
          8. Leaf blades broadest at about the middle........ *Ficus lapathifolia*

KEY F

Trees with yellow or orange latex

1. Latex yellow, sticky. Flowers with definite petals.
  2. Flowers blood red. Fruits brown. Lower trunk often with stilt roots........ *Symphonia globulifera*
  2. Flowers whitish. Fruits yellow or orange. Lower trunk without stilt roots........ *Rheedia edulis*
1. Latex pale creamy orange. Flowers without petals........ *Brosimum terrabanum*

EVERGREEN LOWLAND FOREST

KEY G

TREES WITH RED OR PINK SAP

1. Leaves pinnate.
  2. Flowers yellow. Fruits nearly circular in outline. Trunks prominently buttressed .... *Pterocarpus officinalis*
  2. Flowers pinkish lavender or purple. Fruits elongate. Trunks not prominently buttressed .... *Lonchocarpus sericeus* var. *glabrescens*
1. Leaves simple.
  2. Lower surface of the leaves brown or tan, contrasting strongly with the green upper surface.
    3. Leaf blades with 12–15 pairs of lateral nerves, the apex tapering gradually to a long point .... *Virola sebifera*
    3. Leaf blades with 20–30 pairs of lateral nerves, the apex abruptly acute or acuminate .... *Virola Koschnyi*
  2. Lower surface of the leaves not brown or tan.
    3. Leaf blades with 14–21 pairs of lateral nerves; very long and narrow in outline, tapering gradually to an acute tip.
      4. Seeds about 1″ long .... *Virola guatemalensis*
      4. Seeds about ¾″ long .... *Virola surinamensis*
    3. Leaf blades with about 30–35 pairs of lateral nerves, abruptly short acuminate at the tip .... *Virola nobilis*

KEY H

TREES WITH BLACK SAP OR GUM

1. Leaves pinnate, with four leaflets (two pairs) .... *Prioria copaifera*
1. Leaves simple.
  2. Leaves peltate, nearly circular in outline, deeply lobed. Branches hollow, infested with stinging ants.
    3. Trunks and branches white or green, the margins of the rings prominently elevated.
      4. Fruiting strands 8–16″ long. Leaves green on the lower surface .... *Cecropia obtusifolia*
      4. Fruiting strands 3–4″ long. Leaves white on the lower surface .... *Cecropia peltata*
    3. Trunks and especially branches black, the margins of the rings not elevated or conspicuous .... *Cecropia Sandersoniana*
  2. Leaves not peltate, circular in outline or lobed. Large trees, usually with a few conspicuous red leaves scattered through the crown .... *Hieronyma tectissima*

KEY I

LEAVES DIGITATELY COMPOUND

1. Leaves opposite.
  2. Leaflets 3. Flowers blue .... *Vitex Cooperi*

2. Leaflets 5.
   3. Flowers pink .... *Tabebuia pentaphylla*
   3. Flowers yellow .... *Tabebuia chrysantha*
1. Leaves alternate.
   2. Leaflets 3 .... *Crataeva tapia*
   2. Leaflets 5–7.
      3. Trees very large, with very prominently developed buttresses. Flowers less than 1½″ long. Pods filled with kapok .... *Ceiba pentandra*
      3. Trees of moderate size, seldom exceeding about 50 ft. Lower trunk never prominently buttressed. Flowers about 8″ long, with showy dark-red stamens. Fruits very large, filled with angular, fleshy seeds .... *Pachira aquatica*

KEY J

Leaves pinnately compound

1. Leaves bipinnate (having a central nonwoody axis and lateral pinnae).
   2. Margins of the individual leaflets serrate .... *Dipterodendron costaricense*
   2. Margins of the individual leaflets not serrate.
      3. Fruit an elongate, pendulous, red, beanlike pod .... *Pithecolobium austrinum*
      3. Fruit a dark brown, coiled, earlike pod .... *Enterolobium cyclocarpum*
1. Leaves once pinnate (having a central nonwoody axis and lateral leaflets).
   2. Leaflets 2 (a single pair) .... *Cynometra hemitomophylla*
   2. Leaflets more than 2.
      3. Leaflets 3. Flowers red .... *Erythrina costaricensis*
      3. Leaflets usually more than 3. Flowers not red.
         4. Small trees, usually less than 15 ft. in height, nearly always with multiple trunks. Fruits nearly globose, red, often fused in pairs .... *Pseudima costaricensis*
         4. Trees of varying size, never with multiple trunks. Fruits never fused in pairs.
            5. Bark and wood bitter.
               6. Small trees, with slender, axillary, unbranched, pendulous inflorescences. Fruits red .... *Picramnia latifolia*
               6. Large trees, with branching inflorescences.
                  7. Leaflets 9–20, leathery, broadly rounded at the apex. Fruits olive-like, 1-seeded .... *Simarouba glauca*
                  7. Leaflets 23–35, not leathery, acute at the apex. Fruit a woody capsule, with many winged seeds .... *Cedrela mexicana*
            5. Bark and wood not bitter, or unknown.
               6. Flowers mimosa-like. Fruit a legume.
                  7. Individual leaflets with a small gland at the base. Individual flower heads about 2″ in diameter, on long, slender, pendulous stems .... *Pithecolobium glanduligerum*
                  7. Leaf rachis with a conspicuous gland at the base of each pair of leaflets. Individual flower heads less than 2″ in diameter, never on long, pendulous stems.

8. Pinnae with 2–3 pairs of leaflets.
 9. Leaf rachis narrowly winged..*Inga sapindoides*
 9. Leaf rachis not winged..*Inga quaternata*
8. Pinnae with more than 3 pairs of leaflets.
 9. Leaflets and rachis usually with a minute but conspicuous reddish puberulence. Floral calyx usually more than ¼″ long..*Inga multijuga*
 9. Leaflets and rachis nearly glabrous. Floral calyx about ⅛″ long...*Inga Ruiziana*

6. Flowers pea- or beanlike, lavender or purple.
 7. Individual leaflets less than 2″ long..............*Lonchocarpus minimiflorus*
 7. Individual leaflets more than 2″ long.
  8. Inflorescences terminal, usually conspicuously exceeding the pinnae in length. Fruits nearly globose........................*Andira inermis*
  8. Inflorescences usually axillary, about equaling the pinnae in length. Fruits flat...*Lonchocarpus guatemalensis*

6. Flowers not pea- or beanlike, usually greenish white or yellow.
 7. Leaflets broadly rounded at the apex.
  8. Mature leaflets usually less than 4″ long. Fruits olive-like, 1-seeded...*Simarouba glauca*
  8. Mature leaflets usually more than 6″ long. Fruits very large and woody, several seeded...*Carapa Slateri*
 7. Leaflets pointed at the apex.
  8. Leaflets 3–8 in number.
   9. Inflorescences slender, elongate, often pendulous unbranched spikes.
    10. Flowers bright yellow. Pods usually more than 6″ long, with very large seeds...*Swartzia panamensis*
    10. Flowers not yellow. Pods much less than 6″ long.
     11. Lower leaflets alternate on the rachis. Flowers minute. Fruits scarlet...*Picramnia latifolia*
     11. Lower leaflets opposite on the rachis. Flowers more than ¼″ long. Fruits brown..........................*Guarea Hoffmanniana*
   9. Inflorescences short, compact, branching panicles...*Trichilia montana*
  8. Leaflets 10–35 in number.
   9. Leaflets 10–18. Fruits plumlike, 1-seeded...............*Spondias mombin*
   9. Leaflets 23–35. Fruits woody, with many winged seeds...*Cedrela mexicana*

KEY K

Leaves simple, opposite

1. Trees with multiple trunks and stilt roots....................................*Bravaisia integerrima*
1. Trees with a solitary trunk and no stilt roots.
 2. Flowers white.

3. Large trees, 50–90 ft. in height, with prominently developed buttresses. Wood bright orange. Flowers in broad, axillary cymes............... *Chimarrhis latifolia*
3. Small trees, 10–30 ft. in height, the trunks not buttressed. Wood not orange.
   4. Lower, unexpanded portion of the flower contracted into a slender tube.
      5. Flowers solitary or in sessile, few-flowered fascicles at the ends of the leafy branches.
         6. Leaves 4–16″ long. Outer surface of the flowers glabrous........................ .... *Randia grandifolia*
         6. Leaves 1½–8″ long. Outer surface of the flowers densely covered with a fine, silky tomentum... *Anisomeris Recordii*
      5. Flowers usually many, in branching cymes or panicles........................ .... *Ixora nicaraguensis*
   4. Lower part of the flower not contracted into a tube......... *Tovomitopsis glauca*
2. Flowers yellow or greenish yellow.
   3. Leaves broadest at about the middle. Flowers very small, pale yellow, in short, unbranched spikes.. *Rinorea pubipes*
   3. Leaves broadest above the middle.
      4. Leaf veins and petioles red. Flowers rich yellow, in terminal cymes. Shrubs or small trees, to about 18 ft... *Hamelia magnifolia*
      4. Leaf veins and petioles not red. Flowers greenish yellow. Trees, 25–50 ft. in height.. *Sickingia Maxonii*

KEY L

Leaves simple, alternate

1. Leaves palmately lobed.
   2. Slender, unbranched, papaya-like plants, to about 15 ft. Leaves very deeply lobed.. *Carica pennata*
   2. Trees not as above. Leaves shallowly lobed.
      3. Large trees, often with prominently developed buttresses. Flowers yellowish, covered with dark-red, velvety hairs.. *Sterculia apetala*
      3. Trees of varying size. Flowers white.
         4. Flowers tubular, more than 4″ long. Seeds surrounded by kapok-like fluff...... .... *Ochroma lagopus*
         4. Flowers about 1½″ long. Seeds without kapok-like fluff... *Hampea platanifolia*
1. Leaves not palmately lobed.
   2. Flowers dark pink or red, Hibiscus-like.................................. *Malvaviscus arboreus*
   2. Flowers not as above.
      3. Flowers yellow or orange.
         4. Flowers fleshy, nearly sessile in the leaf axils. Fruits Annona-like................ .... *Rollinia Jimenezii*
         4. Flowers not fleshy. Fruits not Annona-like.
            5. Flowers bright orange, distributed in dense, sessile clusters along the branches. Fruits cacao-like.. *Theobroma angustifolium*
            5. Flowers yellow. Fruits not cacao-like.
               6. Leaves large and leathery. Flowers bright yellow............ *Ouratea Valerii*
               6. Leaves not leathery. Flowers pale yellow.
                  7. Leaves with 3 longitudinal nerves.......................... *Goethalsia meiantha*

7. Leaves with a single longitudinal nerve.
  8. Flowers in erect, relatively slender panicles. Fruits not red........ *Ocotea Ira*
  8. Flowers in broad, pendulous cymes. Fruits red.
    9. Branches and veins of the lower surfaces of the leaves densely ferruginous-tomentulose........ *Neea elegans*
    9. Branches and veins of the lower leaf surfaces glabrous........ *Neea Popenoei*

3. Flowers not yellow or orange.
  4. Leaf margins serrate or crenate.
    5. Undersurfaces of the leaves brown........ *Luehea Seemannii*
    5. Undersurfaces of the leaves not brown.
      6. Branchlets somewhat scurfy. Leaves broadest above the middle, the base narrowly wedge-shaped........ *Saurauia yasicae*
      6. Branchlets not scurfy. Leaves broadest at about the middle.
        7. Flowers in dense, nearly sessile, globose heads........ *Chlorophora tinctoria* (Pistillate flowers)
        7. Flowers never in sessile, globose heads.
          8. Flowers in unbranched spikes or racemes.
            9. Flowers minute, in catkin-like spikes, not equaling the leaves in length........ *Chlorophora tinctoria* (Staminate flowers)
            9. Flowers about ½" in diameter, the spikes not catkin-like, equaling or exceeding the leaves in length........ *Homalium eurypetalum*
          8. Flowers in short, compact, branching cymes or panicles.
            9. Leaf margins very coarsely serrate........ *Xylosma excelsum*
            9. Leaf margins minutely sinuate-serrate........ *Gloeospermum diversipetalum*
  4. Leaf margins entire.
    5. Leaves in two ranks along the branches, conspicuously pendulous. Flowers fleshy........ *Guatteria chiriquensis*
    5. Leaves not as above. Flowers not fleshy.
      6. Flowers and fruits subtended by a conspicuous red, circular calyx.
        7. Fruits black at maturity. Calyx undivided........ *Heisteria longipes*
        7. Fruits white at maturity. Calyx often deeply lobed........ *Heisteria concinna*
      6. Flowers and fruits without a conspicuous red calyx.
        7. Individual flowers relatively large, about 1–1½" in diameter.
          8. Flowers sessile in the leaf axils. Calyx green........ *Quararibea guatemalteca*
          8. Flowers in terminal or axillary racemes. Calyx pale brown........ *Capparis Sinclairii*
        7. Individual flowers very small, ½" or less in diameter.
          8. Leaves broadest above the middle.
            9. Rachis of the inflorescence angular, dark red. Flowers green. Fruits not kidney-shaped........ *Ocotea rivularis*
            9. Rachis of the inflorescence not angular, pale greenish white. Flowers white, aging pink. Fruits kidney-shaped........ *Anacardium excelsum*

8. Leaves broadest at or below the middle.
  9. Flowers subtended by many, conspicuous, greenish-white bracts...... ...... *Hernandia didymantha*
  9. Flowers not subtended by conspicuous bracts.
    10. Flowers in dense, nearly sessile, globose heads...... ...... *Chlorophora tinctoria* (Pistillate form)
    10. Flowers in spikes or panicles.
      11. Flowers in slender, unbranched spikes.
        12. Branches often armed with spines. Wood yellow...... ...... *Chlorophora tinctoria* (Staminate form)
        12. Branches never armed with spines.
          13. Leaves with long, acuminate apices and narrow, wedge-shaped bases...... *Trophis chorizantha*
          13. Leaves with acute or shortly acuminate apices and rounded bases...... *Croton glabellus*
        12. Flowers in broad, axillary panicles...... ...... *Casearia banquitana* var. *laevis*

(2)—LITTORAL WOODLAND

The following lists and keys are of the characteristic species found above high tide along the varied shore lines of Golfo Dulce and the adjacent coast. Here shadowy, blue, receding ranks of steep forested headlands alternate with sandy beaches, where, at certain times, a silvery mist forms over the long, smooth combers breaking at the base of picturesque groves of coconut palms. These, together with *Hibiscus tiliaceus, Gynerium sagittatum,* and *Conocarpus erecta,* form the dominant elements for miles along the beaches, particularly from the delta of the Río Coto to Banco Point. The formation probably never exceeds about a hundred yards in width, and is usually much less, merging imperceptibly near river mouths with the Mangrove Woodland, and inland with either the Evergreen Lowland Forest or the Lower Montane Forest, depending on the nature of the terrain. The great majority of the species will be found to have leathery leaves, and to be evergreen. Rocky shores below the high, forested headlands are considerably more varied, but tend to be dominated by *Protium costaricense, Swartzia simplex, Bombax barrigon* (Pavón Bay), *Nectandra latifolia, Apeiba Tibourbou, Enallagma latifolia,* and *Conocarpus erecta.*

Sandy Beaches

Alibertia edulis
Bactris balanoidea—Spiny palms.

LITTORAL WOODLAND

Bombax barrigon—Bulging trunks, with vertical green and gray stripes.
Citharexylum viride
Coccoloba padiformis
Cocos nucifera
Conocarpus erecta
Ficus lapathifolia—Milky latex.
Hibiscus tiliaceus—Broad, heart-shaped leaves. Showy yellow or reddish-brown flowers.
Hippomane mancinella—Leathery leaves, with a gland at the base of the blade. Poisonous fruits and milky, caustic latex.
Ximenia americana

KEY

1. Palms.
  2. Multiple, intensely spiny trunks .... *Bactris balanoidea*
  2. Solitary, unarmed trunks .... *Cocos nucifera*
1. Not palms.
  2. Trunks conspicuously bulging near the base, with vertical green, yellow, and gray stripes. Leaves digitately compound .... *Bombax barrigon*
  2. Trunks not as above. Leaves not compound.
    3. Leaves opposite.
      4. Fruits yellow, globose, usually solitary, about 1" in diameter .... *Alibertia edulis*
      4. Fruits bright orange, in long, showy, terminal racemes. Individual fruits about ¼" in diameter .... *Citharexylum viride*
    3. Leaves alternate.
      4. Trees with milky latex.
        5. Leaf petiole with a gland at the base of the blade. Fruits superficially like crab apples but poisonous. Latex caustic .... *Hippomane mancinella*
        5. Petiole without a gland at the base of the blade. Fruits hollow, figlike. Latex not caustic .... *Ficus lapathifolia*
      4. Trees without milky latex.
        5. Basal portion of the leaf petiole enveloping the stem .... *Coccoloba padiformis*
        5. Basal portion of the leaf petiole not as above.
          6. Leaves broadly heart-shaped. Flowers large, showy, yellow, aging reddish brown .... *Hibiscus tiliaceus*
          6. Leaves not heart-shaped.
            7. Bark reddish. Branches spiny .... *Ximenia americana*
            7. Bark and branches not as above .... *Conocarpus erecta*

Rocky Seashores

Apeiba Tibourbou—Spiny, sea urchin-like fruits.
Calophyllum braziliense var. *Rekoi*—Milky latex.
Enallagma latifolia—Fruits like small calabashes.
Guettarda macrosperma
Inga multijuga
Nectandra latifolia
Nectandra perdubia
Protium copal var. *glabrum*
Protium costaricense
Swartzia simplex

LOWER MONTANE RAIN FOREST

KEY

1. Leaves pinnate.
  2. Leaf rachis with a conspicuous disklike gland at the base of each pair of leaflets. Flowers mimosa-like........ *Inga multijuga*
  2. Leaf rachis not as above.
    3. Leaflets broadly elliptic, usually more than 4″ long... *Protium copal* var. *glabrum*
    3. Leaflets narrowly elliptic-lanceolate, usually less than 4″ long........ *Protium costaricense*
1. Leaves simple.
  2. Leaves opposite.
    3. Shrubs or small trees, to about 40 ft. in height. Plants without milky latex........ *Guettarda macrosperma*
    3. Trees, usually more than 40 ft. in height, with milky latex........ *Calophyllum braziliense* var. *Rekoi*
  2. Leaves alternate.
    3. Leaves with minutely serrate margins, the blades rough to the touch. Fruits spiny........ *Apeiba Tibourbou*
    3. Leaves with entire margins, the blades smooth to the touch. Fruits not spiny.
      4. Flowers bright yellow. Fruit a bright-orange pod........ *Swartzia simplex*
      4. Flowers not yellow. Fruits not orange.
        5. Flowers relatively large, tubular, greenish white. Fruits gourdlike........ *Enallagma latifolia*
        5. Flowers very small, not tubular. Fruits small, black, 1-seeded drupes.
          6. Axils of the veins of the lower leaf surface with small but distinct tufts of hairs........ *Nectandra perdubia*
          6. Axils without tufts of hairs........ *Nectandra latifolia*

(3)—LOWER MONTANE RAIN FOREST

Hillsides below about 2,000 ft. have approximately the same rainfall as the adjacent floodplains, but the clay ridges and rocky slopes support a much richer arborescent cover. As would be expected, local conditions of soil, slope, drainage, exposure to wind, past history, etc., will often be reflected in the vegetation, which is characterized by a number of quite distinctive consociations and associations. It is well to note, however, that superficially identical situations may have entirely different populations, so that it seems probable that they may sometimes be the result of fortuitous circumstance rather than any basic ecological condition. In some cases it appears to be questionable, or even unlikely, that some of these associations will indefinitely perpetuate themselves on a given piece of terrain. This is particularly true of colonies of *Schizolobium* or *Vochysia* which have invaded artificial clearings, or normally infrequent elements which have become established in unusual concentrations

because of favorable conditions provided by the fall of some forest giant. In general, the more difficult the situation, the more likely that it will support a highly specialized or even locally unique association or consociation.

Most of our hillside forests seem to be much denser than those of the floodplains, at least in the sense that the crowns of the dominant elements usually meet, often forming a nearly continuous cover. A somewhat higher percentage of the emergent species are deciduous, and for somewhat longer periods, though there are never enough of them present at any one time to give the forest the wintry appearance frequently seen on rocky outcrops in a drier climate. Understory species consist of palms and other elements, all of which are evergreen. While there is about the same percentage of woody understory species as in the lowlands, there are far fewer of the giant herbs, such as the aroids, *Calatheas, Heliconias,* and others, that elsewhere congest the forest floor. Many of the ridges, in particular, are quite open, so that one can wander almost at will between the majestic columns that rise like the pillars of some immemorial temple. A walk in the early days of the dry season through these arcaded halls, when the air is perfumed with the fragrance of the great *Espavels* or *Guacimo colorados* can be a delightful experience. High spurs, particularly above Palmar Norte, provide far vistas of Caño Island and of the rolling mountains blanketed in unbroken forest, with the silvery expanse of Golfo Dulce flanked by low, blue hills. Many slopes tend to be steep, often rocky walls, down which the streams plunge in long, feathery cascades. Access to the summits is to be had only by following the ridges, and attempts at descent via stream courses can be extremely dangerous in some localities.

No single species or association of species is universally dominant, the larger trees forming completely mixed stands on all but the specialized situations. Eighty-nine known species form the canopy, which may vary from about 90 to 150 ft. in height. Two hundred and seven understory species have been collected, which range from 15 to approximately 90 ft. in height, but which, in any case, seldom reach the canopy excepting on very steep slopes. These species could be separated into three more or less distinct tiers, according to size, though these tend in actual practice to be obscured by the varying stages of development of species which will eventually form part of the dominant cover. Below the arborescent layers there are shrubby and herbaceous levels, so that in effect a 5-layered forest exists over much of the formation. The more conspicuous associations, consociations, and faciations are listed as follows:

Brosimum utile—Tetragastris panamensis—Rheedia madruno—Lacmellea panamensis association.—Characteristic of steep ridges leading to the summit of the Cerros de Retinto near Palmar Norte, at 1,000–1,500 ft. elevation.

Brosimum utile consociation.—Extremely high, sharply drained slopes and ridges may support a nearly pure stand of this species with hundreds of individuals whose crowns join to form the canopy. Particularly striking examples may be seen at about 2,500 ft. above Rey Curre, and in the hills back of Jalaca in the Esquinas District.

Iriartea gigantea consociation.—Pure stands of this magnificent stilt palm are to be found in damp coves in the hills near El Cedral, at about 2,000–2,500 ft.

Coccoloba roseiflora consociation.—Typical of many of the nearly perpendicular rocky walls above Palmar Norte, at 2,000–2,500 ft. The red flushes of new leaves are quite conspicuous from a distance in late January and early February.

Brosimum utile—Vantanea Barbourii faciation.—An excellent example of this faciation may be seen in the low clay hills back of the citrus orchard in Palmar. Typical of well-drained ridges to about 2,500 ft.

Peltogyne purpurea—Brosimum utile—Vantanea Barbourii association.—The presence of *Peltogyne* in this association seems to be determined by (1) the presence of red clay, and (2) an evenly distributed rainfall in excess of 200″. Very good examples may be seen in the Esquinas Forest Preserve, and in the hills near the delta of the Río Esquinas at 100–350 ft. elevation.

Calycophyllum candidissimum—Maytenus pallidifolius—Eugenia palmarum association.—Apparently limited to slightly sloping areas of restricted extent in the hills above Palmar Norte, at 1,000–1,500 ft.

Lafoensia punicifolia consociation.—Pure stands of this species have been reported from the vicinity of Buenos Aires and Cañas Gordas, but have not been seen by the writer.

Schizolobium parahybum—Sterculia mexicana—Hieronyma tectissima association.—Not so well defined as most of the other associations, sometimes tending to be an almost pure consociation of *Schizolobium*, and probably the result of coincidence. Several good examples may be seen on the ridges above Palmar Norte, and in the hills near Golfito, usually at about 1,000 ft. It has been my observation that *Schizolobium* in particular requires abundant light in its early stages of development, and it appears probable that many if not all of the conspicuous stands of this species represent old clearings.

Vochysia consociation or association.—The relative frequence and distribution of the three species of this genus present in the area are not well understood, but it appears likely that all three form nearly pure stands, extremely conspicuous during the flowering season, which may extend from March until June. Large colonies may be seen, with their characteristic orange flowers, in the hills near Piedras Blancas and near Golfito. It seems at least possible that these centers may be very old clearings made a generation or more ago, since all species of the genus in Central America have the habit of invading cutover land. Very striking examples of this peculiarity may be seen in the low hills back of Bluefields in eastern Nicaragua, and to a more limited extent in small clearings on the shores of Lake Yojoa, in Honduras.

Goethalsia meiantha—Didymopanax Morototoni association.—Typical of well-developed second growth on old clearings. A good example may be seen in the low hills near the Palmar citrus grove.

Goethalsia meiantha consociation.—Usually limited to a few dozen individuals in dense groves on old clearings.

Large trees, which reach the canopy layer—90–150 ft.

Anacardium excelsum—Rather infrequent above the floodplains. Flowers with a clovelike fragrance.
Andira inermis—Purple or lavender flowers in terminal panicles.
Aspidosperma megalocarpon—Papery, circular seeds, about 2″ in diameter.
Astronium graveolens—Crushed leaves have a sharp, distinctive odor.
Basiloxylon excelsum
Batocarpus costaricense—Osage orange-like fruits.
Billia colombiana
Brosimum sapiifolium—Very narrow leaves and milky latex.
Brosimum terrabanum—Milky latex.
Brosimum utile—Milky latex.
Byrsonima densa
Calocarpum borucanum—Sapote-like fruits.
Calophyllum braziliense var. *Rekoi*—Milky latex.
Calycophyllum candidissimum—Flowers subtended by a showy white bract.
Carapa Slateri—Flushes of bright-red new leaves from the ends of the branches.
Caryocar costaricense—Sap has a vinegar-like odor.
Cedrela fissilis—Bitter bark and wood.
Cedrela mexicana—Bitter bark and wood.
Ceiba pentandra—Mature specimens with tremendous buttresses.
Chimarrhis latifolia—Wood orange.
Chrysophyllum mexicanum—Undersides of leaves reddish brown. Milky latex.
Coccoloba roseiflora
Couratari panamensis—Prominent buttresses. Showy pink flowers.

Cynometra hemitomophylla
Dialium guianense
Dipterodendron costaricense—Fernlike foliage.
Dussia macrophyllata—Sap dries pink when exposed to the air.
Dussia mexicana
Enterolobium cyclocarpum—Pods dark brown, coiled, earlike.
Euterpe panamensis—Slender, single-trunked palm.
Ficus Bullenei—Milky latex.
Ficus costaricana—Milky latex.
Ficus lapathifolia—Milky latex.
Ficus Werckleana—Milky latex.
Goethalsia meiantha
Guarea longipetiola
Hieronyma tectissima—Black sap.
Huberodendron Allenii—Tremendous plank buttresses.
Hura crepitans—Intensely spiny trunks.
Iriartea gigantea—Palm, with low, stilt roots.
Jacaranda copaia—Showy blue flowers.
Ladenbergia Brenesii
Ladenbergia chariantha
Lafoensia punicifolia
Licania arborea
Licania operculipetala
Luehea Seemannii—Leaves brownish on the lower surface.
Lysiloma guanacastense—Shaggy bark and fernlike leaves.
Minquartia guianensis
Mortoniodendron anisophyllum—Prominent buttresses. Fruit a 5-celled capsule.
Myroxylon balsamum var. *Pereirae*—Sap with an agreeable, balsamic odor.
Ormosia panamensis—Seeds bright red, with a large black blotch.
Peltogyne purpurea—Leaflets 2. Wood purple.
Pentaplaris Doroteae—Calyx with 5 wings.
Persea americana—Wild avocado.
Phyllocarpus septentrionalis—Showy red flowers.
Pithecolobium austrinum—Fernlike foliage.
Pithecolobium macradenium
Platymiscium pinnatum
Pourouma aspera—Palmately lobed leaves. Black sap.
Pouteria chiricana—Milky latex.
Pouteria heterodoxa—Milky latex.
Pouteria neglecta—Deeply furrowed trunks. Milky latex.
Prioria copaifera—Black gum.
Sacoglottis excelsa
Scheelea rostrata—Tall palm, with very large pinnate fronds.
Schizolobium parahybum—Fernlike leaves and showy yellow flowers.
Simarouba glauca—Bitter bark.
Sloanea laurifolia
Sloanea picapica—Fruits covered with purple, needle-like spines.
Sterculia mexicana—Digitately compound leaves.

Sweetia panamensis
Tabebuia chrysantha—Showy yellow flowers.
Tabebuia pentaphylla—Showy pink flowers.
Tachigalia versicolor—Tremendous buttresses. Showy pinkish-lavender flowers.
Terminalia amazonia
Terminalia lucida—Guava-like bark.
Tetragastris panamensis
Vantanea Barbourii—Freshly cut wood smells like crushed sugar cane.
Virola guatemalensis—Red sap. Nutmeg-like fruits.
Virola Koschnyi—Red sap. Nutmeg-like fruits.
Virola nobilis—Red sap. Nutmeg-like fruits.
Virola sebifera—Red sap. Nutmeg-like fruits.
Virola surinamensis—Red sap. Nutmeg-like fruits.
Vitex Cooperi—Brownish-green foliage. Flowers blue.
Vochysia Allenii—Showy dark-yellow flowers.
Vochysia ferruginea—Showy orange flowers.
Vochysia hondurensis—Showy orange flowers.

Understory species—15–90 ft.

Aegiphila costaricensis
Aegiphila martinicensis
Albizzia filicina—Scaling bark and fernlike foliage.
Albizzia longepedata
Alchornea glandulosa var. *Pittieri*
Alibertia edulis
Andira inermis—Purple or lavender flowers in showy panicles.
Annona Pittieri
Apeiba aspera—Spiny, sea urchin-like fruits.
Apeiba Tibourbou—Spiny, sea urchin-like fruits.
Ardisia Cutteri
Ardisia Dodgei
Ardisia Dunlapiana
Ardisia revoluta
Ardisia Standleyana
Asterogyne Martiana—Dwarf, unarmed palm, with bifid fronds.
Astrocaryum alatum—Palm, with compact clusters of spiny fruits.
Astrocaryum Standleyanum—Large palm, with intensely spiny trunks and showy, pendulous clusters of unarmed orange fruits.
Bactris Baileyana—Multiple-stemmed, intensely spiny palms with red fruits.
Bactris balanoidea—Multiple-stemmed, intensely spiny palms with purple fruits.
Bactris divisicupula—Multiple-stemmed palms with whitish flat spines.
Bactris sp.—*Allen 6765*—Fronds entire, bifid at the apex.
Bellucia costaricensis
Belotia macrantha
Belotia reticulata
Bombax barrigon—Swollen trunk with vertical green and gray stripes.

Bombax sessile
Bravaisia integerrima—Stilt roots.
Bursera simaruba—Conspicuous red, peeling bark.
Calliandra grandifolia
Carpotroche platyptera—Flowers and fruits produced from the trunk and branches, as in cacao.
Casearia arborea
Cassipourea podantha
Castilla fallax—Milky latex. Large, conspicuously drooping leaves.
Cecropia obtusifolia—Nearly circular peltate leaves. Sap black.
Cecropia peltata—Nearly circular peltate leaves. Sap black.
Cecropia Sandersoniana—Nearly circular peltate leaves. Sap black.
Cespedesia macrophylla—Showy yellow flowers. Very large leaves, clustered at the ends of the branches.
Chamaedorea Wendlandiana—Slender, unarmed palms.
Chamaedorea Woodsoniana—Slender, unarmed palms.
Chamaedorea sp.—*Allen 6742*—Slender, unarmed palms.
Chamaedorea sp.—*Moore 6527*—Slender, unarmed palms.
Chimarrhis latifolia—Wood orange.
Chione costaricensis
Chrysophyllum panamense—Milky latex.
Clarisia mexicana
Coccoloba roseiflora—Bases of the leaf petioles enveloping the stems. Flowers pink.
Coccoloba Standleyana—Leaves very large.
Compsoneura Sprucei—Red sap.
Condaminea corymbosa—Small tree, with paired leaves to 30″ long.
Cordia alliodora—Branches with hollow, gall-like swellings.
Cordia gerascanthus—Branches with hollow, gall-like swellings.
Cordia protracta
Coussapoa parviceps
Croton glabellus
Croton panamensis
Croton xalapensis
Cryosophila guagara—Palm with spiny trunk and fan-shaped leaves.
Cupania guatemalensis
Cupania largifolia
Cupania macrophylla
Dialyanthera otoba
Didymopanax Morototoni—Leaves digitately compound, the undersides pale brown.
Diospyros ebenaster—Black, deeply furrowed trunks.
Dipterodendron costaricense—Fernlike foliage.
Dracaena americana—Very long, strap-shaped leaves, clustered at the ends of the branches.
Duguetia panamensis
Erythrina costaricensis—Leaflets 3. Flowers bright red.
Erythroxylon lucidum
Eschweilera calyculata—Fruits woody, with a boxlike lid.
Eugenia palmarum—Smooth, conspicuous white bark.

Eugenia sp.—*Allen 5979*
Euterpe panamensis—Slender, single-trunked, unarmed palm.
Faramea sessifolia
Faramea suerrensis
Ficus Bullenei—Milky latex. Leaves brownish, velvety.
Genipa Caruto
Geonoma binervia—Unarmed palm.
Geonoma congesta—Unarmed palm.
Geonoma sp.—*Allen 6750*—Unarmed palm.
Godmania aesculifolia—Digitately compound, opposite leaves.
Goethalsia meiantha
Guarea aligera—Leaf rachis winged.
Guarea guara
Guatteria aeruginosa—Leaves and stems conspicuously velvety.
Guatteria amplifolia
Guatteria lucens
Gustavia angustifolia—Small trees, with very large spatulate leaves, clustered at the ends of the branches.
Hamelia magnifolia—Leaf petioles and veins red. Shrub or small tree.
Hampea Allenii
Hasseltia floribunda—Leaves with 3 longitudinal nerves.
Hasseltia quinquenervia—Leaves with 5 longitudinal nerves.
Heisteria concinna
Heisteria longipes—Fruits subtended by a collar-like red calyx.
Heliocarpus appendiculatus—Showy pink fruits.
Henriettella tuberculosa
Herrania purpurea—Leaves digitately compound. Flowers and fruits produced from the trunk, as in cacao.
Hirtella americana
Hirtella triandra
Hyospathe Lehmannii—Unarmed palm, with green, multiple trunks.
Inga laurina
Iriartea gigantea—Single-trunked, unarmed palm, with low stilt roots.
Jacaranda lasiogyne—Flowers purple.
Jacaratia dolichaula
Lacmellea panamensis—Trunk with scattered, broad-based pyramidal spines. Sweet, milky latex.
Ladenbergia Brenesii
Ladenbergia chariantha
Lafoensia punicifolia
Licaria Cervantesii—Acorn-like fruits.
Licaria Cufodontisii—Acorn-like fruits.
Linociera panamensis—Showy white flowers.
Lonchocarpus latifolius
Macrocnemum glabrescens—Opposite leaves and showy pink flowers.
Malvaviscus arboreus—Pink, hibiscus-like flowers.
Marila pluricostata
Mayna echinata

Maytenus pallidifolius—Twisted, deeply furrowed trunks.
Meliosma Allenii
Meliosma anisophylla
Meliosma longipetiola
Miconia argentea—Leaves white or pale tan on the lower surface.
Miconia caudata
Miconia dodecandra
Miconia elata
Miconia hondurensis
Miconia impetiolaris
Miconia laevigata
Miconia Matthaei
Miconia pteropoda
Miconia rubiginosa
Miconia Schlimii
Miconia scorpioides
Miconia sp.—*Allen 5814*
Mortoniodendron guatemalense
Muntingia calabura
Myriocarpa longipes—Small trees, with minute flowers in long threadlike spikes.
Nectandra latifolia
Nectandra perdubia
Neea elegans
Neea Popenoei
Neonicholsonia Watsonii—Unarmed, dwarf, single-stemmed palm.
Ochroma lagopus—Large, tubular, white flowers. Wood very light.
Ocotea Ira
Ocotea mollifolia
Ocotea pergamentacea
Ocotea veraguensis
Ocotea Williamsii
Oenocarpus panamanus—Unarmed palms, with multiple trunks.
Olmedia falcifolia—Milky latex.
Oreopanax capitatus
Parmentiera macrophylla
Pentagonia gymnopoda—Small tree, with very large, opposite, pinnatisect leaves.
Perebea molliflora—Milky latex.
Perebea trophophylla—Milky latex.
Persea pallida
Persea Skutchii
Picramnia latifolia—Small tree, with pinnate leaves. Bark very bitter.
Pithecolobium glanduligerum—Showy white, mimosa-like flowers on long, pendulous stems.
Plumeria rubra var. *acutifolia*—Showy white flowers and milky latex.
Pourouma aspera—Broad, palmately lobed leaves and black sap.
Pouteria chiricana—Milky latex.
Pouteria subrotata—Milky latex.
Pouteria triplarifolia—Milky latex.

Protium copal var. *glabrum*
Protium costaricense
Protium neglectum var. *sessiliflorum*
Pseudolmedia spuria—Milky latex.
Psychotria chiapensis
Quassia amara—Leaf rachis winged. Bark intensely bitter.
Rheedia edulis—Yellow latex. Fruits smooth.
Rheedia madruno—Yellow latex. Fruits rough.
Rondeletia urophylla
Roupala complicata
Scheelea rostrata—Single-trunked, robust palms, with very large fronds.
Socratea durissima—Single-trunked palms, with very conspicuous stilt roots.
Stemmadenia grandiflora—Milky latex. Flowers yellow.
Stemmadenia nervosa—Milky latex.
Swartzia picramnoides—Bright yellow, orchid-like flowers.
Swartzia simplex—Bright yellow, orchid-like flowers.
Sweetia panamensis
Symphonia globulifera—Yellow, sticky latex. Flowers blood red.
Synechanthus angustifolius—Slender, unarmed palm.
Tabernaemontana longipes—Milky latex.
Talisia nervosa—Slender trees, with very large pinnate leaves, which arc clustered at the ends of the branches.
Tetrathylacium costaricense—Leaves large, conspicuously drooping. Dark-red flowers in pendulous clusters.
Theobroma simiarum—Flowers and fruits produced directly from the trunk.
Tocoyena Pittieri
Tovomitopsis costaricana
Tovomitopsis glauca
Tovomitopsis grandifolia
Trema micrantha
Trichilia acutanthera
Trichilia montana
Trichilia Skutchii
Trichilia tuberculata
Triplaris melaenodendron—Showy pink, 3-winged fruits.
Unonopsis Pittieri
Urera alceifolia
Urera caracasana
Vismia ferruginea—Orange sap, which dries blood red.
Vismia guianensis—Orange sap, which dries blood red.
Vitex Cooperi—Brownish-green leaves. Flowers blue.
Warscewiczia coccinea—Shrub or small tree, with brilliant red, poinsettia-like flowers.
Welfia Georgii—Tall, unarmed, single-trunked palms.
Ximenia americana
Xylopia sericophylla
Xylosma excelsum—Trunks usually armed with dense clusters of branching, needle-like spines.
Zanthoxylum procerum—Leaves pinnate. Trunks with short, broad-based spines.

*MASTER KEY**

Palms .... KEY A, p. 34
Not Palms.
Flowers and fruits produced directly from the bark, as in cacao .... KEY B, p. 35
Flowers and fruits not produced as above.
Leaves compound .... KEY C, p. 36
Leaves simple, opposite on the stems .... KEY D, p. 42
Leaves simple, alternate on the stems .... KEY E, p. 45
Species with white, milky latex .... KEY F, p. 56
Species with red or pink sap .... KEY G, p. 57
Species with yellow or orange latex .... KEY H, p. 57
Species with black sap or gum .... KEY I, p. 58
Species with prominent buttresses .... KEY J, p. 58
Species with thorny or spiny trunks or branches .... KEY K, p. 60

KEY A

PALMS

1. Fronds nearly circular in outline, fan-shaped .... *Cryosophila Guagara*
1. Fronds not fan-shaped.
  2. Trunks solitary.
    3. Individual leaflets conspicuously wedge-shaped, broadest at the apex. Lower trunk usually with stilt roots.
      4. Stilt roots often more than 6 ft. in height .... *Socratea durissima*
      4. Stilt roots seldom more than 2–3 ft. in height, sometimes lacking in young specimens .... *Iriartea gigantea*
    3. Individual leaflets never wedge-shaped, or broadest at the apex.
      4. Trunks and/or fronds armed with spines.
        5. Trunks intensely spiny, about 6–8″ in diameter. Fruits orange, not spiny, in showy pendulous clusters .... *Astrocaryum Standleyanum*
        5. Trunks about 4″ in diameter. Fruits intensely spiny, in short, compact clusters .... *Astrocaryum alatum*
      4. Trunks and fronds not spiny.
        5. Trees small or dwarf, never exceeding 20 ft. in height.
          6. Fronds undivided, the apex conspicuously forked .... *Asterogyne Martiana*
          6. Fronds pinnate.
            7. Plants dwarf, the trunks usually less than 1 ft. in height.
              8. Inflorescences spicate, undivided .... *Neonicholsonia Watsonii*
              8. Inflorescence branching .... *Chamaedorea* sp.—*Allen 6742*
            7. Plants with trunks more than 3 ft. in height.
              8. Staminate and pistillate flowers produced on separate scapes.
                9. Fruits obovoid, about ⅜″ long .... *Chamaedorea* sp.—*Moore 6527*

*Species that fall in more than one category, such as trees with both spiny trunks and milky latex, will be found in both sections of the key.

9. Fruits oblong, about 5/8" long .... *Chamaedorea Wendlandiana*
8. Staminate and pistillate flowers produced on the same scape.
9. Flowers produced in deep pits on the rachis. Fruits dark purple or black .... *Geonoma binervia*
9. Flowers not produced in pits. Fruits yellow, orange, or red .... *Synechanthus angustifolius*
5. Tall trees, usually 40–75 ft. in height.
6. Strands of the inflorescence octagonal in cross section. Flowers produced in deep pits. Fruits laterally compressed, almond-like, deep purple .... *Welfia Georgii*
6. Strands of the inflorescence not octagonal in cross section. Flowers not in pits. Fruits not laterally compressed; deep yellow and conspicuously beaked at the apex .... *Scheelea rostrata*
2. Trunks multiple.
3. Trunks and/or fronds spiny.
4. Fronds undivided .... *Bactris* sp.—*Allen 6765*
4. Fronds pinnatisect.
5. Spines pale tan in color .... *Bactris divisicupula*
5. Spines dark brown or black.
6. Spathes intensely spiny. Fruits deep purple .... *Bactris balanoidea*
6. Spathes merely woolly. Fruits red .... *Bactris Baileyana*
3. Trunks and fronds not spiny.
4. Plants tall, usually more than 40 ft. in height.
5. Crownshafts purplish. Fruits about 3/4" in diameter .... *Oenocarpus panamanus*
5. Crownshafts green. Fruits less than 1/2" in diameter .... *Euterpe panamensis*
4. Plants usually less than 40 ft. in height.
5. Staminate and pistillate flowers produced on separate scapes .... *Chamaedorea Woodsoniana*
5. Staminate and pistillate flowers produced on the same scape.
6. Spadix broomlike, the slender, rodlike basal part much longer than the fruiting strands .... *Synechanthus angustifolius*
6. Spadix not broomlike, the basal portion much shorter than the fruiting strands.
7. Flowers produced in deep pits on the rachis.
8. Canes about 2" in diameter. Fruits about 1/2" in diameter .... *Geonoma congesta*
8. Canes less than 1/2" in diameter. Fruits about 1/4" in diameter .... *Geonoma* sp.—*Allen 6750*
7. Flowers not produced in pits .... *Hyospathe Lehmannii*

KEY B

FLOWERS AND FRUITS PRODUCED DIRECTLY FROM THE BARK, AS IN CACAO

1. Leaves compound.
2. Leaves pinnate .... *Parmentiera macrophylla*
2. Leaves digitately compound .... *Herrania purpurea*

LOWER MONTANE RAIN FOREST

1. Leaves simple.
   2. Medium-sized trees, to about 60 ft. in height. Pods cacao-like, about 1 ft. long. Flowers red .... *Theobroma simiarum*
   2. Small trees, 12–15 ft. in height. Fruits nearly globose, with 8 thin, green, longitudinal ribs on the surface. Flowers white .... *Carpotroche platyptera*

KEY C

Trees with compound leaves

1. Leaves digitately compound.
   2. Leaves opposite.
      3. Leaflets 5–9.
         4. Flowers about ½″ long .... *Godmania aesculifolia*
         4. Flowers about 3″ long.
            5. Flowers pink .... *Tabebuia pentaphylla*
            5. Flowers yellow .... *Tabebuia chrysantha*
      3. Leaflets 3.
         4. Inflorescences axillary. Flowers blue .... *Vitex Cooperi*
         4. Inflorescences terminal. Flowers not blue.
            5. Flowers pale yellow. Sap with a vinegar-like odor .... *Caryocar costaricense*
            5. Flowers white, aging red. Sap without a vinegar-like odor ... *Billia colombiana*
   2. Leaves alternate.
      3. Flowers and fruits produced directly from the bark of the trunk and larger branches, as in cacao. Small trees, to about 12 ft. in height .... *Herrania purpurea*
      3. Flowers and fruits not produced directly from the bark. Large trees.
         4. Undersides of the leaves pale brown .... *Didymopanax Morototoni*
         4. Undersides of the leaves not brown.
            5. Trunks armed with stout, conical spines .... *Ceiba pentandra* (Young specimens)
            5. Trunks unarmed.
               6. Flowers bright yellow, showy. (Leaves usually opposite, but sometimes alternate on young trees) .... *Tabebuia chrysantha*
               6. Flowers not bright yellow.
                  7. Leaflets 3–5. Lower part of the flower contracted into a slender tube .... *Jacaratia dolichaula*
                  7. Leaflets 5–9. Lower part of the flower not contracted into a tube.
                     8. Flowers without petals .... *Sterculia mexicana*
                     8. Flowers with definite petals.
                        9. Trunks conspicuously swollen, with vertical green, yellow, and gray stripes .... *Bombax barrigon*
                        9. Trunks not as above.
                           10. Flowers about 6″ long. Trunks not strongly buttressed .... *Bombax sessile*
                           10. Flowers about 1½″ long. Trunks very strongly buttressed .... *Ceiba pentandra*
1. Leaves pinnately compound.

2. Leaves twice pinnate (the central axis of the frond with lateral pinnae).
 3. Fronds opposite.
  4. Inflorescences terminal .... *Jacaranda copaia*
  4. Inflorescences axillary .... *Jacaranda lasiogyne*
 3. Fronds alternate.
  4. Flowers bright yellow, showy. Trunks often prominently buttressed .... *Schizolobium parahybum*
  4. Flowers not bright yellow or showy. Trunks not prominently buttressed.
   5. Flowers mimosa-like (a puff ball). Fruit a more or less beanlike pod.
    6. Leaf petiole with a conspicuous, cup-shaped gland, which is more than ½″ in length .... *Pithecolobium macradenium*
    6. Leaf petiole not as above.
     7. Leaflets more than ½″ wide. Flowers cream or pale tan .... *Albizzia longepedata*
     7. Individual leaflets less than ½″ wide. Flowers white.
      8. Seed pods flat, straight, with parallel margins.
       9. Individual leaflets about ⅛″ long. Pods dark brown, nearly 2″ wide .... *Lysiloma guanacastense*
       9. Individual leaflets more than ⅛″ long. Pods about 1″ wide .... *Albizzia filicina*
      8. Seed pods terete, with undulant margins, or coiled, never with parallel margins.
       9. Fruit a long, red, terete, pendulous pod .... *Pithecolobium austrinum*
       9. Fruit a dark-brown, coiled, earlike pod .... *Enterolobium cyclocarpum*
   5. Flowers not mimosa-like. Fruit a short, 3-celled capsule .... *Dipterodendron costaricense*
2. Leaves once pinnate (the central axis with lateral leaflets).
 3. Pinnae opposite on the stems .... *Platymiscium pinnatum*
 3. Pinnae alternate on the stems.
  4. Leaflets 2 or 3.
   5. Small trees, less than 50 ft. in height.
    6. Leaflets 2.
     7. Flowers red, mimosa-like .... *Calliandra grandifolia*
     7. Flowers not red, not mimosa-like .... *Cupania macrophylla*
    6. Leaflets 3.
     7. Flowers red. Branches often somewhat thorny .... *Erythrina costaricensis*
     7. Flowers pale yellow or greenish yellow.
      8. Inflorescences more than 2″ long .... *Tetragastris panamensis*
      8. Inflorescences less than 2″ long .... *Trichilia montana*
   5. Very large trees, 90–140 ft. in height.
    6. Leaflets 2.
     7. Leaflets 2–3″ long. Wood purple .... *Peltogyne purpurea*
     7. Leaflets 4–6″ long. Wood not purple .... *Cynometra hemitomophylla*
    6. Leaflets more than 2.
     7. Bark usually red, thin, and peeling .... *Bursera simaruba*
     7. Bark not as above.
      8. Trunks armed with short, broad-based spines .... *Zanthoxylum procerum*

8. Trunks not armed with spines.
 9. Flowers and fruits produced directly from the trunk..*Parmentiera macrophylla*
 9. Flowers and fruits not produced from the trunk.
  10. Trees tall, often exceeding 80 ft. in height. Key based on flowers. (See also key based on fruits.)
   11. Flowers conspicuous from a distance.
    12. Flowers blood red..................*Phyllocarpus septentrionalis*
    12. Flowers not red.
     13. Flowers pink, or pinkish lavender.
      14. Leaflets 7–11, alternate on the rachis...*Dussia mexicana*
      14. Leaflets 12–14, opposite on the rachis...*Tachigalia versicolor*
     13. Flowers white, purple, or dark lavender.
      14. Flowers purple or dark lavender. Leaflets more than 4.
       15. Individual leaflets more than 1½″ wide...*Lonchocarpus latifolius*
       15. Individual leaflets less than 1½″ wide.
        16. Margin of the calyx entire or obscurely lobed.................................*Andira inermis*
        16. Margin of the calyx with long, pointed teeth...............................*Ormosia panamensis*
      14. Flowers white. Leaflets 4.............*Prioria copaifera*
   11. Flowers not conspicuous from a distance.
    12. Flowers olive green or pinkish lavender.
     13. Flowers olive green.........................*Dialium guianense*
     13. Flowers pinkish lavender.........*Dussia macrophyllata*
    12. Flowers yellow or white.
     13. Flowers pale yellow or greenish yellow.
      14. Petals united......................*Tetragastris panamensis*
      14. Petals not united.......................*Protium*, all species
     13. Flowers white or greenish white.
      14. Leaflets 16–35.
       15. Leaflets densely pubescent beneath (lens!). Capsules about 1⅛″ long............*Cedrela fissilis*
       15. Leaflets glabrous beneath. Capsules about 1½″ long..*Cedrela mexicana*
      14. Leaflets 4–15.
       15. Margins of leaflets serrate...*Astronium graveolens*
       15. Margins of leaflets not serrate.
        16. Largest mature leaflets more than 6″ long.
         17. Leaflets broadly rounded at the apex. Inflorescences subterminal...*Carapa Slateri*

17. Leaflets broadly acute at the apex. Inflorescent axillary or lateral.. *Guarea longipetiola*
16. Largest mature leaflets less than 4″ long.
17. Leaflets broadly rounded at the apex, paler on the lower surface.. *Simarouba glauca*
17. Leaflets acute or acuminate at the apex.
18. Inflorescences spicate or racemose, unbranched.
19. Leaf rachis with a small, disk-like gland at the base of each pair of leaflets.................... *Inga laurina*
19. Leaf rachis without disklike glands.. *Myroxylon balsamum* var. *Pereirae*
18. Inflorescences slender, branching panicles.............. *Sweetia panamensis*

10. Trees tall, often exceeding 80 ft. in height. (Key based on fruits.)
11. Fruits more or less woody.
12. Fruits very large, 5–6″ in diameter, with 8 large, fleshy seeds.. *Carapa Slateri*
12. Fruits less than 4″ in diameter.
13. Fruits 1-seeded.
14. Leaflets 4.. *Prioria copaifera*
14. Leaflets 7–11.
15. Seeds bright red.
16. Sap red........................ *Dussia macrophyllata*
16. Sap not red........................ *Dussia mexicana*
15. Seeds not red.............................. *Andira inermis*
13. Fruits with more than 1 seed.
14. Seeds winged.
15. Leaflets densely pubescent beneath (lens!). Capsules about 1⅛″ long............ *Cedrela fissilis*
15. Leaflets glabrous beneath. Capsules about 1½″ long.............................. *Cedrela mexicana*
14. Seeds not winged.
15. Fruits fig-shaped, reddish... *Guarea longipetiola*
15. Fruits not fig-shaped, usually yellow.
16. Sap red........................ *Dussia macrophyllata*
16. Sap not red........................ *Dussia mexicana*
11. Fruits not woody.
12. Fruits thin, flat, not opening at maturity.
13. Fruit a samara (a single seed at the end of a long, papery wing). Sap with a strong, agreeable balsamic scent...................... *Myroxylon balsamum* var. *Pereirae*

13. Fruit not a samara. Sap not as above.
    14. Leaflets subacute at the apex, with a small but distinct terminal cleft. Fruits usually less than 4" long .... *Sweetia panamensis*
    14. Leaflets never cleft at the apex.
        15. Winged seeds with a thin, brittle, dark-brown outer covering which separates at maturity .... *Tachigalia versicolor*
        15. Winged seeds without an outer covering.
            16. Seed wing conspicuously reticulate veined, with a single longitudinal nerve .... *Phyllocarpus septentrionalis*
            16. Seed pod not reticulate veined or with a longitudinal nerve ... *Lonchocarpus latifolius*

12. Fruits opening at maturity.
    13. Seeds bright red, with a conspicuous black blotch .... *Ormosia panamensis*
    13. Seeds not bright red.
        14. Fruit a flat, beanlike pod .... *Inga laurina*
        14. Fruit not a flat, beanlike pod.
            15. Margins of the leaflets serrate .... *Astronium graveolens*
            15. Margins of the leaflets not serrate.
                16. Leaflets broadly rounded at the apex, paler in color on the lower surface .... *Simarouba glauca*
                16. Leaflets acute or acuminate, not paler on the lower surface.
                    17. Fruits 1-seeded, broadest above the middle .... *Dialium guianense*
                    17. Fruits 3–4 seeded (sometimes 1-seeded by abortion), broadest below the middle.
                        18. Fruits usually somewhat oblique .... *Protium*, all species
                        18. Fruits symmetrical .... *Tetragastris panamensis*

10. Trees of medium size or small, seldom exceeding 50 ft. in height.
    11. Leaf rachis winged.
        12. Flowers pink. Bark intensely bitter .... *Quassia amara*
        12. Flowers not pink. Bark not bitter.
            13. Leaflets 5. Fruits orange, with black seeds. Small trees, to about 15 ft. .... *Swartzia picramnoides*
            13. Leaflets more than 5. Fruits dull red, with scarlet seeds. Larger trees, often to 50 or more feet .... *Guarea aligera*
    11. Leaf rachis not winged.
        12. Flowers purple or lavender.

13. Leaflets 5–9 in number, more than 1½" wide .... *Lonchocarpus latifolius*
13. Leaflets 9–11 in number, less than 1½" wide .... *Andira inermis*

12. Flowers white or pale yellow.
13. Flowers pale yellow.
14. Leaflets 8–20 .... *Guarea guara*
14. Leaflets less than 8.
15. Leaflets with a distinct glandular thickening at the base of the blade.
16. Petals united .... *Tetragastris panamensis*
16. Petals distinct .... *Protium*, all species
15. Leaflets without a glandular thickening at the base of the blade.
16. Inflorescences very short. Fruits not tuberculate .... *Trichilia montana*
16. Inflorescences about ½ the length of the leaves. Fruits tuberculate .... *Trichilia tuberculata*
13. Flowers white.
14. Flowers mimosa-like.
15. Flowers in elongate, slender spikes .... *Inga laurina*
15. Flowers in dense, globose heads .... *Pithecolobium glanduligerum*
14. Flowers not mimosa-like.
15. Flowers pea or beanlike. Fruit thin and flat .... *Sweetia panamensis*
15. Flowers not as above. Fruits not thin or flat.
16. Leaves very large, averaging about 3 ft. in length. Slender trees, with extremely hard wood .... *Talisia nervosa*
16. Leaves not as above.
17. Leaflets broadly rounded at the apex, paler in color on the lower surface .... *Simarouba glauca*
17. Leaflets acute or acuminate at the apex, not paler in color on the lower surface.
18. Inflorescences terminal. Fruits broadest near the apex.
19. Leaflets 6–10 .... *Cupania guatemalensis*
19. Leaflets 2–5.
20. Leaflets 2 or 4. Capsules glabrous .... *Cupania macrophylla*
20. Leaflets 5. Capsules brown, velvety .... *Cupania largifolia*
18. Inflorescences axillary.

19. Inflorescences slender, pendulous spikes. Fruits scarlet. Bark bitter.................. *Picramnia latifolia*
19. Inflorescences and bark not as above.
    20. Terminal leaflets 2.................. *Guarea guara*
    20. Terminal leaflet 1.
        21. Fruits tuberculate.................. *Trichilia tuberculata*
        21. Fruits not tuberculate.................. *Trichilia acutanthera*

KEY D

Trees with simple, opposite leaves

1. Sap colored or milky white.
    2. Leaves brownish on the lower surface. Sap reddish orange, drying blood red.
        3. Petals pale orange.................. *Vismia guianensis*
        3. Petals white.................. *Vismia ferruginea*
    2. Leaves not brownish on the lower surface. Latex yellow or milky white.
        3. Latex yellow.
            4. Flowers blood red. Lower trunk often with stilt roots...... *Symphonia globulifera*
            4. Flowers white. Lower trunk without stilt roots.
                5. Leaves broadly oblong-elliptic, the tips shortly and abruptly acute. Fruits very rough.................. *Rheedia madruno*
                5. Leaves narrowly lanceolate, gradually tapering to an acute or acuminate tip. Fruits smooth.................. *Rheedia edulis*
        3. Latex milky white.
            4. Trunk armed with low, broad-based spines.................. *Lacmellea panamensis*
            4. Trunk not armed with spines.
                5. Flowers reddish tan, in pendulous cymes which about equal the leaves in length.................. *Tabernaemontana longipes*
                5. Flowers not reddish tan, not in pendulous cymes.
                    6. Flowers white. Large trees with leathery leaves.................. *Calophyllum braziliense* var. *Rekoi*
                    6. Flowers not white. Small trees, the leaves not leathery.
                        7. Flowers bright yellow, with a conspicuous whitish calyx.................. *Stemmadenia grandiflora*
                        7. Flowers yellowish pink, without a conspicuous whitish calyx.................. *Stemmadenia nervosa*
1. Sap not milky white, or colored.
    2. Trees with conspicuous stilt roots.................. *Bravaisia integerrima*
    2. Trees without stilt roots.
        3. Leaves about 2 ft. long, the blade deeply cut into pinnate lobes.................. *Pentagonia gymnopoda*
        3. Leaves not as above.

4. Flowers small, but subtended by a very showy red or white expanded calyx lobe.
   5. Calyx lobes white. Large trees, to about 75 ft. .... *Calycophyllum candidissimum*
   5. Calyx lobes blood red. Inflorescences superficially resemble those of the poinsettia .... *Warscewiczia coccinea*
4. Flowers without a showy expanded calyx lobe.
   5. Leaf blade with 3 or more longitudinal nerves.
      6. Inflorescences axillary.
         7. Leaves broadly ovate, the tips rounded or very abruptly acuminate .... *Bellucia costaricensis*
         7. Leaves lanceolate, acuminate .... *Henriettella tuberculosa*
      6. Inflorescences terminal.
         7. Flowers blue, the lower part contracted into a tube ... *Faramea suerrensis*
         7. Flowers not as above .... *Miconia*, all species
   5. Leaf blade with a single longitudinal nerve.
      6. Bark smooth, white, conspicuous. Small trees, to about 40 ft. in height .... *Eugenia palmarum*
      6. Bark not white.
         7. Leaves very large, about 30″ long and 12–14″ wide. Small trees, to about 20 ft. in height .... *Condaminea corymbosa*
         7. Leaves not as above.
            8. Flowers blue or pink.
               9. Flowers blue. Shrubs or small trees .... *Faramea sessifolia*
               9. Flowers pink. Trees, 40–75 ft. .... *Macrocnemum glabrescens*
            8. Flowers red, orange, yellow, or white.
               9. Flowers dark red, orange, or yellow.
                  10. Small trees, less than 35 ft. in height.
                     11. Leaf veins and petioles red .... *Hamelia magnifolia*
                     11. Leaf veins and petioles not red.
                        12. Leaves leathery, elliptic-oblong. Heartwood dark red .... *Tovomitopsis grandifolia*
                        12. Leaves lanceolate, not leathery. Heartwood not red .... *Aegiphila martinicensis*
                  10. Larger trees, mostly 40–100 ft. in height.
                     11. Flowers orange or dark yellow, conspicuous from a distance.
                        12. Leaves obtuse or emarginate at the apex, in whorls of 3 at each node .... *Vochysia hondurensis*
                        12. Leaves acute at the apex, in pairs at each node.
                           13. Leaves densely brown-tomentose on the lower surface .... *Vochysia ferruginea*
                           13. Leaves glabrous on the lower surface .... *Vochysia Allenii*
                     11. Flowers not conspicuous from a distance.
                        12. Flowers large, woody, bell-shaped, with many prominent, dark-red stamens .... *Lafoensia punicifolia*
                        12. Flowers not woody or bell-shaped.

13. Leaves usually more than 6″ long. Lower part of the flower contracted into a very long, slender tube..*Tocoyena Pittieri*
13. Leaves usually less than 6″ long. Flowers not as above..*Byrsonima densa*

9. Flowers white.
10. Inflorescences axillary.
11. Leaves less than 3″ long.
12. Leaves with many tiny translucent dots when seen against the light (lens!). Flowers in short racemes..*Eugenia,* all species
12. Leaves without translucent dots. Flowers in nearly sessile clusters..*Cassipourea podantha*
11. Leaves more than 3″ long.
12. Inflorescence a simple, pendulous raceme.
13. Leaf blade with more than 30 pairs of secondary veins...*Marila pluricostata*
13. Leaf blade with about 15 pairs of secondary veins.
14. Flowers in very dense, compact clusters. Wood orange. Trees more than 50 ft. in height..*Chimarrhis latifolia*
14. Flowers in slender, open clusters. Wood not orange. Trees seldom more than 25 ft. in height...*Aegiphila costaricensis*
10. Inflorescences terminal or pseudoterminal.
11. Small trees, usually less than 30 ft.
12. Flowers more than 1″ long.
13. Flowers in dense terminal cymes, the slender tubular flowers prominently exserted from a compact cluster of overlapping bracts..................*Psychotria chiapensis*
13. Flowers solitary or in sessile fascicles, never subtended by a conspicuous cluster of bracts..*Alibertia edulis*
12. Flowers less than 1″ long.
13. Leaves less than 3″ long. Racemes about 1″ long..*Eugenia,* all species
13. Leaves more than 3″ long. Racemes or panicles much more than 1″ in length.
14. Leaf axils with small, but distinct stipules. Leaves caudate-acuminate..................*Rondeletia urophylla*
14. Leaf axils without stipules. Leaves not caudate-acuminate.
15. Floral pedicels very long and slender, usually exceeding 1″ in length...*Aegiphila costaricensis*
15. Floral pedicels ¼″ long or less.
16. Flowers about ¼″ in diameter..*Tovomitopsis costaricana*

16. Flowers more than ¼″ in diameter........ *Tovomitopsis glauca*

11. Larger trees, 40–90 ft. in height.
12. Flowers with conspicuous, very slender, petaloid bracteoles........ *Linociera panamensis*
12. Flowers not as above.
13. Lower part of the flower contracted into an elongate tube. Flowers in terminal panicles.
14. Corolla tube stout, about 1½ times as long as the lobes........ *Ladenbergia Brenesii*
14. Corolla tube slender, about 3–4 times as long as the lobes........ *Ladenbergia chariantha*
13. Floral tube very short. Flowers in broad terminal cymes........ *Chione costaricensis*

KEY E

Trees with simple, alternate leaves

1. Leaves palmately lobed.
2. Leaves peltate, deeply lobed, with 7–13 divisions.
3. Trunks and branches black, not prominently ringed. Fruiting clusters about 4″ long, yellow........ *Cecropia Sandersoniana*
3. Trunks and branches green or white, conspicuously ringed.
4. Undersides of leaves green. Fruiting clusters 8–16″ long... *Cecropia obtusifolia*
4. Undersides of leaves white. Fruiting clusters about 3–4″ long........ *Cecropia peltata*
2. Leaves not peltate.
3. Trees with black sap. Flowers inconspicuous. Fruits juicy........ *Pourouma aspera*
3. Trees without black sap. Flowers large, white, showy. Fruits filled with fluffy, kapok-like fiber........ *Ochroma lagopus*
1. Leaves not palmately lobed.
2. Trees with colored or milky latex.
3. Latex milky white.
4. Flowers with definite petals.
5. Leaves usually more than 8″ long.
6. Leaves conspicuously clustered at the ends of the branches.
7. Inflorescences terminal. Flowers more than 1″ in diameter........ *Plumeria rubra* forma *acutifolia*
7. Inflorescences axillary. Flowers less than 1″ in diameter.
8. Fruits globose, densely covered with thick, brownish-orange, woolly processes........ *Pouteria neglecta*
8. Fruits ellipsoid, glabrous, pale gray........ *Calocarpum borucanum*
6. Leaves not conspicuously clustered at the ends of the branches........ *Pouteria triplarifolia*
5. Leaves usually less than 6″ long.
6. Leaves conspicuously reddish brown on the lower surface........ *Chrysophyllum mexicanum*

6. Leaves not reddish brown on the lower surface.
   7. Scapes up to 1″ or more in length, the flowers in dense heads at the ends of the branches........ *Pouteria heterodoxa*
   7. Flowers in sessile, or nearly sessile, fascicles.
      8. Leaf petioles about 1″ long. Blades rounded or subacute at the apex........ *Pouteria subrotata*
      8. Leaf petioles about ½″ long. Blades acute or acuminate at the apex.
         9. Flowers completely sessile........ *Pouteria chiricana*
         9. Flowers on slender peduncles about ⅛″ long........ *Chrysophyllum panamense*

4. Flowers without definite petals.
   5. Leaves large, rough to the touch, conspicuously pendulus........ *Castilla fallax*
   5. Leaves not large, pendulous or rough to the touch.
      6. Fruits multiple, in sessile clusters, red at maturity.
         7. Fruits glabrous, leaves shortly acute or acuminate........ *Perebea trophophylla*
         7. Fruits velvety, leaves abruptly caudate-acuminate........ *Perebea molliflora*
      6. Fruits solitary or in pairs, not red.
         7. Leaves with serrate or crenate margins.
            8. Leaves broadly elliptic, acute or shortly acuminate. Secondary nerves 10–12 pairs. Fruits like a small Osage orange........ *Batocarpus costaricense*
            8. Leaves narrowly lanceolate, caudate-acuminate. Secondary nerves 16–18 pairs........ *Olmedia falcifolia*
         7. Leaf margins not serrate or crenate.
            8. Male and female flowers produced on separate receptacles........ *Pseudolmedia spuria*
            8. Male and female flowers produced on the same receptacle.
               9. Flowers borne on the inside of a hollow, often globose receptacle. Fruits more or less figlike, with many small seeds.
                  10. Leaves brownish, velvety........ *Ficus Bullenei*
                  10. Leaves smooth, leathery.
                      11. Leaves usually more than 10″ long........ *Ficus Werckleana*
                      11. Leaves usually less than 10″ long.
                          12. Leaf blades broadest above the middle........ *Ficus costaricana*
                          12. Leaf blades broadest at about the middle........ *Ficus lapathifolia*
               9. Flowers borne on the outer surface of a solid, spherical, fleshy receptacle. Fruits with 1 or 2 relatively large seeds.
                  10. Leaves broadly oblong-elliptic, with abruptly acute or acuminate tips and rounded bases. Fruit about 1″ in diameter........ *Brosimum utile*
                  10. Leaves narrowly lanceolate, with gradually acuminate tips and more or less wedge-shaped bases. Fruit about ½″ in diameter........ *Brosimum sapiifolium*

3. Latex or sap red, orange, or yellow.
   4. Sap red.

5. Leaf blades with a pair of lateral longitudinal nerves arising from the base of the principal vein. Fruit a very small, 3-celled capsule.
  6. Leaves usually densely stellate-tomentose, particularly on the lower surface, the blades typically not broadly cordate at the base..*Croton xalapensis*
  6. Leaves usually sparsely stellate-tomentose on the lower surface, the blades typically broadly cordate at the base..............................*Croton panamensis*
5. Leaf blades without lateral longitudinal nerves at the base. Fruits nutmeg-like, the dark-brown seed covered by a red, fleshy, macelike aril.
  6. Aril surrounding the seed undivided, or divided only at the tip. Tertiary nerves of the leaves conspicuous..................................*Compsoneura Sprucei*
  6. Aril surrounding the seed divided for more than ½ its length. Tertiary nerves of the leaves not conspicuous.
    7. Lower surface of the leaves brown or tan, contrasting strongly with the upper surface.
      8. Leaf blades with 12–15 pairs of lateral nerves, the upper part tapering gradually to a long point..*Virola sebifera*
      8. Leaf blades with 20–30 pairs of lateral nerves, the apex abruptly acute or acuminate...*Virola Koschnyi*
    7. Lower surface of the leaves not brown or tan.
      8. Leaf blades with 14–21 pairs of lateral nerves; very long and narrow, tapering gradually to an acute tip.
        9. Seeds about 1″ long...*Virola guatemalensis*
        9. Seeds about ¾″ long..*Virola surinamensis*
      8. Leaf blades with 30–35 pairs of lateral nerves, the apex abruptly short-acuminate...*Virola nobilis*

4. Sap or latex orange or yellow.
  5. Fruit large, woody, laterally compressed, much broader above the middle, filled with many circular, papery seeds..............*Aspidosperma megalocarpon*
  5. Fruit rough, globose, fleshy, resembling a small Osage orange..*Batocarpus costaricense*
  5. Fruit not as above...*Brosimum terrabanum*

2. Sap or latex not colored or milky.
  3. Leaves very large, conspicuously clustered at the ends of the branches.
    4. Leaves very long, strap-shaped, with parallel margins......*Dracaena americana*
    4. Leaves broadly ovate, the margins not parallel.
      5. Relatively large trees, to about 75 ft. Flowers bright yellow, in erect, terminal panicles..*Cespedesia macrophylla*
      5. Trees usually less than 30 ft.
        6. Flowers sessile or nearly sessile, in terminal clusters.
          7. Leaves usually more than 18″ long...........................*Gustavia angustifolia*
          7. Leaves usually less than 9″ long...*Genipa Caruto*
  3. Leaves not conspicuously clustered at the ends of the branches.
    4. Flowers and fruits subtended by a showy red, circular calyx about ⅝″ in diameter.
      5. Fruits black at maturity..*Heisteria longipes*
      5. Fruits white at maturity...*Heisteria concinna*

4. Flowers and fruits not subtended by a showy red calyx.
    5. Flowers and fruits minute, in conspicuous, pendulous, threadlike strands up to 15″ or more in length..*Myriocarpa longipes*
    5. Flowers and fruits never in pendulous, threadlike strands.
        6. Trunks or branches thorny or spiny.
            7. Trunks with dense clusters of long, branching, needle-like spines..*Xylosma excelsum*
            7. Spines not long and needle-like, never in dense branching clusters.
                8. Large trees, the trunks entirely covered with short, sharp thorns. Flowers red, but inconspicuous from a distance............*Hura crepitans*
                8. Small trees, with smooth trunks and spiny branches. Flowers white..*Ximenia americana*
        6. Trunks or branches not thorny or spiny.
            7. Large trees, with very conspicuously developed buttresses.
                8. Leaf blades with 3 longitudinal nerves.
                    9. Lower surface of the leaves brownish. Flowers white, followed by angular capsules which are star-shaped in cross section..*Luehea Seemannii*
                    9. Lower surface of the leaves not brownish. Flowers pale yellow, fruits of 3 thin, oblong, winged seeds, fused along the central axis..*Goethalsia meiantha*
                8. Leaf blades with a single longitudinal nerve.
                    9. Bark thin, often conspicuously peeling, pale tan or nearly white, guava-like in superficial appearance........................*Terminalia lucida*
                    9. Bark not as above.
                        10. Fruit a 5-valved capsule, often woody.
                            11. Fruits densely covered with purple, needle-like spines..*Sloanea picapica*
                            11. Fruits not spiny.
                                12. Fruits very large, about 9″ long. Trees to 150 ft., with tremendous plank buttresses.........*Huberodendron Allenii*
                                12. Fruits less than 3″ long.
                                    13. Leaves glossy, somewhat leathery, acute at the apex, usually less than 6″ long. Valves of the fruit very thick..*Sloanea laurifolia*
                                    13. Leaves not glossy or leathery, acuminate at the apex, usually more than 6″ long. Valves of the fruit thin..*Mortoniodendron anisophyllum*
                        10. Fruit not a 5-valved capsule.
                            11. Fruit a cylindric, woody, boxlike pod, to about 4″ long. Flowers pink, conspicuous from a distance..*Couratari panamensis*
                            11. Fruits papery, winged, about ¾″ long. Flowers greenish tan, inconspicuous........................*Terminalia amazonia*
            7. Trees of varying size, without conspicuously developed buttresses.
                8. Leaves with palmately 5–7 veined blades. Wood very light..*Ochroma lagopus*
                8. Leaves not palmately 5–7 veined.

9. Flowers and fruits produced directly from the trunk and larger branches, as in cacao.
  10. Flowers fleshy, yellowish .... *Guatteria amplifolia*
  10. Flowers not fleshy; white or red.
    11. Small trees rarely exceeding 15 ft. in height. Flowers white. Fruits nearly globose, with 8 thin, longitudinal vanes on the surface .... *Carpotroche platyptera*
    11. Trees 40–60 ft. in height. Flowers red. Fruit a slender, velvety pod about 1 ft. long .... *Theobroma simiarum*
9. Flowers and fruits not produced directly from the trunk and larger branches.
  10. Fruits bristly or spiny.
    11. Fruits globose, bright yellow. Seeds red .... *Mayna echinata*
    11. Fruits sea urchin-like. Seeds not red.
      12. Spines short and hard .... *Apeiba aspera*
      12. Bristles longer, pliant .... *Apeiba Tibourbou*
  10. Fruits not bristly or spiny.
    11. Fruit a large, strongly asymmetrical, woody pod, filled with brown, winged seeds .... *Basiloxylon excelsum*
    11. Fruit not as above.
      12. Inflorescences terminal.
        13. Flowers or fruits red, pink, lavender, or purple.
          14. Trees more than 40 ft. in height.
            15. Mature fruits enveloped in a showy, dark-pink, 3- or 5-lobed calyx.
              16. Calyx lobes 3 .... *Triplaris melaenodendron*
              16. Calyx lobes 5 .... *Pentaplaris Doroteae*
            15. Fruiting calyx not as above.
              16. Flowers dark pink, conspicuous .... *Belotia macrantha*
              16. Flowers pale lavender .... *Belotia reticulata*
          14. Small trees, less than 40 ft. in height.
            15. Fruits dark pink, with a marginal fringe of slender bristles. Fruits showy from a distance .... *Heliocarpus appendiculatus*
            15. Fruits not as above, not conspicuous.
              16. Flowers pink.
                17. Flowers hibiscus-like .... *Malvaviscus arboreus*
                17. Flowers not hibiscus-like.
                  18. Trees 18 ft. in height or less.
                    19. Inflorescences very short and compact, subtended by numerous broad, overlapping bracts .... *Ardisia Dodgei*
                    19. Inflorescences relatively long racemes, the flowers not subtended by bracts.

20. Leaves leathery .... *Ardisia revoluta*
20. Leaves not leathery .... *Belotia macrantha*
16. Flowers purple or lavender.
17. Flowers purple, in simple racemes or narrow panicles.
18. Leaf blades acute at the base .... *Hirtella triandra*
18. Leaf blades obtuse at the base .... *Hirtella americana*
17. Flowers very pale lavender, in spreading panicles .... *Belotia reticulata*
13. Flowers white, tan, or yellow.
14. Flowers white.
15. Leaves with 3–5 longitudinal nerves.
16. Undersides of leaves brownish .... *Luehea Seemannii*
16. Undersides of leaves not brownish.
17. Leaf blades with 3 longitudinal nerves, the margins serrate or dentate .... *Hasseltia floribunda*
17. Leaf blades with 5 longitudinal nerves, the margins not serrate or dentate .... *Hasseltia quinquenervia*
15. Leaves with a single longitudinal nerve.
16. Forks of the young branches with hollow, gall-like swellings, inhabited by ants.
17. Flowers about ¾″ in diameter .... *Cordia gerascanthus*
17. Flowers less than ¾″ in diameter .... *Cordia alliodora*
16. Forks of the young branches without hollow, gall-like swellings.
17. Leaf blades less than 2½″ long, the margins slightly serrate .... *Sacoglottis excelsa*
17. Leaf blades more than 2½″ long, the margins entire.
18. Leaves clustered at the ends of the branches. Rachis of the inflorescence conspicuously whitish.
19. Large trees, to 120 ft. Leaves with very short petioles. Fruits kidney-shaped .... *Anacardium excelsum*
19. Small trees, usually less than 40 ft. Leaves with very long slender petioles .... *Oreopanax capitatus*
18. Leaves not clustered at the ends of the branches. Rachis of the inflorescence not conspicuously whitish.
19. Leaves acute or acuminate at the apex.
20. Flowers in unbranched spikes .... *Croton glabellus*
20. Flowers in branching panicles.
21. Leaves usually more than 6″ long .... *Ardisia Standleyana*
21. Leaves usually less than 6″ long .... *Ocotea veraguensis*
19. Leaves rounded or retuse at the apex. Freshly cut wood with an odor like crushed sugar cane .... *Vantanea Barbourii*
14. Flowers tan, yellow, or greenish yellow.

15. Staminate and pistillate flowers produced on separate inflorescences.......*Alchornea glandulosa* var. *Pittieri*
15. Flowers perfect, with both staminate and pistillate elements present.
 16. Flowers bright yellow, orchid-like.......*Swartzia simplex*
 16. Flowers not bright yellow or orchid-like.
  17. Leaf blades with 3 longitudinal nerves.......*Goethalsia meiantha*
  17. Leaf blades with a single longitudinal nerve.
   18. Leaves of the flowering branches usually less than 3″ long. Flowers greenish tan, in clusters of slender spikes. Fruits winged, papery.......*Terminalia amazonia*
   18. Leaves of the flowering branches more than 3″ long. Flowers not in clusters of slender spikes. Fruits not papery or winged.
    19. Flowers fleshy, more than 1″ in diameter. Fruits woody, with a boxlike lid...*Eschweilera calyculata*
    19. Flowers and fruits not as above.
     20. Leaf blades rounded at the base and apex, paler on the lower surface.......*Licania arborea*
     20. Leaf blades acute or acuminate, not paler on the lower surface.
      21. Flowers pale pinkish tan.......*Ardisia Dunlapiana*
      21. Flowers not pale pinkish tan.
       22. Small trees, usually less than 18 ft. in height. Inflorescences about 1″ long.......*Mortoniodendron guatemalense*
       22. Trees more than 20 ft. in height. Inflorescences more than 1″ long.
        23. New growth and lower leaf surface conspicuously ferruginous-tomentose.......*Neea elegans*
        23. New growth and lower leaf surface glabrous.
         24. Inflorescences elongate, pendulous cymes, longer than the leaves.......*Neea Popenoei*
         24. Inflorescences not as above.
          25. Leaf blades more than 7″ long.......*Ocotea Ira*
          25. Leaf blades less than 7″ long.
           26. Fruits with a subtending cup, which may be greater or less than the diameter of the fruit.
            27. Cup with a double rim.

Fruit acorn-like in superficial appearance.

28. Outer rim of cup flaring and crisped... *Licaria Cufodontisii*

28. Outer rim of cup not as above... *Licaria Cervantesii*

27. Cup with a single rim, less than the fruit in diameter... *Ocotea Williamsii*

26. Fruits without a subtending cup... *Licania operculipetala*

12. Inflorescences axillary.

13. Small trees, less than 30 ft. in height.

14. Leaves with 3 or 5 longitudinal nerves.

15. Leaf blades broadly ovate, with entire margins and symmetrical bases. Petioles more than 2" long... *Hampea Allenii*

15. Leaf blades narrowly lanceolate, usually with serrate margins and strongly asymmetrical bases. Petioles less than ½" long.

16. Flowers white, about ¾" in diameter... *Muntingia calabura*

16. Flowers green, minute, less than ⅛" in diameter... *Trema micrantha*

14. Leaves with a single longitudinal nerve.

15. Leaves with serrate margins.

16. Leaves very large, often as much as 1 ft. in length, 2-ranked and pendulous on the branches. Flowers dark red, clusters pendulous... *Tetrathylacium costaricense*

16. Leaves less than 6" long. Flowers white... *Casearia arborea*

15. Leaf margins not serrate.

16. Leaves more than 1 ft. long.

17. Bases of the leaf petioles entirely enveloping the stems... *Coccoloba Standleyana*

17. Bases of the leaf petioles not enveloping the stems.

18. Very slender trees, usually less than 15 ft. in height. Leaves conspicuously caudate-acuminate... *Cordia protracta*

18. Trees more than 15 ft. in height.

19. Leaf petioles not more than 1" long.

20. Leaves broadest at the middle. Veins of the lower surface brownish and finely velvety... *Ocotea mollifolia*

20. Leaves broadest above the middle. Veins not brownish or velvety... *Meliosma anisophylla*

19. Leaf petioles more than 2" long... *Meliosma Allenii*

16. Leaves less than 1 ft. long.

17. Flowers solitary, or in very short, few-flowered clusters less than 2″ long.
    18. Leaves less than 3″ long. Small trees, usually less than 15 ft. in height........................*Erythroxylon lucidum*
    18. Leaves more than 3″ long.
        19. Flowers minute, less than ¼″ in diameter.
            20. Inflorescences about ½″ long. Flowers white...*Trichilia Skutchii*
            20. Inflorescences more than 1″ long. Flowers greenish tan..........................*Coussapoa parviceps*
        19. Flowers ½″ in diameter or larger.
            20. Leaves more than 3″ wide...*Guatteria amplifolia*
            20. Leaves less than 3″ wide.
                21. Branchlets and veins of the lower leaf surface densely velvety.......*Guatteria aeruginosa*
                21. Branchlets and veins not as above.
                    22. Leaves narrowly oblong, with abruptly acuminate tips. Fruits not in multiple clusters..............................*Annona Pittieri*
                    22. Leaves elliptic-lanceolate, acute.
                        23. Leaves leathery. Flowers fleshy. Multiple yellow fruits, in long pediceled clusters.......................*Guatteria lucens*
                        23. Leaves not leathery. Flowers not fleshy. Fruits Annona-like...*Duguetia panamensis*
17. Flowers in well-developed panicles or racemes.
    18. Flowers purple.
        19. Leaf blades acute at the base.......*Hirtella triandra*
        19. Leaf blades obtuse at the base...*Hirtella americana*
    18. Flowers white, yellow, or greenish yellow.
        19. Flowers white.
            20. Leaf petioles nearly as long the the blade..*Roupala complicata*
            20. Leaf petiole much shorter than the blade.
                21. Flowers minute, less than ⅛″ in diameter. Fruit a 3-celled capsule.......*Croton glabellus*
                21. Flowers nearly ½″ in diameter when fresh. Fruit a 1-seeded drupe.........*Persea pallida*
        19. Flowers yellow or greenish yellow.
            20. Leaf petioles nearly as long as the blades..*Roupala complicata*
            20. Leaf petioles much shorter than the blades.
                21. Leaves 9–12″ long.......*Meliosma longipetiola*
                21. Leaves less than 9″ long.
                    22. Flowers bright yellow, orchid-like...*Swartzia simplex*
                    22. Flowers not as above.......*Persea Skutchii*

13. Trees more than 40 ft. in height.
  14. Leaves with 3 or 5 longitudinal veins.
    15. Leaf blades broadly ovate, with entire margins. Petioles more than 2″ long .... *Hampea Allenii*
    15. Leaf blades narrowly lanceolate, usually with serrate margins. Petioles less than ½″ long .... *Trema micrantha*
  14. Leaves with a single longitudinal vein.
    15. Leaf petiole nearly as long as the blade .... *Roupala complicata*
    15. Leaf petiole much shorter than the blade.
      16. Trunks deeply and conspicuously furrowed.
        17. Leaves more than 12″ long. Flowers orange .... *Meliosma Allenii*
        17. Leaves less than 12″ long. Flowers essentially white.
          18. Leaves about 4 times as long as broad. Fruits about 3″ in diameter .... *Diospyros ebenaster*
          18. Leaves about twice as long as broad .... *Maytenus pallidifolius*
      16. Trunks not deeply furrowed.
        17. Flowers bright yellow, orchid-like .... *Swartzia simplex*
        17. Flowers not bright yellow or orchid-like.
          18. Basal part of the leaf petiole enveloping the stem .... *Coccoloba roseiflora*
          18. Basal part of the leaf petiole not enveloping the stem.
            19. Flowers solitary, or in very short, few-flowered clusters less than 2″ long.
              20. Leaves less than 1″ wide, the lower surface silky .... *Xylopia sericophylla*
              20. Leaves more than 1″ wide.
                21. Individual flowers either completely staminate or completely pistillate.
                  22. Pistillate flowers sessile, solitary or in pairs .... *Clarisia mexicana*
                  22. Pistillate flowers in slender racemes or panicles.
                    23. Leaves acuminate at the apex. Flowers in unbranched racemes .... *Croton glabellus*
                    23. Leaves acute at the apex. Flowers in branching panicles .... *Coussapoa parviceps*
                21. Individual flowers with both staminate and pistillate elements .... *Guatteria lucens*
            19. Flowers never solitary or in few-flowered fascicles. Inflorescences more than 2″ long.
              20. Flowers in unbranched spikes or racemes.
                21. Floral pedicel with a pair of conspicuous linear bracts .... *Persea pallida*
                21. Floral pedicel without conspicuous bracts.

22. Leaf blades obtuse or subacute at the apex, paler in color on the lower surface.............................*Dialyanthera otoba*
22. Leaf blades acute or acuminate at the apex, not paler in color on the lower surface.
    23. Racemes solitary in the leaf axils...... ....*Minquartia guianensis*
    23. Racemes in fascicles of 2–4 together in the leaf axils.........*Croton glabellus*

20. Flowers in branching cymes or panicles.
  21. Flowers white.
    22. Flowers more than 1/4″ in diameter, the pedicels with a pair of conspicuous linear bracts.......................*Persea pallida*
    22. Flowers less than 1/4″ in diameter, the pedicels without conspicuous bracts.
      23. Axils of the veins of the lower leaf surface with distinct tufts of hairs (lens!)..................*Nectandra perdubia*
      23. Axils of the veins not as above......... ....*Nectandra latifolia*
  21. Flowers yellow.
    22. Leaf petiole elongate, about 1/2 the length of the blade. Sap black................ ....*Hieronyma tectissima*
    22. Leaf petioles much less than 1/2 the length of the blade.
      23. Leaf blades more than 4 times as long as broad...*Ocotea pergamentacea*
      23. Leaf blades not more than 3 times as long as broad.
        24. Individual flowers either completely staminate or completely pistillate.
          25. Branchlets and lower leaf surface conspicuously ferruginous-tomentose.......*Neea elegans*
          25. Branchlets and lower leaf surfaces essentially glabrous......... ....*Neea Popenoei*
        24. Individual flowers with both staminate and pistillate elements present.
          25. Calyx segments unequal, the outer series shorter. Fruits more than 3″ in diameter......... ....*Persea americana*

25. Outer and inner series of calyx segments of about equal length. Fruits about ¾″ in diameter......*Ocotea Williamsii*

KEY F

Trees with white, milky latex

1. Leaves opposite.
  2. Large trees with leathery leaves. Leaf tips never acuminate...*Calophyllum braziliense* var. *Rekoi*
  2. Small trees with thin leaves. Leaf tips acuminate.
    3. Leaves more than 7″ long. Inflorescences pendulous...*Tabernaemontana longipes*
    3. Leaves less than 7″ long. Inflorescences not as above.
      4. Flowers bright yellow..................*Stemmadenia grandiflora*
      4. Flowers yellowish pink..................*Stemmadenia nervosa*
1. Leaves alternate.
  2. Leaves conspicuously clustered at the ends of the branches.
    3. Leaves very leathery, broadest at about the middle. Flowers white, more than 1″ in diameter, in terminal panicles or cymes......*Plumeria rubra* forma *acutifolia*
    3. Leaves not leathery, broadest near the apex. Flowers less than 1″ in diameter, nearly sessile in the leaf axils or on the defoliate nodes below the leaves.
      4. Fruits nearly globose, woolly..................*Pouteria neglecta*
      4. Fruits ellipsoidal, smooth..................*Calocarpum borucanum*
  2. Leaves not conspicuously clustered at the ends of the branches.
    3. Lower leaf surface conspicuously reddish brown.
      4. Fruits hollow..................*Ficus Bullenii*
      4. Fruits not hollow..................*Chrysophyllum mexicanum*
    3. Lower surface of the leaves not reddish brown.
      4. Flowers with definite petals.
        5. Leaves more than 8″ long..................*Pouteria triplarifolia*
        5. Leaves less than 6″ long.
          6. Flowers in small, dense heads on slender, often branching scapes, which are usually 1″ or more in length..................*Pouteria heterodoxa*
          6. Flowers in sessile or nearly sessile fascicles.
            7. Leaf petioles about 1″ long, the blades rounded or subacute at the apex..................*Pouteria subrotata*
            7. Leaf petioles about ½″ long, the blades acute or acuminate at the apex.
              8. Flowers completely sessile..................*Pouteria chiricana*
              8. Flowers on slender peduncles about ⅛″ long.......................................*Chrysophyllum panamense*
      4. Flowers minute, without definite petals.
        5. Leaves very large, conspicuously 2-ranked and pendulous on the branches, rough to the touch..................*Castilla fallax*
        5. Leaves not as above.
          6. Fruits multiple, in sessile clusters, bright red at maturity.

7. Fruits glabrous. Leaves shortly acute or acuminate........ *Perebea trophophylla*
7. Fruits velvety. Leaves abruptly caudate-acuminate... *Perebea molliflora*
6. Fruits solitary or in pairs, not red at maturity.
7. Leaves with serrate or crenate margins.
8. Leaves broadly elliptic, acute or shortly acuminate, with 10–12 pairs of secondary nerves. Fruit like a small Osage orange........ *Batocarpus costaricense*
8. Leaves narrowly lanceolate, caudate-acuminate. Secondary nerves 16–18 pairs. Fruit a small, 1-seeded drupe........ *Olmedia falcifolia*
7. Leaf margins not serrate or crenate.
8. Staminate and pistillate flowers produced on separate receptacles........ *Pseudolmedia spuria*
8. Staminate and pistillate flowers produced on the same receptacle.
9. Fruits hollow........ *Ficus*, all species
9. Fruits not hollow........ *Brosimum*, all species

KEY G

Trees with red or pink sap

1. Leaves pinnate........ *Dussia macrophyllata*
1. Leaves simple.
2. Leaves opposite.
3. Petals white........ *Vismia ferruginea*
3. Petals pale orange........ *Vismia guianensis*
2. Leaves alternate.
3. Inflorescences terminal. Fruit a small, 3-celled capsule........ *Croton*
3. Inflorescences axillary. Fruit nutmeg-like, the single seed covered with a fleshy, red, macelike aril.
4. Aril undivided, or divided only at the tip. Tertiary nerves of the leaves conspicuous........ *Compsoneura Sprucei*
4. Aril divided for more than half its length. Teritary nerves of the leaves not conspicuous........ *Virola*, all species

KEY H

Trees with yellow or orange latex

1. Leaves alternate.
2. Fruit a large, woody, asymmetrical, laterally compressed pod, with many circular, thin, transparent seeds........ *Aspidosperma megalocarpon*
2. Fruit globose.
3. Fruit more than 1" in diameter........ *Batocarpus costaricense*
3. Fruit less than 1" in diameter........ *Brosimum terrabanum*
1. Leaves opposite.
2. Lower leaf surface brownish. Sap reddish orange, drying blood red........ *Vismia*, both species

2. Lower leaf surface green. Sap bright yellow.
   3. Flowers red. Lower trunk often with stilt roots .... *Symphonia globulifera*
   3. Flowers white. Lower trunk never with stilt roots.
      4. Leaves narrow, linear-lanceolate, with acute or acuminate tips. Tall trees, to 80–100 ft. Fruits smooth, orange or yellow .... *Rheedia edulis*
      4. Leaves broadly oblong-elliptic, with an acute tip. Fruits very rough, bright yellow .... *Rheedia madruno*

KEY I

Trees with black sap or gum

1. Leaves pinnate .... *Prioria copaifera*
1. Leaves not pinnate.
   2. Leaves palmately lobed.
      3. Leaves 3–5 lobed. Flowers in paniculate cymes. Trunks and branches not hollow or infested with stinging ants .... *Pourouma aspera*
      3. Leaves peltate, with 7–13 lobes. Flowers in slender, often pendulous spikes. Trunks and branches hollow, inhabited by colonies of stinging ants.
         4. Trunks and young branches green or white, conspicuously ringed.
            5. Fruiting clusters 8–16″ long. Leaves green on the lower surface .... *Cecropia obtusifolia*
            5. Fruiting clusters 3–4″ long. Leaves white on the lower surface .... *Cecropia peltata*
         4. Trunks and young branches black, not conspicuously ringed .... *Cecropia Sandersoniana*
   2. Leaves simple .... *Hieronyma tectissima*

KEY J

Trees with prominent buttresses

1. Trees with milky latex.
   2. Fruits hollow.
      3. Usually strangler epiphytes, developing arched, flying buttresses in old specimens .... *Ficus lapathifolia*
      3. Not strangler epiphytes. Tall, handsome, white-trunked trees, the lower trunk buttressed for about 8–10 ft. .... *Ficus Werckleana*
   2. Fruits not hollow .... *Brosimum utile*
1. Trees without milky latex.
   2. Bark pale tan, guava-like in superficial appearance .... *Terminalia lucida*
   2. Bark not guava-like.
      3. Trunks and branches hollow, inhabited by stinging ants. Leaves peltate, conspicuously palmately lobed .... *Cecropia Sandersoniana*
      3. Trunks and branches not as above. Leaves not peltate.
         4. Leaves compound.
            5. Leaves digitately compound.

6. Leaflets 7–9, distinctly sticky and resinous when young. Flowers less than ½″ in diameter. Fruit a cluster of 5 woody, podlike carpels.. *Sterculia mexicana*
6. Leaflets 5–7, not sticky or resinous. Flowers about 1″ in diameter. Fruit an oblong capsule.. *Ceiba pentandra*

5. Leaves pinnate.
6. Leaves bipinnate (having a central nonwoody axis and 2 or more lateral pinnae).
7. Flowers bright yellow, conspicuous from a distance.. *Schizolobium parahybum*
7. Flowers white, not conspicuous.................. *Enterolobium cyclocarpum*
6. Leaves once pinnate (the central nonwoody axis with lateral leaflets).
7. Leaflets 2.
8. Leaflets 2–3″ long. Seed pods flat. Wood purple.. *Peltogyne purpurea*
8. Leaflets 4–6″ long. Seed pods not flat. Wood not purple.. *Cynometra hemitomophylla*
7. Leaflets more than 2.
8. Margins of the leaflets serrate. Fresh bark and crushed leaves with a strong, characteristic odor.......................... *Astronium graveolens*
8. Margins of the leaflets not serrate. Bark and leaves without a strong, distinctive odor.
9. Leaflets alternate on the rachis.
10. Leaflets 7–11. Flowers pea or beanlike. Seeds bright red.
11. Floral calyx divided nearly to the base. Flowers pinkish lavender, inconspicuous...................... *Dussia macrophyllata*
11. Floral calyx tubular, with short, distinct marginal teeth. Flowers pink, conspicuous from a distance.. *Dussia mexicana*
10. Leaflets 5–7. Flowers olive green, not pealike. Seeds not red. .... *Dialium guianense*
9. Leaflets opposite on the rachis.
10. Leaves not leathery, the apices acute. Flowers pinkish lavender, conspicuous from a distance. Fruits flat, broadly winged.. *Tachigalia versicolor*
10. Leaves leathery, broadly rounded at the apex. Flowers greenish white, inconspicuous. Fruits large, woody, nearly globose.. *Carapa Slateri*

4. Leaves simple.
5. Leaves opposite.. *Chimarrhis latifolia*
5. Leaves alternate.
6. Sap black. Crown with scattered red leaves.............. *Hieronyma tectissima*
6. Sap not black. Crown without scattered red leaves.
7. Undersurfaces of leaves brownish.............................. *Luehea Seemannii*
7. Undersurfaces of leaves not brownish.
8. Fruits densely covered with purple, needle-like spines.. *Sloanea picapica*

8. Fruits not spiny.
 9. Fruits 1-seeded.
  10. Inflorescences terminal .... *Licania operculipetala*
  10 Inflorescences axillary .... *Minquartia guianensis*
 9. Fruits with more than 1 seed.
  10. Leaves with 3 longitudinal nerves .... *Goethalsia meiantha*
  10. Leaves with a single longitudinal nerve.
   11. Leaf petioles about equaling the blades in length. Giant trees, with tremendous plank buttresses, ascending the trunk for 40–60 ft. .... *Huberodendron Allenii*
   11. Leaf petioles much shorter than the blades. Buttresses much less conspicuously developed.
    12. Fruit an elongate, woody, tubular pod, with a triangular detractable plug. Seeds winged .... *Couratari panamensis*
    12. Fruit a 5-valved capsule.
     13. Leaves acuminate. Fruits not woody .... *Mortoniodendron anisophyllum*
     13. Leaves acute. Fruits woody .... *Sloanea laurifolia*

KEY K

TREES WITH THORNY OR SPINY TRUNKS OR BRANCHES

1. Leaves compound.
 2. Leaves digitately compound .... *Ceiba pentandra* (Young specimens)
 2. Leaves pinnate.
  3. Leaflets 3. Flowers red .... *Erythrina costaricensis*
  3. Leaflets 5–17. Flowers white .... *Zanthoxylum procerum*
1. Leaves simple.
 2. Leaves opposite on the stems. Latex sweet, milky .... *Lacmellea panamensis*
 2. Leaves alternate on the stems. Plants without milky latex.
  3. Spines long and needle-like, in dense branching clusters at intervals along the trunk .... *Xylosma excelsum*
  3. Spines short, never in branching clusters.
   4. Trunks intensely spiny. Leaves with 2 prominent glands at the base of the blade. Large trees .... *Hura crepitans*
   4. Trunks sparingly spiny or unarmed. Leaves without glands at the base of the blade. Small trees .... *Ximenia americana*

(4)—UPPER MONTANE RAIN FOREST

In this formation are to be found those species that occur in high valleys and on steep ridges and mountain walls above about 2,500 ft. Separation from the Lower Montane Forest is not sharp, and many of the species of that formation will still be found in favorable situations, a

particularly notable example being *Brosimum utile,* which occurs in nearly pure stands on many of the upper slopes and ridges, to well over 4,000 ft. Sharp ridges and nearly perpendicular slopes tend, as would be expected, to be poor in species, most of which are found in either nearly pure stands or in a very limited association of three or four species. Deep, well-drained soils on the highest levels, particularly near Cañas Gordas and Agua Buena, support great climax forests dominated by various species of *Quercus* and *Cedrela Tonduzii.* These stands are relatively limited in our area, though they cover extensive tracts in adjacent territory, both in Costa Rica and Panama. Because of the difficult and precipitous nature of much of the terrain and the almost complete absence of access trails our knowledge of this formation is relatively poor. Some of the more obvious associations and consociations are listed below, though it is to be expected that further studies in the area may show some of these to have been the result of fortuitous circumstances, or to have been based on insufficient observation.

Quercus—Cedrela Tonduzii association.—Typical of deep, well-drained areas above about 4,000 ft., good examples being found near Cañas Gordas and Agua Buena.

Iriartea gigantea consociation.—This extremely handsome stilt palm occurs in nearly pure stands of several acres in extent in small, sheltered coves in the hills at about 2,500–3,000 ft. elevation, several excellent examples being found near El Cedral on the trail between Palmar Norte and Buenos Aires.

Brosimum utile consociation.—Very good examples of these nearly pure stands may be seen at 2,500–4,000 ft. elevation in the hills between Palmar Norte and Boruca, particularly on the heights above the settlement of Rey Curre.

Coccoloba roseiflora consociation.—Typical of very steep, rocky walls, at about 3,000 ft. elevation, particularly in the Cerros de Retinto, slightly west of Palmar Norte.

Ladenbergia Brenesii consociation.—Typical of deep soils on hill crests at about 2,500–3,000 ft. above Palmar Norte.

Euterpe panamensis—Cespedesia macrophylla association.—Characteristic of the uppermost slopes and crests of the high, very steep ridges at 3,000–4,000 ft. elevation between Palmar Norte and Boruca.

Preliminary list of arborescent species

Brosimum utile—Large trees with sweet, milky latex.
Carpotroche platyptera—Small trees, flowers and fruits produced directly from the trunk and larger branches.
Cedrela Tonduzii—A large tree, with shaggy bark and pinnate leaves.
Cespedesia macrophylla—Very large leaves, clustered at the ends of the branches.
Coccoloba roseiflora—Conspicuous flushes of new, pink leaves in February.
Coussapoa parviceps
Cymbopetalum costaricense
Euterpe panamensis—Slender, unarmed palm.
Guatteria aeruginosa
Hedyosmum mexicanum
Iriartea gigantea—Stilt palms, often forming nearly pure stands.
Ladenbergia Brenesii
Maurea glauca
Miconia Matthaei
Nectandra salicifolia
Oreopanax capitatus
Oreopanax xalapense
Persea pallida
Phoebe costaricana
Pourouma aspera
Quercus sp.
Stemmadenia Alfari

Our present knowledge of this formation is far too fragmentary for the preparation of field keys.

EDAPHIC FORMATIONS

(5)—SWAMP FOREST

Extensive areas of swamp forest are found in the lower Río Coto and in parts of the Terraba floodplain, particularly between the Laguna de Sierpe and the Mangrove Woodlands of the lower delta. All of this considerable tract is characterized by its low relief and poor drainage. The floristic cover varies conspicuously from place to place, and probably reflects soil differences and degrees of relative salinity, as well as possibly other factors. In some instances, the break between the Swamp Forest and adjacent areas of either Palm Swamp or Herbaceous Swamp is very abrupt and striking, while in other cases the stands merge imperceptibly with the margins of the Evergreen Lowland Forest so that the boundaries are not distinguishable when seen from the air. The streams, particularly in the Sierpe area are very sluggish, and are filled with brilliant mineral

green masses of floating vegetation, mostly of *Pistia stratiotes,* which contrast handsomely with the darker surrounding foliage. The outstanding associations, consociations, and faciations may be listed as follows:

Pterocarpus officinalis—Carapa Slateri association.—This association is found on the extreme margins of the Swamp formation, adjoining the Evergreen Lowland Forest, and to a certain degree merging with it. *Pterocarpus* is seldom, if ever, found above the swamp, but *Carapa Slateri* frequently grows on well-drained clay ridges, though never in dense stands, and usually as unbuttressed specimens.

Pterocarpus officinalis—Mora oleifera association.—This association is typical of areas on the margins of Mangrove Woodland, very good examples being found along the road to the inner pastures near Golfito, in the delta of the Río Esquinas and near Farms 19 and 20 in the lower delta of the Río Terraba.

Symphonia globulifera consociation.—Extensive, rather open stands of this species occur on the margins of Herbaceous Swamp in the Sierpe area, particularly back of Jalaca Farm.

Principal arborescent species

Anacardium excelsum
Astrocaryum alatum
Astrocaryum Standleyanum
Bactris balanoidea
Bactris militaris
Carapa Slateri
Corozo oleifera
Cryosophila guagara
Grias Fendleri
Hampea platanifolia
Herrania purpurea
Hieronyma tectissima
Homalium eurypetalum
Hura crepitans
Inga marginata
Inga multijuga
Inga punctata
Inga Ruiziana
Ixora nicaraguensis
Luehea Seemannii
Mora oleifera
Neonicholsonia Watsonii
Ouratea Valerii
Pachira aquatica
Prioria copaifera
Pterocarpus officinalis
Quararibea guatemalteca
Raphia taedigera
Sapium jamaicense
Scheelea rostrata
Sickingia Maxonii
Socratea durissima
Symphonia globulifera
Tetrathylacium costaricense
Trichilia montana

KEY

1. Palms.
  2. Leaves fan-shaped.. *Cryosophila guagara*
  2. Leaves not fan-shaped.
    3. Trunks solitary.

4. Lower trunk with very conspicuously developed stilt roots .... *Socratea durissima*
4. Trunk without stilt roots.
   5. Trunk and/or fronds armed with many long, needle-like spines.
      6. Tall palms, 40–65 ft. in height. Fruits orange, unarmed, in long showy pendulous clusters .... *Astrocaryum Standleyanum*
      6. Small palms, 12–25 ft. in height. Fruits spiny, in short, compact clusters .... *Astrocaryum alatum*
   5. Trunk and fronds unarmed.
      6. Plants dwarf, the trunk less than 1 ft. in height. Inflorescences erect unbranched spikes .... *Neonicholsonia Watsonii*
      6. Trees not dwarf, the trunks more than 1 ft. in height.
         7. Trunks massive, often creeping. Fruiting clusters sessile in the leaf axils, like those of the African oil palm .... *Corozo oleifera*
         7. Trunks tall, never creeping. Fruiting clusters conspicuously pendulous .... *Scheelea rostrata*

3. Trunks multiple.
   4. Trunks and/or fronds intensely spiny. Small palms. Fruits not covered with overlapping scales.
      5. Fronds with many lateral pinnae .... *Bactris balanoidea*
      5. Fronds undivided .... *Bactris militaris*
   4. Trunks not spiny. Large palms. Fruits covered with closely overlapping scales .... *Raphia taedigera*

1. Not palms.
   2. Trunks completely covered with short, sharp spines .... *Hura crepitans*
   2. Trunks unarmed.
      3. Leaves compound.
         4. Leaves digitately compound.
            5. Small trees less than 15 ft. in height. Flowers and fruits produced directly from the trunk .... *Herrania purpurea*
            5. Trees more than 30 ft. in height. Flowers and fruits produced from the ends of the branches .... *Pachira aquatica*
         4. Leaves pinnate.
            5. Leaflets 4.
               6. Leaf rachis winged .... *Inga marginata*
               6. Leaf rachis not winged.
                  7. Basal pair of leaflets about ½ the size of the terminal pair .... *Inga punctata*
                  7. Basal pair of leaflets about the same size as the terminal pair.
                     8. Fruits nearly circular in outline, 1-seeded, about 3″ in diameter. Trunks exude a copious black gum when cut .... *Prioria copaifera*
                     8. Fruits longer than broad, often several seeded. Seeds beanlike, about 6″ long. Trunks without black gum .... *Mora oleifera*
            5. Leaflets more than 4.
               6. Leaf rachis with a small but distinct disklike gland at the base of each pair of leaflets.
                  7. Leaflets and rachis with a minute but conspicuous brownish puberu-

lence. Calyx usually more than ¼″ long........................*Inga multijuga*

7. Leaflets and rachis nearly glabrous. Calyx about ⅛″ long..*Inga Ruiziana*

6. Leaf rachis without disklike glands.

7. Flowers pea- or beanlike, yellow or orange, conspicuous from a distance. Trunk with red sap..............................*Pterocarpus officinalis*

7. Flowers not pea- or beanlike, not conspicuous from a distance. Trunk without red sap.

8. Leaflets broadly rounded at the apex. Large trees, to more than 100 ft..*Carapa Slateri*

8. Leaflets acuminate at the apex. Small trees............*Trichilia montana*

3. Leaves simple.

4. Leaves very large, conspicuously clustered at the ends of the branches. Flowers and fruits produced directly from the trunk.................*Grias Fendleri*

4. Leaves not conspicuously clustered at the ends of the branches. Flowers and fruits not produced directly from the trunk and larger branches.

5. Leaves opposite on the stems.

6. Trees with bright-yellow, sticky latex. Flowers red..*Symphonia globulifera*

6. Trees without yellow latex. Flowers not red.

7. Leaves very large, usually more than 6″ wide............*Sickingia Maxonii*

7. Leaves small, usually less than 2″ wide..................*Ixora nicaraguensis*

5. Leaves alternate on the stems.

6. Trees with milky white latex...*Sapium jamaicense*

6. Trees without milky latex.

7. Leaves palmately lobed..*Hampea platanifolia*

7. Leaves not palmately lobed.

8. Leaf blades with 3 longitudinal nerves, the lower surface conspicuously brownish..*Luehea Seemannii*

8. Leaf blades with a single longitudinal nerve, the lower surface not brownish.

9. Trees with black sap. Crowns usually with a few conspicuous red, senescent leaves...*Hieronyma tectissima*

9. Trees without black sap. Leaves not as above.

10. Leaves conspicuously rounded at the apex..*Anacardium excelsum*

10. Leaves acute or acuminate at the apex.

11. Flowers yellow..*Ouratea Valerii*

11. Flowers not yellow.

12. Flowers red, in conspicuous pendulous clusters..*Tetrathylacium costaricense*

12. Flowers white or greenish white, never in pendulous clusters.

13. Flowers in elongate racemes or panicles..*Homalium eurypetalum*

13. Flowers in nearly sessile fascicles in the leaf axils..*Quararibea guatemalteca*

(6)—PALM SWAMP

A nearly pure consociation of *Raphia taedigera* occurs in the Sierpe area in somewhat discontinuous stands, each several hundreds of acres in extent. These are noteworthy in being the only known occurrence of the genus on the Pacific coast of Central America. The groves are very attractive when seen from the air, the brown, older leaves contrasting handsomely with the darker green of the feathery mature fronds.

Symphonia globulifera—Raphia taedigera faciation.—There are considerable areas near the Laguna de Sierpe where these two species are found in mixed stands of about equal numbers, which are quite conspicuous when seen from the air.

Principal species

Corozo oleifera
Pterocarpus officinalis
Raphia taedigera
Scheelea rostrata
Symphonia globulifera

KEY

1. Palms.
   2. Trunks multiple. Fruits covered with conspicuous, overlapping scales. The dominant species of the formation.................*Raphia taedigera*
   2. Trunks solitary. Fruits not covered with overlapping scales.
      3. Trunks short and massive, often repent. Fruiting clusters sessile in the frond axils.................*Corozo oleifera*
      3. Trunks tall, never repent. Fruiting clusters conspicuously pendulous.................*Scheelea rostrata*
1. Not palms.
   2. Trees with sticky, yellow latex. Lower trunk often with stilt roots.................*Symphonia globulifera*
   2. Trees with red sap. Lower trunks prominently buttressed.................*Pterocarpus officinalis*

(7)—MANGROVE WOODLAND

Great areas in the deltas of the Río Coto and Río Terraba, and to a lesser extent in the lower course of the Río Esquinas and fringing the landlocked harbor of Golfito are dominated by trees of the Mangrove association, whose principal species is *Rhizophora mangle*. Mature stands, when seen from a distance or at extreme high tide, bear a considerable

superficial resemblance to groves of willows or poplars, because of their straight trunks and remarkable uniformity of height and diameter. Younger colonies on mud bars or river banks have a denser, more rounded profile, rising in a gradual curve to the largest trees in the center of the stand, or nearest the shore, as the case may be. "Mangrove Swamp" is the usually accepted term for such a locality, but they are not swamps in the ordinary sense of the word (that is, the drainage is not impeded) but are rather tidal forests whose distribution in river deltas and estuaries is determined by the high degree of tolerance of the individual species to salt or strongly brackish water. Such areas are in direct contradiction to the usual rule of extreme diversity so typical of most tropical forests, since comparatively few species, mostly of completely unrelated families, have been able to meet the exacting conditions which the habitat presents. Many ecologists, notably Schimper (1903) and Warming (1925), have recognized the role of mangroves in extending coasts and building islands. The roots of the Red Mangrove *(Rhizophora mangle)* are masterpieces of plant engineering, being made up of complicated arching tiers of bracing stilts, admirably suited to the dual purpose of supporting the trunk and leaves above the shifting currents and checking the tidal flow, causing the precipitation of mud and debris. The soil level is thus built up, year after year, preparing the ground for subsequent, less tolerant types of vegetation. Thousands of acres have been thus reclaimed from the sea. Red Mangrove may range from about 15 to over 100 ft. in height, the size of the trees and the character of their interlocking roots depending on the nature of the underlying soil and the tidal range. On exposed coasts, mature trees may average about 30 ft. in height, with an open, branching habit, while old specimens in protected situations, such as the inner delta of the Río Esquinas, have fine, straight, unbranched trunks and comparatively few stilt roots at the base. The bark of this species is utilized to a certain extent in the local tanneries, particularly near Golfito, producing, when used alone, a rather brittle, dark-red leather. Some of the larger trees furnish timbers for construction and shipbuilding, but the greatest part of that cut is burned in the swamps, to produce a very high-grade charcoal. The small, leathery fruits, which are of an inverted pear-shape, are borne in profusion near the tips of the branches, among the opposite, leathery leaves. These fruits are viviparous, the seeds germinating before they fall, and are commonly to be seen in all stages of development, the roots growing downward as long, green, pencil- or spindle-shaped rods, which may be nearly 1 ft. in length before they drop to the mud below.

Higher ground in the center of the grove is typically covered with a mature growth of the Black Mangrove (*Avicennia nitida*). These are often found in nearly pure stands of straight-trunked, unbranched specimens averaging about 1 ft. in diameter, and 60 ft. or more in height. They lack the stilt roots characteristic of Rhizophora, but have in their place thousands of curious finger-like breathing organs, or pneumatophores, which project in myriads from the mud flats beneath the trees. In some situations, particularly above all but the highest tides, these trees may be greatly reduced in size, sometimes forming low, much-branched bushes which lack the typical breathing roots. The leaves have the peculiarity at times of being covered with salt crystals, giving rise to the common Spanish name of *Palo de sal*. The flowers are small and white, and are produced in axillary or terminal clusters. They usually bloom at the end of the dry season, and have a delightful fragrance, reminiscent of warm honey, attracting swarms of bees, so that the entire grove often hums like a huge apiary. Further aquaintance will reveal other interesting trees, particularly *Laguncularia racemosa* and *Pelliciera rhizophorae*, mature specimens of the latter species with very curious, and highly characteristic buttressed bases. Upper reaches of the estuaries, and the landward margins of the Mangrove association merge imperceptibly with the *Pterocarpus officinalis—Mora oleifera* association of the Swamp Forests, or with the Littoral Woodland, as the case may be.

Principal species

Avicennia nitida	*Mora oleifera*
Conocarpus erecta	*Pelliciera rhizophorae*
Corozo oleifera	*Pterocarpus officinalis*
Hibiscus tiliaceus	*Raphia taedigera*
Laguncularia racemosa	*Rhizophora mangle*

KEY

1. Palms.
  2. Trunks multiple. Fruits covered with conspicuous, overlapping scales .... *Raphia taedigera*
  2. Trunks solitary. Fruits not covered with scales .... *Corozo oleifera*
1. Not palms.
  2. Trees with conspicuous stilt roots .... *Rhizophora mangle*
  2. Trees without stilt roots.
    3. Leaves pinnate.
      4. Leaflets 4 .... *Mora oleifera*
      4. Leaflets 6 to 9 .... *Pterocarpus officinalis*
    3. Leaves simple.

4. Leaves opposite on the stems.
   5. Leaf petiole with 2 conspicuous glands. Base of the tree without finger-like pneumatophores........ *Laguncularia racemosa*
   5. Leaf petiole without glands. Base of the tree with dense colonies of erect, finger-like pneumatophores........ *Avicennia nitida*
4. Leaves alternate on the stems.
   5. Leaves broadly cordate. Flowers yellow, turning reddish brown with age........ *Hibiscus tiliaceus*
   5. Leaves not broadly cordate. Flowers pink or white.
      6. Flowers large and attractive, followed by large, woody, brown fruits. Base of tree curiously buttressed........ *Pelliciera rhizophorae*
      6. Flowers small and inconspicuous, followed by small, conelike fruits. Base of the tree not buttressed........ *Conocarpus erecta*

(8)—HERBACEOUS SWAMP

Large areas near the Laguna de Sierpe are dominated by varying kinds of herbaceous plants, in sharp contrast to the surrounding arborescent cover. It seems at least possible that many of these open tracts may represent old lagoons that have been entirely filled by the encroaching vegetation, particularly by grasses and sedges, such as *Paspalum fasciculatum* and *Cyperus canus,* or broad-leaved plants such as *Heliconia imbricata, Heliconia latispatha, Heliconia bihai, Calathea insignis,* or *Calathea lutea.* Individual species often form definite colonies, which are frequently nearly circular in outline and quite conspicuous when seen from the air. These areas tend to be slowly invaded by *Symphonia globulifera,* and to a lesser extent by *Pachira aquatica.*

Brackish herbaceous swamp, on the margins of Mangrove Woodland, is usually dominated by a consociation of a robust fern, *Acrostichum aureum,* though other woody and herbaceous elements, such as *Rhabdadenia biflora* and *Pancratium littorale,* are usually present.

(9)—GALLERY WOODLAND

In this category are to be found the arborescent species peculiar to riverbanks and gravel bars of the larger streams which drain the area. This is seldom listed as a distinct formation, but the species found in such locations are either not duplicated in the adjacent forest, or occur in such concentrations as to give them an entirely different aspect. It is suspected that the gravel bars in particular are the source of many of the species which invade old clearings and make up the bulk of the second-growth thickets, which are so profoundly different in composition from the original forest cover. A number of very distinctive faciations

have been noted, which seem to be consistent enough to be worth recording.

Pithecolobium longifolium faciation.—Peculiar to the rocky banks of fast-flowing streams above the influence of the highest tides. Good examples may be seen along the Río Terraba above Palmar Norte.

Tessaria mucronata consociation.—Typical of gravel bars in fast-flowing streams above the influence of the tides. Very good examples may be seen in the Río Piedras Blancas, and in the Río Esquinas near Kilometer 42. The thickets are quite reminiscent of groves of young willows when seen from a distance.

Albizzia filicina consociation.—This species forms flat-topped groves of dozens of shaggy-barked individuals on the gravelly banks of fast-flowing streams, good examples being found in the Esquinas pastures, and in the Río Esquinas near Kilometer 42.

Gynerium sagittatum consociation.—Thickets of the common, robust arrow cane form on mudbanks along the lower courses of small, meandering streams. Several good examples may be seen in Coto, and along the railroad line between Lagarto Station and Golfito.

Cecropia—Ochroma association.—Typical of old, relatively stable gravel bars throughout the area.

TRANSITIONAL FORMATIONS

(10)—SAVANNA

As has been previously noted, it seems quite doubtful that the present extensive grasslands near Potrero Grande and San Andrés are a natural formation. Although there is some evidence that this area may be in a rain shadow, this may simply have made annual burnings easier, and hastened the extension of the pasture lands and the progressive destruction of the forest. A few ridges, especially near Boruca, support a typical open woodland of stunted trees of *Curatella americana* and *Byrsonima crassifolia,* characteristic of dry areas; but ravines, stream courses, and all inaccessible tracts are still in heavy forest. It is believed that trees would gradually reinvade the savanna if fire were controlled.

(11)—PASTURES

Since all the pastures in the area have been artificially created, it is of course to be expected that many of the larger trees found in such loca-

tions will be chance remnants of the original forest cover, and should, in the majority of cases be sought in the keys to the Evergreen Lowland Forest or Lower Montane Forest, depending on the nature of the terrain. There are, however, a number of such species that are so common in many of the pastures that they have been entered in the present list to facilitate the tracing of the more conspicuous elements. Besides these chance remnants, there are quite a number of characteristic things, usually of small or medium size, that seem to be nearly confined to these clearings and are seldom seen elsewhere.

Principal Species

Acacia costaricensis—Small trees, with paired, hollow thorns, inhabited by stinging ants.
Acacia farnesiana—Spines pale in color, not hollow.
Acacia melanoceras—Very large, hollow, bull's-horn thorns and stinging ants.
Acacia spadicigera—Hollow thorns and ants.
Acrocomia vinifera—Large, single-trunked, spiny palms, common near Potrero Grande.
Aegiphila martinicensis
Albizzia filicina—Trees form small, flat-topped groves along fast flowing streams.
Anacardium excelsum—Large trees, remnants of the original forest.
Anacardium occidentale—A small, introduced species, persistent near old house sites.
Astrocaryum Standleyanum—Tall, intensely spiny palms, left standing because the trunks are too hard to cut.
Bravaisia integerrima—A bushy tree, with multiple trunks and stilt roots.
Byrsonima crassifolia—"Nance." A small tree, with edible yellow fruits.
Cassia grandis—Medium-sized tree with attractive pink flowers.
Cassia reticulata—Small, round-topped tree, with bright-yellow flowers.
Cassia spectabilis—Small, usually flat-topped tree, with bright-yellow flowers.
Cedrela mexicana
Ceiba pentandra—Often very large, with strongly developed buttresses.
Chlorophora tinctoria—Rather bushy, round-topped trees, frequently with spiny branches along the lower trunk.
Citharexylum viride
Cordia alliodora—Rather showy white flowers produced in February.
Cordia collococca—Showy red fruits.
Cordia gerascanthus—Like a larger flowered *alliodora*. Apparently limited in our area to the Peninsula de Osa.
Cordia lasiocalyx
Corozo oleifera—Short palms with massive trunks.
Crescentia cujete—An introduced species, rather frequently found near old house sites and in pastures near Palmar Norte. The gourdlike fruits are produced from the trunk and larger branches.
Croton glabellus
Curatella americana—Typical of dry pastures and grassy ridges near Boruca.

PASTURES

Diphysa robinioides—Mostly grown as living fence posts, but also occasional near old house sites.
Dracaena americana—Long, strap-shaped leaves, clustered at the ends of the branches.
Enterolobium cyclocarpum—Usually very large, spreading trees, with fine, fernlike foliage.
Ficus Tonduzii—Milky latex.
Ficus Werckleana—Tall, white-trunked trees, with milky latex.
Guazuma ulmifolia
Hamelia patens
Hura crepitans—Large or medium-sized trees, with very spiny trunks.
Inga marginata
Inga portobellensis
Inga punctata
Inga quaternata
Inga Ruiziana
Inga sapindoides
Inga spectabilis
Luehea Seemannii—Leaves with 3 longitudinal nerves, the lower surfaces conspicuously brownish.
Mangifera indica—An introduced species, occasionally persistent near old house sites.
Miconia argentea—Undersides of leaves pale tan or white.
Muntingia calabura
Nectandra salicifolia
Persea Skutchii
Posoqueria latifolia
Psidium Friedrichsthalianum
Psidium Guajava
Pterocarpus Hayesii—Showy orange flowers.
Sapindus saponaria
Sapium thelocarpum—Milky latex.
Scheelea rostrata—Robust, single-trunked palm. Fruits in large, pendulous clusters. Common near Puerto Cortes in pastures.
Solanum verbascifolium var. *adulterinum*
Spondias mombin—Yellow, plumlike fruits.
Stemmadenia Donnell-Smithii—Milky latex.
Sterculia apetala—Large, palmately lobed leaves.
Swartzia panamensis—Pendulous racemes of attractive yellow flowers. Seeds very large.
Tabebuia chrysantha—Showy yellow flowers.
Terminalia amazonia
Terminalia bucidioides
Terminalia lucida—Pale, guava-like bark.
Theobroma bicolor—Lower surfaces of the leaves nearly white.
Theobroma cacao—Persistent near old house sites.
Trema micrantha
Vismia ferruginea—Red or orange sap.
Vismia guianensis—Red or orange sap.
Vitex Cooperi—Digitately compound leaves. Small blue flowers.
Vochysia ferruginea—Showy orange flowers.

PASTURES

Zanthoxylum procerum—Sparsely thorny trunks and pinnate leaves.

Zexmenia frutescens—Small trees, with bright-yellow flowers. Common in brushy pastures near Boruca.

KEY

1. Palms. (See also keys to Evergreen Lowland Forest and Lower Montane Forest.)
 2. Trunks and fronds intensely spiny.
  3. Trunks stout, usually more or less covered by the persistent bases of the old fronds. Fruits globose, greenish yellow at maturity .... *Acrocomia vinifera*
  3. Trunks relatively slender, never covered by the persistent bases of the old fronds. Fruits ellipsoid, deep orange at maturity .... *Astrocaryum Standleyanum*
 2. Trunks and fronds not spiny.
  3. Trunks short and massive, often creeping. Fruit clusters sessile in the frond axils .... *Corozo oleifera*
  3. Trunks tall, never creeping. Fruit clusters pendulous .... *Scheelea rostrata*
1. Not palms (See also keys to Evergreen Lowland Forest and Lower Montane Forest).
 2. Trunks or lower branches armed with spines.
  3. Trunks armed with short, sharp spines.
   4. Leaves simple. Flowers red. Usually large trees .... *Hura crepitans*
   4. Leaves digitately compound .... *Ceiba pentandra* (Young specimens)
  3. Trunks unarmed. Lower branches spiny.
   4. Leaves simple .... *Chlorophora tinctoria*
   4. Leaves bipinnate.
    5. Thorns hollow, inhabited by stinging ants.
     6. Flowers in globose heads .... *Acacia melanoceras*
     6. Flowers in elongate spikes.
      7. Fruits dehiscent at maturity .... *Acacia costaricensis*
      7. Fruits indehiscent .... *Acacia spadicigera*
    5. Thorns not hollow, not inhabited by ants .... *Acacia farnesiana*
 2. Trunks or branches not armed with spines.
  3. Flowers and fruits produced directly from the trunk and larger branches.
   4. Leaves in dense, sessile fascicles from the nodes. Fruits smooth, green, gourd-like .... *Crescentia cujete*
   4. Leaves alternate on the stems. Fruits more or less ridged, usually yellow, purplish, or reddish when mature .... *Theobroma cacao*
  3. Flowers and fruits not produced directly from the trunk or larger branches.
   4. Trees with multiple trunks and conspicuous stilt roots .... *Bravaisia integerrima*
   4. Trees without multiple trunks or stilt roots.
    5. Leaves compound.
     6. Leaves digitately compound.
      7. Leaflets 3. Flowers blue .... *Vitex Cooperi*
      7. Leaflets 5–7. Flowers not blue.
       8. Lower trunks usually very prominently buttressed. Flowers pink or pinkish lavender .... *Ceiba pentandra*
       8. Lower trunk not prominently buttressed. Flowers bright yellow .... *Tabebuia chrysantha*

6. Leaves pinnate.
  7. Leaves bipinnate .... *Enterolobium cyclocarpum*
  7. Leaves once pinnate.
    8. Flowers pink, yellow, or orange.
      9. Flowers pink .... *Cassia grandis*
      9. Flowers yellow or orange.
        10. Flowers in elongate, pendulous racemes .... *Swartzia panamensis*
        10. Flowers never in pendulous racemes.
          11. Flowers pea- or beanlike.
            12. Small trees, usually less than 40 ft. in height. Flowers yellow. Common in fence rows and occasional near old house sites .... *Diphysa robinioides*
            12. Larger trees, usually more than 60 ft. in height. Flowers orange, very showy .... *Pterocarpus Hayesii*
          11. Flowers not pea- or beanlike.
            12. Leaflets broadly rounded at the apex .... *Cassia reticulata*
            12. Leaflets acute or acuminate at the apex .... *Cassia spectabilis*
    8. Flowers white, greenish white, or pale tan.
      9. Flowers mimosa-like. Fruit a legume.
        10. Leaf rachis more or less winged.
          11. Leaflets 3 pairs .... *Inga sapindoides*
          11. Leaflets 2 pairs.
            12. Flowers in slender spikes .... *Inga marginata*
            12. Flowers in pedunculate heads .... *Inga portobellensis*
        10. Leaf rachis not winged.
          11. Leaflets usually 6 pairs .... *Inga Ruiziana*
          11. Leaflets 2 or 3 pairs.
            12. Floral corolla more than ½″ long. Fruits more than 8″ long .... *Inga spectabilis*
            12. Floral corolla less than ½″ long. Fruits less than 8″ long.
              13. Leaflets 2 pairs .... *Inga punctata*
              13. Leaflets 3 pairs .... *Inga quaternata*
      9. Flowers not mimosa-like. Fruit not a legume.
        10. Leaflets more than 20. Wood bitter .... *Cedrela mexicana*
        10. Leaflets less than 20. Wood not bitter.
          11. Bark rough, strongly corrugated. Fruits plumlike, edible .... *Spondias mombin*
          11. Bark not strongly corrugated. Fruits waxy, not edible .... *Sapindus saponaria*

5. Leaves simple.
  6. Leaves opposite, or in whorls of 4 at the nodes.
    7. Relatively large trees, usually more than 40 ft. in height.
      8. Trees with milky latex. Leaves green on the lower surface .... *Stemmadenia Donnell-Smithii*
      8. Trees without milky latex. Leaves brownish on the lower surface .... *Vochysia ferruginea*
    7. Relatively small trees, usually less than 40 ft. in height.

8. Trees with orange sap, which dries blood red.
 9. Petals white .... *Vismia ferruginea*
 9. Petals pale orange .... *Vismia guianensis*
8. Trees without distinctive sap.
 9. Inflorescences axillary.
  10. Veins of the lower leaf surface elevated and conspicuous. Fruits sweet .... *Psidium Guajava*
  10. Veins of the lower leaf surface inconspicuous. Fruits acid .... *Psidium Friedrichsthalianum*
 9. Inflorescences terminal.
  10. Lower portion of the flower contracted into a slender tube 3″ or more in length .... *Posoqueria latifolia*
  10. Flowers not as above.
   11. Inflorescence an unbranched raceme.
    12. Leaves brownish on the lower surface .... *Byrsonima crassifolia*
    12. Leaves green on the lower surface ... *Citharexylum viride*
   11. Inflorescence a branching cyme or panicle.
    12. Leaves in whorls of 4 at each node. Flowers reddish orange .... *Hamelia patens*
    12. Leaves in pairs at each node. Flowers yellow or white.
     13. Flowers bright yellow, daisy-like. Common on waste land near Boruca .... *Zexmenia frutescens*
     13. Flowers not daisy-like.
      14. Leaves pale tan on the lower surface. Flowers white .... *Miconia argentea*
      14. Leaves not pale tan on the lower surface. Flowers pale yellow .... *Aegiphila martinicensis*

6. Leaves alternate on the stems.
 7. Leaves strap-shaped, conspicuously clustered at the ends of the branches .... *Dracaena americana*
 7. Leaves not conspicuously strap-shaped or clustered at the ends of the branches.
  8. Trees with white, milky latex.
   9. Leaves with 2 small glands at the base of the blade .... *Sapium thelocarpum*
   9. Leaves without glands at the base of the blade.
    10. Leaf blades with about 10 pairs of lateral nerves .... *Ficus Tonduzii*
    10. Leaf blades with 20–25 pairs of lateral nerves .... *Ficus Werckleana*
  8. Trees without white, milky latex.
   9. Leaves large, conspicuously 3–5 lobed .... *Sterculia apetala*
   9. Leaves not conspicuously lobed.
    10. Leaves with 3–5 longitudinal nerves.
     11. Leaves with 3 longitudinal nerves.
      12. Leaves conspicuously brownish on the lower surface. Inflorescences terminal .... *Luehea Seemannii*

12. Leaves not conspicuously brownish on the lower surface. Inflorescences axillary............................*Trema micrantha*

11. Leaves with at least 5 longitudinal nerves, the blades nearly white on the lower surface......................*Theobroma bicolor*

10. Leaves with a single longitudinal nerve.

11. Bark thin, pale tan, or nearly white, often peeling in large irregular patches, guava-like in superficial appearance...... ....*Terminalia lucida*

11. Bark not as above.

12. Leaves very rough (sandpapery) to the touch.................. ....*Curatella americana*

12. Leaves not sandpapery to the touch.

13. Terminal 1 or 2" of the leafy twigs conspicuously thickened, with many scars marking the location of the bases of former leaves.........*Terminalia bucidioides*

13. Leafy twigs not conspicuously thickened.

14. Inflorescences axillary.

15. Flowers relatively large, about 1" in diameter when fresh..........................*Muntingia calabura*

15. Flowers much less than 1" in diameter.

16. Leaf margins serrate.

17. Staminate and pistillate flowers produced on separate trees. Wood yellow, hard, and heavy...*Chlorophora tinctoria*

17. Flowers perfect. Wood not yellow...... ....*Guazuma ulmifolia*

16. Leaf margins not serrate.

17. Individual flowers either completely pistillate or completely staminate.

18. Inflorescences either catkin-like (male) or globose (female), produced on separate trees..................... ....*Chlorophora tinctoria*

18. Inflorescences racemose, both male and female flowers found on the same branch............*Croton glabellus*

17. Individual flowers perfect, with both staminate and pistillate elements present.

18. Leaves broadly elliptic, very abruptly short acuminate at the apex. Floral calyx shallowly lobed............... ....*Cordia lasiocalyx*

18. Leaves lanceolate, tapering gradually to an acute or acuminate apex. Floral calyx cleft to the base.

19. Flowers white.. *Nectandra salicifolia*
19. Flowers yellow...*Persea Skutchii*

14. Inflorescences terminal.
15. Leaves conspicuously rounded at the apex.
16. Large trees, to over 100 ft. in height. Leaves usually more than 10″ long. Common...... *Anacardium excelsum*
16. Small trees, seldom exceeding 20 ft. in height. Leaves usually less than 8″ long. Occasional near old house sites...................... *Anacardium occidentale*
15. Leaves more or less pointed at the apex, never conspicuously rounded.
16. Base of the leaf petiole conspicuously thickened where it joins the stem. Cultivated tree, occasional near old house sites... *Mangifera indica*
16. Base of the leaf petiole not conspicuously thickened. Native species.
17. Forks of the young branches usually with hollow, gall-like swellings inhabited by ants. Flowers white, conspicuous.
18. Flowers about ¾″ in diameter......... *Cordia gerascanthus*
18. Flowers less than ½″ in diameter...... *Cordia alliodora*
17. Forks of the young branches without hollow gall-like swellings. Flowers not conspicuous.
18. Branchlets rough to the touch...*Solanum verbascifolium* var. *adulterinum*
18. Branchlets not rough to the touch.
19. Leaves conspicuously broadest above the middle........................ *Terminalia amazonia*
19. Leaves broadest at about the middle............... *Nectandra salicifolia*

(12)—FENCE ROWS AND TRAILSIDE THICKETS

A few quite distinctive species are commonly found as living fence posts, and a fair number of other things are most frequently encountered in brushy, trailside thickets. While this is not an ecological formation in the usual sense of the word, the separate listing of such species may serve as an aid in their identification.

FENCE ROWS AND TRAILSIDE THICKETS

Principal species

Aegiphila martinicensis
Bursera simaruba
Carica papaya
Casearia aculeata
Casearia arguta
Casearia guianensis
Casearia myriantha
Casearia sylvestris
Cecropia obtusifolia
Cecropia peltata
Cecropia Sandersoniana
Citharexylum viride
Coccoloba acuminata
Cochlospermum vitifolium
Croton glabellus
Cupania largifolia
Cyphomandra costaricensis
Dicraspidia Donnell-Smithii
Didymopanax Morototoni
Diphysa robinioides
Erythrina costaricensis
Ficus Goldmanii
Gliricidia sepium
Guazuma ulmifolia
Hamelia patens
Hampea Allenii
Hampea platanifolia
Malvaviscus arboreus var. *penduliflorus*
Miconia argentea
Muntingia calabura
Myriocarpa longipes
Ochroma lagopus
Phyllanthus acuminatus
Piper arboreum
Piper reticulatum
Psidium Friedrichsthalianum
Psidium Guajava
Randia armata
Ricinus communis
Rollinia Jimenezii
Roupala complicata
Sapindus saponaria
Sapium thelocarpum
Simaba cedron
Siparuna patelliformis
Solanum verbascifolium var. *adulterinum*
Spondias mombin
Spondias purpurea
Tabebuia chrysantha
Trema micrantha
Vismia ferruginea
Vismia guianensis
Vitex Cooperi
Zexmenia frutescens

KEY

1. Living fence posts (that is, deliberately planted from stakes).
  2. Leaves simple. Trees with milky latex........ *Ficus Goldmanii*
  2. Leaves compound.
    3. Leaves digitately compound........ *Tabebuia chrysantha*
    3. Leaves pinnate.
      4. Bark thin, often coppery red........ *Bursera simaruba*
      4. Bark not coppery red.
        5. Leaflets 3........ *Erythrina costaricensis*
        5. Leaflets more than 3.
          6. Leaflets conspicuously rounded at the apex. Flowers red or yellow.
            7. Leaflets usually more than 10, sour to the taste. Flowers minute, red, followed by red or yellow, plumlike fruits........ *Spondias purpurea*
            7. Leaflets usually less than 10, not sour to the taste. Flowers yellow, conspicuous, followed by an inflated, papery pod........ *Diphysa robinioides*

6. Leaflets conspicuously pointed at the apex.
   7. Flowers pink, showy, followed by a long, beanlike pod .... *Gliricidia sepium*
   7. Flowers greenish or yellowish, followed by small 3-angled fruits .... *Bursera simaruba*

1. Not living fence posts, or at least not deliberately planted for that purpose from stakes.
  2. Leaves compound.
    3. Leaves digitately compound.
      4. Leaves opposite. Leaflets 3 .... *Vitex Cooperi*
      4. Leaves alternate. Leaflets 5–10.
        5. Leaflets 5 .... *Tabebuia chrysantha*
        5. Leaflets 7–10 .... *Didymopanax Morototoni*
    3. Leaves pinnately compound.
      4. Leaflets 3. Flowers red .... *Erythrina costaricensis*
      4. Leaflets more than 3. Flowers not red.
        5. Bark thin, often coppery red .... *Bursera simaruba*
        5. Bark not coppery red.
          6. Inflorescences axillary.
            7. Individual leaflets about 1″ long .... *Phyllanthus acuminatus*
            7. Individual leaflets more than 1″ long .... *Bursera simaruba*
          6. Inflorescences terminal.
            7. Bark rough, strongly corrugated. Fruits plumlike, yellow .... *Spondias mombin*
            7. Bark not rough or strongly corrugated. Fruits not plumlike.
              8. Pinnae very large, often more than 2 ft. long .... *Simaba cedron*
              8. Pinnae much less than 2 ft. long.
                9. Leaflets 5 .... *Cupania largifolia*
                9. Leaflets more than 5 .... *Sapindus saponaria*
  2. Leaves simple.
    3. Leaves in pairs or in whorls of 4 at the nodes.
      4. Leaves in whorls of 4 at each node .... *Hamelia patens*
      4. Leaves in pairs at each node.
        5. Leaf blade with 3 or more longitudinal nerves .... *Miconia argentea*
        5. Leaf blade with a single longitudinal nerve.
          6. Sap orange, drying blood red.
            7. Petals white .... *Vismia ferruginea*
            7. Petals pale orange .... *Vismia guianensis*
          6. Sap not distinctively colored.
            7. Inflorescences or flowers axillary.
              8. Flowers white.
                9. Veins of the lower leaf surface elevated and conspicuous. Fruits sweet .... *Psidium Guajava*
                9. Veins of the lower leaf surface inconspicuous. Fruits acid .... *Psidium Friedrichsthalianum*
              8. Flowers yellow .... *Siparuna patelliformis*
            7. Inflorescences terminal.

8. Branches spiny............ *Randia armata*
8. Branches not spiny.
9. Flowers bright yellow, daisy-like............ *Zexmenia frutescens*
9. Flowers not bright yellow or daisy-like.
10. Inflorescences unbranched racemes............ *Citharexylum viride*
10. Inflorescences paniculate cymes............ *Aegiphila martinicensis*
3. Leaves solitary at each node.
4. Leaves palmately lobed.
5. Leaves peltate.
6. Hollow branches inhabited by stinging ants. Inflorescences pendulous.
7. Trunks and young branches green or white, conspicuously ringed.
8. Fruiting spikes 3–4″ long. Undersurfaces of leaves white............ *Cecropia peltata*
8. Fruiting spikes 8–16″ long. Undersurfaces of leaves green............ *Cecropia obtusifolia*
7. Trunks and young branches blackish, not conspicuously ringed............ *Cecropia Sandersoniana*
6. Branches not inhabited by stinging ants. Inflorescences erect............ *Ricinus communis*
5. Leaves not peltate.
6. Flowers bright yellow............ *Cochlospermum vitifolium*
6. Flowers white.
7. Leaves very deeply and conspicuously lobed............ *Carica papaya*
7. Leaves shallowly lobed.
8. Flowers usually more than 4″ long............ *Ochroma lagopus*
8. Flowers less than 2″ long............ *Hampea platanifolia*
4. Leaves not palmately lobed.
5. Flowers or inflorescences terminal.
6. Trees, 40–60 ft. in height, with abundant milky latex............ *Sapium thelocarpum*
6. Shrubs or small trees, 12–18 ft. in height, without milky latex.
7. Flowers red or dark pink.
8. Flowers hibiscus-like, more than 1″ long............ *Malvaviscus arboreus* var. *penduliflorus*
8. Flowers not as above, about ⅛″ long............ *Coccoloba acuminata*
7. Flowers white............ *Solanum verbascifolium* var. *adulterinum*
5. Flowers or inflorescences axillary.
6. Inflorescences slender, erect, unbranched spikes.
7. Leaves with 5 longitudinal nerves............ *Piper reticulatum*
7. Leaves with a single longitudinal nerve.
8. Individual flowers minute, distinguishable only with a lens. Base of the leaf blade very strongly asymmetrical............ *Piper arboreum*
8. Individual flowers ¼″ or more in diameter. Base of the leaf blade not asymmetrical.
9. Crushed leaves with an offensive, skunklike odor............ *Roupala complicata*
9. Crushed leaves not as above............ *Croton glabellus*
6. Inflorescences not erect, unbranched spikes.

7. Inflorescences conspicuous, white, pendulous, threadlike strands, often 2 ft. or more in length........ *Myriocarpa longipes*
7. Inflorescences not as above.
  8. Leaf bases covered by nearly circular, peltate, foliaceous stipules. Flowers bright yellow, attractive........ *Dicraspidia Donnell-Smithii*
  8. Leaf bases not covered by foliaceous stipules. Flowers not bright yellow.
    9. Flowers in pendulous, scorpioid cymes... *Cyphomandra costaricensis*
    9. Flowers not in pendulous scorpioid cymes, mostly in short, sessile fascicles in the leaf axils.
      10. Flowers relatively large, 1″ or more in diameter when fresh.
        11. Leaf blades with a single longitudinal nerve. Flowers more or less fleshy or leathery........ *Rollinia Jimenezii*
        11. Leaf blades with 3 longitudinal nerves.
          12. Leaf petioles about ½ the length of the blades, which are nearly as broad as long........ *Hampea Allenii*
          12. Leaf petioles much less than ½ the length of the blades, which are about 3 times as long as broad........ *Muntingia calabura*
      10. Flowers relatively small, much less than 1″ in diameter when fresh.
        11. Leaves superficially bipinnate in appearance. Fruit a 3-celled capsule........ *Phyllanthus acuminatus*
        11. Leaves not superficially bipinnate. Fruit not a 3-celled capsule.
          12. Flowers in short axillary cymes.
            13. Leaf blades with a single longitudinal nerve........ *Guazuma ulmifolia*
            13. Leaf blades with 3 longitudinal nerves........ *Trema micrantha*
          12. Flowers in sessile axillary fascicles.
            13. Leaf margins not serrate or crenate........ *Casearia sylvestris*
            13. Leaf margins serrate or crenate.
              14. Leaf margins very coarsely and densely serrate........ *Casearia arguta*
              14. Leaf margins finely and minutely serrate, or sinuate-serrate.
                15. Leaf blades usually less than 3″ long, obtuse or narrowly rounded at the apex. Plants often armed with spines........ *Casearia aculeata*
                15. Leaf blades usually more than 4″ long, acute or acuminate at the apex. Plants unarmed.
                  16. Leaves broadest above the middle, the margins minutely serrate...... *Casearia guianensis*
                  16. Leaves broadest at about the middle, with finely sinuate-serrate margins........ *Casearia myriantha*

(13)—SECOND GROWTH

Wherever primary forest is felled, a rank, weedy growth soon springs up, composed in part of robust herbs such as *Neurolaena lobata,* and various of the broad-leaved *Calatheas* and *Heliconias,* but usually quickly dominated by arborescent species which were either rare or entirely absent in the original stand. These changelings are of widely divergent families and genera, but have one thing in common; they thrive on sunlight and, to a lesser degree, on fire and disturbed land. Their origins are rather poorly understood, but it seems apparent that some are derived from riverbank and gravel bar communities, while others are from rare elements in the primary forest. In the former class is *Tessaria mucronata,* which normally grows in pure stands on bars and islands in fast-flowing streams, but which occasionally appears in second-growth scrub, especially along the spoil banks of drainage canals. Balsa *(Ochroma lagopus)* is probably also in this general category, as are *Muntingia calabura* and *Trema micrantha.* The long dormant seeds of Balsa seem to be stimulated by fire, and this may be true also of other species. All the *Cecropias* are found in the climax rain forest as very tall, slender, stilt-rooted specimens, but are very rare, and are probably dependent on chance breaks in the canopy for their development. Artificial clearings, of course, provide ideal conditions for such species. A considerably greater number of species invade the hillside clearings than are found on the flat land, which may reflect the better drainage conditions, or indicate an inherent preference for clay. Hillside clearings in particular seem to go through a long cycle, usually involving a succession of three or more distinct populations before something approximating the original climax association can be re-established. A fairly typical, if somewhat simplified, succession might begin with a rapidly developing herbaceous cover, composed of *Neurolaena* and various *Heliconias* and *Calatheas,* followed for a few years by short-lived trees such as *Ochroma lagopus, Muntingia calabura,* and *Trema micrantha.* These tend to be gradually replaced by *Goethalsia meiantha* and *Didymopanax Morototoni,* or an association of *Schizolobium parahybum* and *Hieronyma tectissima,* or nearly pure consociations of various species of *Vochysia,* or sometimes *Brosimum utile.* These last will probably reach the height of the general canopy, and persist for many years, being slowly invaded thereafter by other less specialized forest elements. If the clearing is taken over by grass, and subjected to annual burning, an artificial savanna will become established.

SECOND GROWTH

Principal species

Apeiba Tibourbou—Hillsides.
Belotia macrantha—Hillsides.
Belotia reticulata—Hillsides.
Carica papaya—Mostly floodplains.
Casearia arborea—Hillsides.
Casearia arguta—Mostly hillsides.
Cecropia obtusifolia—Both hillsides and floodplains.
Cecropia peltata—Both hillsides and floodplains.
Cecropia Sandersoniana—Both hillsides and floodplains.
Cochlospermum vitifolium—Hillsides.
Croton panamensis—Mostly hillsides.
Croton xalapensis—Hillsides.
Cyphomandra costaricensis—Mostly floodplains.
Didymopanax Morototoni—Hillsides.
Goethalsia meiantha—Both hillsides and floodplains.
Hamelia patens—Hillsides.
Hampea platanifolia—Floodplains.
Luehea Seemannii—Both hillsides and floodplains.
Miconia argentea—Mostly hillsides.
Muntingia calabura—Both hillsides and floodplains.
Myriocarpa longipes—Both hillsides and floodplains.
Ochroma lagopus—Both hillsides and floodplains.
Piper arboreum—Both hillsides and floodplains.
Piper reticulatum—Both hillsides and floodplains.
Psidium Guajava—Both hillsides and floodplains.
Rollinia Jimenezii—Both hillsides and floodplains.
Sapium thelocarpum—Both hillsides and floodplains.
Schizolobium parahybum—Hillsides.
Solanum verbascifolium var. *adulterinum*—Both hillsides and floodplains.
Spondias mombin—Both hillsides and floodplains.
Tessaria mucronata—Floodplains.
Trema micrantha—Both hillsides and floodplains.
Vismia ferruginea—Mostly hillsides.
Vismia guianensis—Mostly hillsides.
Vochysia ferruginea—Hillsides.
Zexmenia frutescens—Hillsides.

KEY

1. Leaves compound.
   2. Leaves digitately compound .... *Didymopanax Morototoni*
   2. Leaves pinnately compound.
      3. Leaves bipinnate .... *Schizolobium parahybum*
      3. Leaves once pinnate .... *Spondias mombin*
1. Leaves simple.

2. Leaves opposite or in whorls of 4 at the nodes.
 3. Leaves in whorls of 4 at the nodes........ *Hamelia patens*
 3. Leaves in pairs at each node.
  4. Sap orange, drying blood red.
   5. Leaves coarsely brown tomentose on the lower surface. Petals white........ *Vismia ferruginea*
   5. Leaves pale brown or whitish on the lower surface. Petals pale orange........ *Vismia guianensis*
  4. Sap not distinctively colored.
   5. Leaves with 3 or more longitudinal nerves.
    6. Leaves broadly oblong-ovate, conspicuously whitish or pale tan on the lower surface. Flowers white........ *Miconia argentea*
    6. Leaves not as above. Flowers yellow........ *Zexmenia frutescens*
   5. Leaves with a single longitudinal nerve.
    6. Bark smooth, pale tan, often peeling in irregular patches. Flowers white. Fruits yellow and edible........ *Psidium Guajava*
    6. Bark not as above. Flowers orange. Fruits not edible... *Vochysia ferruginea*
2. Leaves alternate.
 3. Leaf blades palmately lobed.
  4. Leaves peltate. Hollow branches inhabited by stinging ants.
   5. Trunks and young branches green or white, conspicuously ringed.
    6. Leaves white on the lower surface. Fruiting spikes 3–4″ long........ *Cecropia peltata*
    6. Leaves green on the lower surface. Fruiting spikes 8–16″ long........ *Cecropia obtusifolia*
   5. Trunks and young branches blackish, not conspicuously ringed........ *Cecropia Sandersoniana*
  4. Leaves not peltate. Branches not inhabited by stinging ants.
   5. Leaves shallowly lobed.
    6. Flowers 4–6″ long. Leaves usually more than 1 ft. in diameter. Trunks usually solitary........ *Ochroma lagopus*
    6. Flowers less than 2″ long. Leaves less than 1 ft. in diameter. Trunks often multiple........ *Hampea platanifolia*
   5. Leaves deeply and conspicuously lobed.
    6. Leaf blades 1 ft. or more in diameter, the primary lobes also very deeply dissected........ *Carica papaya*
    6. Leaf blades much less than 1 ft. in diameter, the primary lobes not dissected........ *Cochlospermum vitifolium*
 3. Leaf blades not lobed.
  4. Trees with abundant milky latex........ *Sapium thelocarpum*
  4. Trees without milky latex.
   5. Leaf blades with 3–5 longitudinal nerves.
    6. Leaf blades with 5 distinct longitudinal nerves........ *Piper reticulatum*
    6. Leaf blades with 3 longitudinal nerves.
     7. Inflorescences definitely axillary.
      8. Individual flowers minute, less than ⅛″ in diameter. Leaf blades essentially symmetrical at the base, very rough to the touch........ *Trema micrantha*

8. Individual flowers about 1" in diameter when fresh. Leaf blades very strongly asymmetrical at the base, not rough to the touch .... *Muntingia calabura*

7. Inflorescences terminal.

8. Inflorescences erect, unbranched racemes. Sap turns red on exposure to the air.

9. Leaves usually densely stellate-tomentose, particularly on the lower surface, the blades typically not broadly cordate at the base .... *Croton xalapensis*

9. Leaves usually sparsely stellate-tomentose on the lower surface, the blades typically broadly cordate at the base .... *Croton panamensis*

8. Inflorescences branching cymes or panicles. Sap not red.

9. Leaves conspicuously brownish on the lower surface .... *Luehea Seemanii*

9. Leaves not conspicuously brownish on the lower surface.

10. Flowers pale yellow. Individual seeds winged .... *Goethalsia meiantha*

10. Flowers pink, lavender, or white. Individual seeds not winged.

11. Flowers appearing dark pink when seen from a distance. Leaves broadly elliptic-oblong .... *Belotia macrantha*

11. Flowers nearly white when seen from a distance. Leaves narrowly lanceolate .... *Belotia reticulata*

6. Leaf blades with a single longitudinal nerve.

7. Inflorescences terminal.

8. Margins of leaves conspicuously serrate. Flowers yellow, followed by spiny, sea urchin-like fruits .... *Apeiba Tibourbou*

8. Margins of leaves not serrate. Flowers white or lavender. Fruits not as above.

9. Flowers white. Leafy branches densely covered with minute stellate hairs, rough to the touch .... *Solanum verbascifolium* var. *adulterinum*

9. Flowers lavender. Leafy branches not rough to the touch .... *Tessaria mucronata*

7. Inflorescences axillary.

8. Inflorescences conspicuously pendulous.

9. Leaf margins serrate. Flowers and fruits minute, on long, white, threadlike scapes .... *Myriocarpa longipes*

9. Leaf margins entire. Flowers green, with purple stamens, in pendulous scorpioid cymes .... *Cyphomandra costaricensis*

8. Inflorescences not pendulous.

9. Inflorescences essentially sessile in the leaf axils.

10. Leaf margins serrate. Flowers small, white, in dense clusters.

11. Leaf axils of the younger branches with conspicuous linear stipules .... *Casearia arborea*

11. Leaf axils without conspicuous stipules .... *Casearia arguta*

10. Leaf margins not serrate. Flowers rather fleshy, about ¾" in diameter, usually solitary .... *Rollinia Jimenezii*

9. Inflorescences erect, unbranched spikes .... *Piper arboreum*

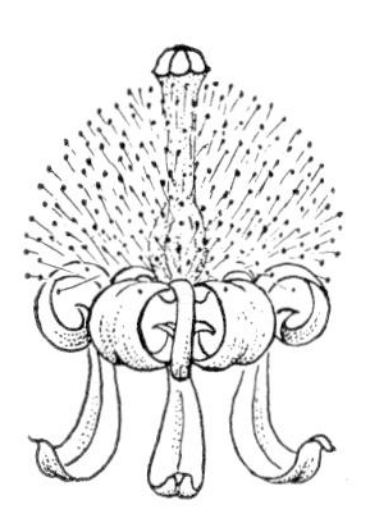

CONSPICUOUS OR DISTINCTIVE SPECIES

CONSPICUOUS FLOWERS

IT WOULD APPEAR that the flowering season of many of our species is determined by periods of rest or growth, which are correlated with the occurrence of dry or rainy weather, a very flexible schedule in actual practice. Trees which normally flower in December or January may be forced into bloom in late November in an abnormally dry year, or one that usually flowers in May or June may appear as early as February after a freakish rainstorm. Some of the *Ingas* flower twice a year, at times of heaviest precipitation, while the exceedingly handsome red *Phyllocarpus* shows itself on the hillsides only after a very dry December.

The species can also be divided into two groups, which might be called EPHEMERALS and DURABLES. The first are composed of genera like *Peltogyne* and *Chimarrhis,* which remain in bloom for only a few days. The second consists of trees such as *Vochysia* and *Macrocnemum,* which persist for several months. Most of the Ephemerals are definitely limited to one season or the other, but some of the Durables, while brought into flower by either dry or rainy weather, last well into the succeeding period.

The following groups include only such species as are showy enough to be noticed from the air, or from a passing motor car. It would be well to note however that one must first be certain that it is the tree itself that is in bloom, and that flowers rather than new leaves have been seen. Red, for example, is a rather uncommon color in the flowers of our local trees, but is very frequently seen in the flushes of new foliage of things like *Carapa Slateri* and *Tetragastris panamensis,* which may be mistaken for inflorescences when seen from a distance. Trees like *Miconia argentea, Luehea Seemannii,* and *Didymopanax Morototoni,* in which the undersurfaces of the leaves are a strongly contrasting white or tan, produce

the illusion of being in bloom when the wind turns the foliage. Some mistletoes, such as *Phoradendron robustissimum,* form conspicuous rounded masses of reddish-brown in the tops of trees, many examples being seen from the air in the hills between the Río Esquinas and Golfito. Epiphytic shrubs, such as *Satyria Warscewiczii,* frequently produce bright-red terminal flushes of new leaves, which are particularly conspicuous in the crowns of some of the larger trees during the month of November. Finally, lianas, which climb to the tree tops, are often very handsome when seen from afar, notable examples being *Cassia spinescens,* whose bright-yellow flowers are frequent in the hills between Farm 18 and Golfito in November and early December, or some of the dark-pink-flowered *Bignoniaceae,* such as *Paragonia pyramidata* or *Arrabidaea candicans,* which may bloom at irregular intervals throughout the year. Once it has been definitely determined that it is a tree that has been seen, the following keys, based on color, may be of some help.

FLOWERS BLUE

1. Leaves digitately compound. Flowering season June to August.......*Vitex Cooperi*
1. Leaves pinnate.
   2. Inflorescences terminal. Secondary leaf petioles not winged.......*Jacaranda copaia*
   2. Inflorescences produced from the axils of the current new growth. Secondary leaf petioles narrowly winged.......*Jacaranda lasiogyne*

FLOWERS PURPLE, LAVENDER, OR PINKISH LAVENDER

1. Leaves simple. Flowers pinkish lavender, October to January.......*Belotia reticulata*
1. Leaves pinnate.
   2. Leaves even-pinnate, with 2 terminal leaflets. Flowers pinkish lavender, July and August.......*Tachigalia versicolor*
   2. Leaves odd-pinnate, with 1 terminal leaflet.
      3. Sap dries blood red. Flowers pinkish lavender, early February.......*Lonchocarpus sericeus* var. *glabrescens*
      3. Sap not red.......*Andira inermis*

FLOWERS OR FRUITS PINK OR PINKISH TAN

1. Flowers or fruits produced during the dry season (December through April).
   2. Leaves simple.
      3. Colored fruits or their covering the conspicuous element.
         4. Individual fruits enveloped in a showy, dark-pink, winged calyx.
            5. Calyx lobes 5. Trees conspicuous in December and January.......*Pentaplaris Doroteae*

5. Calyx lobes 3. Trees conspicuous in February and March........ *Triplaris melaenodendron*
4. Individual fruits with a marginal fringe of bristles........ *Heliocarpus appendiculatus*
3. Flowers the conspicuous element.
4. Leaves with 3 longitudinal nerves.
5. Flowers dark pink........ *Belotia macrantha*
5. Flowers pale pinkish lavender........ *Belotia reticulata*
4. Leaves with a single longitudinal nerve.
5. Leaves alternate. Flowers pale pinkish tan........ *Ardisia Dunlapiana*
5. Leaves opposite. Flowers dark pink, produced from December to April........ *Macrocnemum glabrescens*
2. Leaves compound.
3. Leaves digitately compound........ *Tabebuia pentaphylla*
3. Leaves pinnate.
4. Mature leaflets less than 3″ long.
5. Leaflets rounded at the apex. Flowers dark pink........ *Cassia grandis*
5. Leaflets acute or acuminate. Flowers pale pink........ *Gliricidia sepium*
4. Mature leaflets more than 3″ long.
5. Floral calyx deeply lobed........ *Dussia mexicana*
5. Floral calyx not deeply lobed........ *Lonchocarpus sericeus* var. *glabrescens*
1. Flowers produced during the rainy season (May through November).
2. Flowering season July and August. Individual florets actually yellow when seen close at hand........ *Tachigalia versicolor*
2. Flowering season September and October. Individual flowers dark pink........ *Couratari panamensis*

FLOWERS OR FRUITS RED OR REDDISH BROWN

1. Colored fruits or their covering the conspicuous element.
2. Individual fruits enveloped in a 5-lobed calyx........ *Pentaplaris Doroteae*
2. Individual fruits not as above........ *Casearia banquitana* var. *laevis*
1. Flowers the conspicuous element.
2. Trees confined to sea beaches and margins of mangrove swamp... *Hibiscus tiliaceus*
2. Trees never found near sea beaches.
3. Trees with yellow, sticky latex........ *Symphonia globulifera*
3. Trees without yellow latex.
4. Leaves simple.
5. Inflorescences blood red, superficially poinsettia-like........ *Warscewiczia coccinea*
5. Inflorescences not as above........ *Sterculia apetala*
4. Leaves compound.
5. Leaflets 2........ *Cynometra hemitomophylla*
5. Leaflets more than 2.
6. Leaflets 3. Small trees........ *Erythrina costaricensis*
6. Leaflets more than 3. Tall trees........ *Phyllocarpus septentrionalis*

CONSPICUOUS FLOWERS

FLOWERS ORANGE, YELLOW, OR TAN

1. Flowers produced during the dry season (December through April).
  2. Leaves simple.
    3. Leaves about as broad as long. Trees mostly confined to sea beaches........ *Hibiscus tiliaceus*
    3. Leaves much longer than broad. Trees of hillside forests...... *Vochysia*, all species
  2. Leaves compound.
    3. Leaves digitately compound........ *Tabebuia chrysantha*
    3. Leaves pinnately compound.
      4. Leaves bipinnate (having a central nonwoody axis and 2 or more lateral pinnae)........ *Schizolobium parahybum*
      4. Leaves once-pinnate (having a central axis and 2 or more leaflets)........ *Caryocar costaricense*
        5. Leaflets 3.
        5. Leaflets more than 3.
          6. Pinnate leaves opposite on the stems........ *Platymiscium pinnatum*
          6. Pinnate leaves alternate.
            7. Flowers mimosa-like, white, aging tan........ *Inga sapindoides*
            7. Flowers rich yellow or orange, not mimosa-like.
              8. Leaves even-pinnate, with two terminal leaflets........ *Cassia reticulata*
              8. Leaves odd-pinnate, with a single terminal leaflet........ *Pterocarpus Hayesii*
1. Flowers produced during the rainy season (May through November).
  2. Leaves simple.
    3. Flowering season May to July.
      4. Leaves broadly cordate. Trees mostly confined to sea beaches........ *Hibiscus tiliaceus*
      4. Leaves not cordate. Trees of hillside forests........ *Vochysia*, all species
    3. Flowering season August to October.
      4. Leaves broadly heart-shaped........ *Hibiscus tiliaceus*
      4. Leaves not heart-shaped.
        5. Flowers pale yellow.
          6. Leaves with 3 longitudinal nerves........ *Goethalsia meiantha*
          6. Leaves with a single longitudinal nerve........ *Ocotea Ira*
        5. Flowers dark yellow.
          6. Small, understory trees, usually less than 30 ft. in height... *Ouratea Valerii*
          6. Hillside trees, mostly 50–60 ft. in height........ *Cespedesia macrophylla*
  2. Leaves compound.
    3. Leaves digitately compound........ *Tabebuia chrysantha*
    3. Leaves pinnately compound.
      4. Leaves bipinnate (having a central nonwoody axis and 2 or more lateral pinnae)........ *Schizolobium parahybum*
      4. Leaves once-pinnate (having a central axis and 2 or more lateral leaflets).
        5. Leaves even-pinnate, with 2 terminal leaflets.
          6. Terminal leaflets more than 5″ long........ *Cassia reticulata*
          6. Terminal leaflets less than 3″ long........ *Cassia spectabilis*
        5. Leaves odd-pinnate, with a single terminal leaflet........ *Pterocarpus officinalis*

CONSPICUOUS FLOWERS

FLOWERS WHITE OR GREENISH WHITE

1. Flowers produced during the dry season (December through April).
  2. Trees with conspicuous stilt roots .... *Bravaisia integerrima*
  2. Trees without stilt roots.
    3. Leaves compound.
      4. Leaves digitately compound .... *Billia colombiana*
      4. Leaves pinnately compound.
        5. Leaves bipinnate (having a central nonwoody axis and 2 or more lateral pinnae) .... *Pithecolobium macradenium*
        5. Leaves once-pinnate (having a central axis and 2 or more lateral leaflets).
          6. Flowers mimosa-like.
            7. Leaves with 2–3 pairs of leaflets .... *Inga sapindoides*
            7. Leaves with more than 3 pairs of leaflets .... *Inga Ruiziana*
          6. Flowers not mimosa-like .... *Spondias mombin*
    3. Leaves simple.
      4. Leaves opposite.
        5. Flowers subtended by a broad and conspicuous white calyx lobe .... *Calycophyllum candidissimum*
        5. Flowers not subtended by a broad calyx lobe.
          6. Flowers with conspicuous slender petaloid bracteoles .... *Linociera panamensis*
          6. Flowers not as above .... *Chione costaricense*
      4. Leaves alternate.
        5. Leaves with 5 longitudinal nerves .... *Hasseltia quinquenervia*
        5. Leaves with a single longitudinal nerve.
          6. Forks of young branches usually with gall-like hollow swellings inhabited by ants. Leaves with an onion-like odor when crushed .... *Cordia alliodora*
          6. Forks of young branches not as above. Leaves without onion-like odor.
            7. Flowering season December and January.
              8. Leaves long-acuminate at the apex .... *Nectandra salicifolia*
              8. Leaves recurved or emarginate at the apex .... *Vantanea Barbourii*
            7. Flowering season March and April .... *Homalium eurypetalum*
1. Flowers produced during the rainy season (May through November).
  2. Leaves simple.
    3. Leaves opposite.
      4. Flowers subtended by a broad white petaloid calyx lobe .... *Calycophyllum candidissimum*
      4. Flowers not subtended by a petaloid calyx lobe .... *Chimarrhis latifolia*
    3. Leaves alternate.
      4. Leaves very broad, palmately 5–7 lobed. Flowers large, axillary .... *Ochroma lagopus*
      4. Leaves elliptic-oblong, with 5 longitudinal nerves. Flowers small, in terminal panicles .... *Hasseltia quinquenervia*
  2. Leaves compound.
    3. Leaflets 2. Wood purple .... *Peltogyne purpurea*
    3. Leaflets more than 2. Wood not purple.
      4. Flowers mimosa-like.

5. Leaf rachis winged .... *Inga marginata*
5. Leaf rachis not winged.
  6. Leaflets 2–4 pairs .... *Inga punctata*
  6. Leaflets 5–10 pairs .... *Inga Ruiziana*
4. Flowers not mimosa-like .... *Sapindus saponaria*

CONSPICUOUS VEGETATIVE CHARACTERS

On entering a tropical rain forest, one's first impressions are likely to be of dark, majestic columns, some of which have, upon closer examination, conspicuous characters of trunk, bark, or sap. No matter what their botanical affinities may be, these are usually all that can be seen until the tree is felled or climbed. Dozens of times in the field I have wished that I had some ready means of tracing a species distinguished by some striking feature, such as blood-red sap, or a peculiarly winged leaf petiole, without having to wait years for a lucky find of flowers.

As has been previously noted, native woodsmen make use of conspicuous vegetative characters as a primary means of approach, since these have obvious advantages over more orthodox methods. The following keys are the sort of thing that I wish might have been available to me some twenty years ago when I first began my work in the tropical forests, since they would have helped to solve many a perplexing problem. It was my original intention to have this project cover a considerably greater area, but circumstances beyond my control made this impossible. Most of the species cited however, have very extensive ranges in Central America outside of the Golfo Dulce area, so that it is hoped that the present effort, imperfect and restricted as it is, may be of some aid to those who, through necessity or inclination, spend part of their days in this vast, beautiful, and unbelievably complex wonderland.

TREES WITH VERY LARGE BUTTRESSES

1. Sap milky, or otherwise distinctively colored.
  2. Sap milky.
    3. Fruits hollow.
      4. Leaves more than 10″ long .... *Ficus Werckleana*
      4. Leaves less than 10″ long .... *Ficus lapathifolia*
  2. Sap red or black.
    3. Leaves very large, palmately lobed. Sap turns black on exposure to air .... *Cecropia Sandersoniana*
    3. Leaves not as above.
      4. Leaves simple. Sap black .... *Hieronyma tectissima*
      4. Leaves pinnate. Sap red.

5. Flowers pale orange. Trees mostly in swamps .... *Pterocarpus officinalis*
5. Flowers pink. Trees mostly on hillsides .... *Dussia macrophyllata*

1. Sap not distinctively colored.
  2. Leaves simple.
    3. Trees confined to mangrove swamps .... *Pelliciera rhizophorae*
    3. Trees not found in mangrove swamps.
      4. Leaves very large, palmately lobed .... *Sterculia apetala*
      4. Leaves not as above.
        5. Leaves opposite .... *Chimarrhis latifolia*
        5. Leaves alternate.
          6. Leaves with 3 longitudinal nerves.
            7. Lower leaf surface brownish .... *Luehea Seemannii*
            7. Lower leaf surface not brownish .... *Goethalsia meiantha*
          6. Leaves with a single longitudinal nerve.
            7. Bark thin, pale tan or nearly white, peeling or flaking; guava-like in superficial appearance .... *Terminalia lucida*
            7. Bark not as above.
              8. Fruit a 5-valved capsule, often woody.
                9. Fruits densely covered with purple, needle-like spines .... *Sloanea picapica*
                9. Fruits not spiny.
                  10. Fruits about 9″ long, pendulous .... *Huberodendron Allenii*
                  10. Fruits less than 3″ long, not pendulous.
                    11. Leaves glossy, usually less than 6″ long. Valves of fruit very thick .... *Sloanea laurifolia*
                    11. Leaves not glossy, usually more than 6″ long. Valves of fruit thin .... *Mortoniodendron anisophyllum*
              8. Fruit not a 5-valved capsule .... *Couratari panamensis*
  2. Leaves compound.
    3. Leaves digitately compound.
      4. Flowers about 1½″ long. Pods filled with woolly fiber .... *Ceiba pentandra*
      4. Flowers about ½″ long. Pods not filled with woolly fiber ... *Sterculia mexicana*
    3. Leaves pinnately compound.
      4. Leaves bipinnate (having a central nonwoody axis and 2 or more lateral pinnae).
        5. Flowers white. Fruit a coiled, earlike pod .... *Enterolobium cyclocarpum*
        5. Flowers yellow. Fruit not as above .... *Schizolobium parahybum*
      4. Leaves once-pinnate (having a central axis and 2 or more lateral leaflets).
        5. Leaflets 2.
          6. Leaflets less than 3″ long. Heartwood purple .... *Peltogyne purpurea*
          6. Leaflets more than 4″ long. Heartwood not purple .... *Cynometra hemitomophylla*
        5. Leaflets more than 2.
          6. Flowers conspicuous from a distance.
            7. Flowers pink or pinkish lavender.
              8. Leaflets 7–11, alternate on the rachis .... *Dussia mexicana*
              8. Leaflets 12–14, opposite on the rachis .... *Tachigalia versicolor*

6. Flowers not conspicuous from a distance.
   7. Flowers olive green........ *Dialium guianense*
   7. Flowers white or greenish white.
      8. Margins of leaflets serrate........ *Astronium graveolens*
      8. Margins of leaflets not serrate........ *Carapa Slateri*

TREES WITH RED OR REDDISH-BROWN BARK

1. Plants epiphytic, growing in the tops of other trees........ *Dacryodes epiphytica*
1. Plants not epiphytic.
   2. Leaves simple.
      3. Leaves opposite........ *Calycophyllum candidissimum*
      3. Leaves alternate........ *Ximenia americana*
   2. Leaves pinnately compound.
      3. Trunks prominently buttressed. Bark not thin or peeling........ *Tachigalia versicolor*
      3. Trunks not buttressed. Bark thin, conspicuously peeling........ *Bursera simaruba*

TREES WITH PALE TAN OR NEARLY WHITE, THIN, GUAVA-LIKE BARK

1. Tall trees with buttressed trunks........ *Terminalia lucida*
1. Small trees without buttressed trunks.
   2. Trees of trailside thickets and pastures. Calyx undivided in bud, splitting irregularly on flowering........ *Psidium Guajava*
   2. Trees of hillside forests. Calyx with 4 distinct lobes in bud........ *Eugenia palmarum*

TREES WITH CONSPICUOUSLY SCALING BARK

1. Leaves simple.
   2. Leaves opposite.
      3. Flowers white, conspicuous from a distance. Bark reddish........ *Calycophyllum candidissimum*
      3. Flowers greenish yellow, not conspicuous. Bark not reddish........ *Lafoensia punicifolia*
   2. Leaves alternate.
      3. Leaves more than 3" long. Flowers white........ *Vantanea Barbourii*
      3. Leaves less than 3" long. Flowers greenish tan........ *Terminalia amazonia*
1. Leaves pinnately compound.
   2. Leaves bipinnate (having a central nonwoody axis and 2 or more lateral pinnae)........ *Lysiloma guanacastense*
   2. Leaves once-pinnate (having a central axis and 2 or more lateral leaflets).
      3. Trees found mostly along streams near sea level. Flowers mimosa-like........ *Albizzia filicina*
      3. Trees of highland forests, mostly above 4,000 ft. Flowers not mimosa-like........ *Cedrela Tonduzii*

TREES WITH DISTINCTIVE CHARACTERS

TREES WITH THORNY OR SPINY TRUNKS

1. Palms.
  2. Leaves fan-shaped .... *Cryosophila guagara*
  2. Leaves not fan-shaped.
    3. Trunks solitary.
      4. Trunks more than 8″ in diameter. Fruits globose .... *Acrocomia vinifera*
      4. Trunks less than 8″ in diameter. Fruits not globose.
        5. Individual fruits intensely spiny. Pinnae often in broad, confluent blocks .... *Astrocaryum alatum*
        5. Individual fruits not spiny. Pinnae never in confluent blocks .... *Astrocaryum Standleyanum*
    3. Trunks multiple.
      4. Fronds undivided .... *Bactris militaris*
      4. Fronds pinnatisect.
        5. Spines conspicuously pale tan in color; winged or flattened throughout most of their length .... *Bactris divisicupula*
        5. Spines not pale tan or conspicuously flattened.
          6. Spathes densely woolly, but not armed with spines on the expanded part. Fruits red at maturity .... *Bactris Baileyana*
          6. Spathes intensely spiny on the expanded part. Fruits purple at maturity .... *Bactris balanoidea*
1. Not palms.
  2. Leaves simple.
  3. Plants with milky latex.
    4. Leaf petiole with a pair of glands at the base of the blade .... *Sapium thelocarpum*
    4. Leaf petiole without glands .... *Lacmellea panamensis*
  3. Plants without milky latex.
    4. Trunks irregularly armed with dense fascicles of long, needle-like spines .... *Xylosma excelsum*
    4. Trunks uniformly covered with short, conical spines .... *Hura crepitans*
  2. Leaves compound.
    3. Leaves digitately compound .... *Ceiba pentandra*
    3. Leaves pinnately compound .... *Zanthoxylum procerum*

TREES WITH DEEPLY FURROWED TRUNKS

1. Trees with milky latex.
  2. Fruits globose, woolly .... *Pouteria neglecta*
  2. Fruits ellipsoidal, smooth .... *Calocarpum borucanum*
1. Trees without milky latex.
  2. Leaves opposite .... *Macrocnemum glabrescens*
  2. Leaves alternate.
    3. Flowers orange. Leaves about 20″ long .... *Meliosma Allenii*
    3. Flowers whitish. Leaves less than 8″ long.
      4. Leaves 3–5″ long, obtuse or subacute at the apex .... *Maytenus pallidifolius*
      4. Leaves about 6″ long, acute or acuminate at the apex .... *Diospyros ebenaster*

TREES WITH DISTINCTIVE CHARACTERS

FLOWERS AND FRUITS PRODUCED DIRECTLY FROM THE TRUNK AND BRANCHES

1. Leaves very large, usually more than 3 ft. in length .... *Grias Fendleri*
1. Leaves less than 2 ft. in length.
    2. Leaves produced in sessile fascicles of 3–12 directly from the branches .... *Crescentia cujete*
    2. Leaves not produced in fascicles.
        3. Leaves simple.
            4. Flowers white. Fruits globose .... *Carpotroche platyptera*
            4. Flowers pink or red. Fruits not globose.
                5. Leaves conspicuously paler in color on the lower surface. Flowers red .... *Theobroma simiarum*
                5. Leaves not paler in color on the lower surface. Flowers pink .... *Theobroma cacao*
        3. Leaves compound.
            4. Leaves digitately compound .... *Herrania purpurea*
            4. Leaves pinnate .... *Parmentiera macrophylla*

TREES WITH STILT ROOTS

1. Palms.
    2. Mature specimens with stilt roots more than 6 ft. in height. Stamens more than 50. Seeds with an apical embryo .... *Socratea durissima*
    2. Mature specimens with stilt roots 2–3 ft. in height. Stamens about 14. Seeds with a lateral embryo .... *Iriartea gigantea*
1. Not palms.
    2. Leaves peltate, palmately lobed. Hollow branches inhabited by stinging ants .... *Cecropia*, all species
    2. Leaves not peltate.
        3. Trees with sticky yellow sap .... *Symphonia globulifera*
        3. Trees without yellow sap.
            4. Leaves digitately compound .... *Pachira aquatica*
            4. Leaves simple.
                5. Flowers white, conspicuous from a distance. Trees of pastures and hillside forests below 1,500 ft. .... *Bravaisia integerrima*
                5. Flowers inconspicuous.
                    6. Leaves strongly aromatic when crushed. Trees of highland forests above 2,500 ft. .... *Hedyosmum mexicanum*
                    6. Leaves not aromatic when crushed. Trees of coastal swamps .... *Rhizophora mangle*

TREES WITH MILKY LATEX

1. Leaves opposite.
    2. Trunks spiny .... *Lacmellea panamensis*
    2. Trunks not spiny.

3. Large trees with leathery leaves. Flowers white .... *Calophyllum brasiliense* var. *Rekoi*
3. Small or medium-sized trees with thin leaves. Flowers not white.
    4. Leaves more than 7″ long. Flowers reddish tan, in pendulous cymes .... *Tabernaemontana longipes*
    4. Leaves less than 7″ long. Flowers never in pendulous cymes .... *Stemmadenia*, all species

1. Leaves alternate.
  2. Leaves with two small glands on the petiole .... *Sapium*, both species
  2. Leaves without glands on the petiole.
    3. Leaves clustered at the ends of the branches.
      4. Leaves very leathery. Inflorescences terminal, of showy white flowers .... *Plumeria rubra* forma *acutifolia*
      4. Leaves not leathery. Inflorescences axillary. Flowers not showy.
        5. Fruits woolly, globose .... *Pouteria neglecta*
        5. Fruits elongate, smooth .... *Calocarpum borucanum*
    3. Leaves not clustered at the ends of the branches.
      4. Lower leaf surface reddish brown.
        5. Fruits hollow .... *Ficus Bullenii*
        5. Fruits not hollow .... *Chrysophyllum mexicanum*
      4. Lower leaf surface not reddish brown.
        5. Flowers with definite petals.
          6. Leaves more than 8″ long .... *Pouteria triplarifolia*
          6. Leaves less than 6″ long.
            7. Flowers in dense heads, on slender, often branching scapes up to 1″ or more in length .... *Pouteria heterodoxa*
            7. Flowers in sessile or nearly sessile fascicles.
              8. Leaf petioles about 1″ long, the blades rounded or subacute at the apex .... *Pouteria subrotata*
              8. Leaf petioles about ½″ long, the blades acute or acuminate at the apex.
                9. Flowers completely sessile .... *Pouteria chiricana*
                9. Flowers on slender peduncles about ⅛″ long .... *Chrysophyllum panamense*
        5. Flowers minute, without definite petals.
          6. Leaves very large, conspicuously 2-ranked and pendulous on the branches; rough to the touch .... *Castilla fallax*
          6. Leaves not as above.
            7. Fruits multiple, in sessile clusters, bright red at maturity .... *Perebea*
            7. Fruits solitary or in pairs, not red at maturity.
              8. Leaves with serrate or crenate margins.
                9. Leaves with 10–12 pairs of secondary nerves .... *Batocarpus costaricensis*
                9. Leaves with 16–18 pairs of secondary nerves .... *Olmedia falcifolia*
              8. Leaf margins not serrate or crenate.
                9. Staminate and pistillate flowers produced on separate receptacles .... *Pseudolmedia spuria*

9. Staminate and pistillate flowers produced on the same receptacle.
 10. Fruits hollow........*Ficus*, all species
 10. Fruits not hollow........*Brosimum*, all species

TREES WITH YELLOW OR ORANGE LATEX

1. Leaves alternate.
 2. Fruit a large, woody, laterally compressed, asymmetrical pod, containing many circular, papery, transparent seeds........*Aspidosperma megalocarpon*
 2. Fruit globose, resembling a small Osage orange........*Batocarpus costaricensis*
 2. Fruit not as above........*Brosimum terrabanum*
1. Leaves opposite.
 2. Lower leaf surface brownish. Sap orange, drying blood red......*Vismia*, both species
 2. Lower leaf surface green. Sap bright yellow.
  3. Flowers red. Lower trunk often with stilt roots........*Symphonia globulifera*
  3. Flowers white. Lower trunk never with stilt roots.
   4. Leaves narrowly lanceolate. Fruits smooth........*Rheedia edulis*
   4. Leaves broadly oblong-elliptic. Fruits rough........*Rheedia madruno*

TREES WITH RED OR PINK SAP

1. Leaves pinnate.
 2. Flowers yellow........*Pterocarpus officinalis*
 2. Flowers pink, lavender, or purple.
  3. Stamens free........*Dussia macrophyllata*
  3. Stamens united........*Lonchocarpus sericeus* var. *glabrescens*
1. Leaves simple.
 2. Leaves opposite........*Vismia*, both species
 2. Leaves alternate.
  3. Inflorescences terminal. Fruit a small, 3-celled capsule........*Croton*
  3. Inflorescences axillary. Fruits nutmeg-like, the single seed covered with a fleshy, red, macelike aril.
   4. Aril undivided, or divided only at the apex. Tertiary nerves of the leaves conspicuous........*Compsoneura Sprucei*
   4. Aril divided for more than half its length. Tertiary nerves of the leaves not conspicuous........*Virola*, all species

TREES WITH BLACK SAP OR GUM

1. Leaves pinnate........*Prioria copaifera*
1. Leaves not pinnate.
 2. Leaves palmately lobed.
  3. Leaves 3–5 lobed........*Pourouma aspera*
  3. Leaves 7–13 lobed........*Cecropia*, all species
 2. Leaves simple........*Hieronyma tectissima*

TREES WITH BITTER BARK OR WOOD

1. Leaf petiole winged........*Quassia amara*
1. Leaf petiole not winged.

2. Leaflets subemarginate at the apex.......... *Sweetia panamensis*
2. Leaflets not as above.
 3. Leaflets less than 9.......... *Picramnia latifolia*
 3. Leaflets more than 9.
  4. Trees usually less than 20 ft. in height. Individual leaflets 6–7″ long.......... *Simaba cedron*
  4. Trees usually more than 20 ft. in height.
   5. Leaflets leathery. Fruits olive-like.......... *Simarouba glauca*
   5. Leaflets not leathery. Fruit a capsule, with winged seeds.......... *Cedrela mexicana*

SPECIES WHICH HARBOR STINGING ANTS

1. Leaves pinnate.
 2. Thorns white or pale yellow.......... *Acacia spadicigera*
 2. Thorns dark brown or black.
  3. Flowers in slender spikes.......... *Acacia costaricensis*
  3. Flowers in globose heads.......... *Acacia melanoceras*
1. Leaves not pinnate.
 2. Leaves peltate.......... *Cecropia*, all species
 2. Leaves not peltate.......... *Triplaris melaenodendron*

SPECIES WHICH OCCUR IN CONCENTRATED STANDS

1. Palms.
 2. Leaves fan-shaped.......... *Cryosophila guagara*
 2. Leaves not fan-shaped.
  3. Plants armed with spines (limited to the base of the frond in Corozo).
   4. Trunks massive, solitary.
    5. Tall palms with very spiny trunks and fronds.......... *Acrocomia vinifera*
    5. Short, stout palms. Spines limited to the base of the frond... *Corozo oleifera*
   4. Trunks slender, multiple.
    5. Fronds undivided. Mature fruits red.......... *Bactris militaris*
    5. Fronds pinnatisect. Mature fruits deep purple.......... *Bactris balanoidea*
  3. Plants not armed with spines.
   4. Stilt roots conspicuously developed.......... *Iriartea gigantea*
   4. Plants without stilt roots.
    5. Trunks solitary.
     6. Trunks very short and massive, often reclining. Fruits produced in large, compact clusters deeply seated in the frond axils.......... *Corozo oleifera*
     6. Trunks not as above. Fruits produced in conspicuously pendulous clusters.
      7. Fruits more than 6″ in diameter.......... *Cocos nucifera*
      7. Fruits less than 2″ in diameter.......... *Scheelea rostrata*
    5. Trunks multiple.
     6. Fruits covered with overlapping scales. Species limited to swampy areas.......... *Raphia taedigera*

6. Fruits not scaly. Species found on hillsides............*Oenocarpus panamanus*

1. Not palms.
 2. Leaves simple.
  3. Trees with stilt roots.
   4. Leaves peltate. Hollow branches inhabited by stinging ants..................*Cecropia*
   4. Leaves not peltate. Branches not as above.
    5. Sap yellow..*Symphonia globulifera*
    5. Sap not yellow..*Rhizophora mangle*
  3. Trees without stilt roots.
   4. Leaves peltate.
    5. Hollow branches with stinging ants..*Cecropia*
    5. Branches without stinging ants...*Ricinus communis*
   4. Leaves not peltate.
    5. Sap distinctively colored.
     6. Sap milky...*Brosimum utile*
     6. Sap yellow or red.
      7. Sap yellow..*Symphonia globulifera*
      7. Sap red...*Croton*
    5. Sap not distinctively colored.
     6. Leaves opposite or in whorls of 3 at each node.
      7. Leaves in whorls of 3..*Vochysia hondurensis*
      7. Leaves opposite.
       8. Flowers white.
        9. Inflorescences terminal..*Ixora nicaraguensis*
        9. Inflorescences axillary..*Psidium Guajava*
       8. Flowers orange, yellow, or greenish yellow. (In *Lafoensia* with dark red stamens.)
        9. Forest trees, usually more than 40 ft. in height.
         10. Flowers conspicuous from a distance.
          11. Leaves brownish on the lower surface......*Vochysia ferruginea*
          11. Leaves not as above.......................................*Vochysia Allenii*
         10. Flowers not conspicuous from a distance...*Lafoensia punicifolia*
        9. Trees of pastures and second-growth thickets, usually less than 40 ft. in height.
         10. Leaf margins serrate......................................*Zexmenia frutescens*
         10. Leaf margins not serrate..............................*Byrsonima crassifolia*
     6. Leaves alternate.
      7. Leaves with a single longitudinal nerve.
       8. Trees found only in mangrove swamps..............*Pelliciera rhizophorae*
       8. Trees never in mangrove swamps.
        9. Flowers yellow...*Ouratea Valerii*
        9. Flowers pink, lavender, or white.
         10. Large trees, usually more than 60 ft.
          11. Rachis of inflorescence conspicuously whitish. Fruits kidney-shaped.......................................*Anacardium excelsum*
          11. Rachis of the inflorescence not whitish. Fruits not kidney-shaped.

12. Flowers pink .... *Coccoloba roseiflora*
12. Flowers white .... *Vantanea Barbourii*
10. Small trees, usually less than 30 ft. in height.
11. Leaves broadly rounded at the apex, very rough to the touch .... *Curatella americana*
11. Leaves not rounded at the apex.
12. Flowers whitish .... *Vernonia patens*
12. Flowers pink or lavender.
13. Inflorescences axillary. Flowers dark pink .... *Coccoloba roseiflora*
13. Inflorescences terminal. Flowers lavender .... *Tessaria mucronata*
7. Leaves with 3–7 longitudinal nerves.
8. Leaves at least twice as long as broad. Nerves 3.
9. Flowers pink or lavender .... *Belotia*
9. Flowers white, pale yellow, or green.
10. Inflorescences terminal or subterminal. Flowers pale yellow .... *Goethalsia meiantha*
10. Inflorescences axillary.
11. Flowers white .... *Muntingia calabura*
11. Flowers green .... *Trema micrantha*
8. Leaves about as broad as long. Nerves 5–7.
9. Flowers yellow or reddish brown. Trees found on sea beaches on the margins of mangrove swamp .... *Hibiscus tiliaceus*
9. Flowers white.
10. Flowers usually more than 4" long .... *Ochroma lagopus*
10. Flowers much less than 4" long .... *Hampea platanifolia*
2. Leaves compound.
3. Leaves digitately compound.
4. Leaflets 7–10, brownish on the lower surface .... *Didymopanax Morototoni*
4. Leaflets 5–7, not brownish on the lower surface .... *Pachira aquatica*
3. Leaves pinnately compound.
4. Leaves bipinnate (having a central nonwoody axis and 2 or more lateral pinnae).
5. Flowers yellow .... *Schizolobium parahybum*
5. Flowers white .... *Albizzia filicina*
4. Leaves once-pinnate (having a central axis and 2 or more lateral leaflets).
5. Leaves with 2 terminal leaflets.
6. Leaflets 4.
7. Sap black .... *Prioria copaifera*
7. Sap not black .... *Mora oleifera*
6. Leaflets more than 4 .... *Carapa Slateri*
5. Leaves with a single terminal leaflet.
6. Sap red .... *Pterocarpus officinalis*
6. Sap not red.
7. Inflorescences less than 2" long .... *Trichilia montana*
7. Inflorescences more than 2" long .... *Tetragastris panamensis*

COMMON NAMES

ONE OF THE FUNDAMENTAL ASPECTS of the investigation of any flora is the collection of local common names. These are sometimes highly specific, "Nazareno," for example, being applied only to *Peltogyne purpurea*. Others, such as "Fruta Dorada," may be generic, and used for all species of *Virola;* or, in extreme cases like "Sigua," the name may refer to many species in several genera of the Lauraceae. While such names are unsatisfactory from a technical standpoint, they are important, since they may serve to bridge the gap between the countryman's often intimate knowledge of the local plants and the scientific terms that are the key to the available literature.

The problem in the Golfo Dulce region has been unusually complicated, since the area was almost uninhabited until about 1935. Most of the present population has come from other parts of Costa Rica or from other of the Central American republics, where the plants are only approximately related to ours, and where a different set of common names is in use. A man from Limón, on the Atlantic coast, for example, will know a tree by the name in use there, while a former resident of Chiriquí, Guanacaste, Nicaragua, or Honduras will know the same species by a usually entirely different term. Added to this Spanish, Nahuatl, and Macro-Chibcha tower of Babel are many strictly local names, which do not appear to be known elsewhere, since they are completely absent from the literature. This is understandable, since a relatively high percentage of our trees appear to be endemic, and nearly all closely related to those of the poorly known Atlantic coast. In general, it may be said that countries having a large Indian population, such as Mexico, Guatemala, Salvador, and Honduras, have the most reliable vernacular names, which in many cases have persisted from pre-Columbian times. In countries like Costa Rica and Panama, the names tend to be those of species or genera native to Spain (for example, "Roble," "Cedro," and others), to which the local plants bear some real or fancied resemblance. These usually bristle with qualifying epithets, such as "Macho," "Hembra,"

"Negro," "Colorado," "Silvestre," etc.—all sure signs that the species are not well known. Relatively few of our trees have a single common name of general application throughout the area, and one of the practical local difficulties will be to secure true-to-name material. All the common names known for our species south of Mexico and north of Colombia have been included in the text. It would be well to note however that many of the strictly local terms have been received from a single informant. Subsequent investigation may show some of these to be incorrect. An example of the sort of thing that must be kept constantly in mind might be cited in the cases of *Minquartia guianensis*, a highly durable construction timber, and *Guarea Hoffmanniana*, a nondurable wood suitable for use only in protected situations, both of which are known locally by the name of "Manu." It has been reported that the United Fruit Company purchased a considerable number of railroad ties from a contractor for use in Coto under the vernacular name, expecting to receive *Minquartia*. Ties of *Guarea* were delivered, which proved to be entirely unsuitable, lasting less than a year.

SCIENTIFIC NAMES

SCIENTIFIC NAMES ARE A NECESSARY EVIL, but for the relatively few species that are exploited commercially, should not be too difficult to learn, at least in an age when terms like cyclotron, turbogenerator, aerodynamics, and the technical details of nuclear fission come trippingly from the tongues of high school freshmen. Much is simply a matter of not having the fortitude to try. Who ever heard of anyone who couldn't pronounce *Chrysanthemum?*

This does not imply that difficulties do not exist in scientific terminology. No general catalog of all the Tropical American trees has ever been attempted, and in consequence there are many groups, particularly those whose range extends to the Amazon, in which it is very difficult to arrive at any adequate concept of the relative validity of species. A fair number of local floras are available, but it soon becomes apparent that there has been only a partial and imperfect correlation of the Central and South American names.

Any detailed investigation will also demonstrate that many species, and even genera and families of plants, are in large measure idealized concepts which, as populations, have no actual existence in nature. Each of these artificial entities is represented somewhere in its "true" form, but tends to blend with related concepts as do colors in a spectrum. The most casual examination of specimens in any large herbarium will show families in which whole blocks of such variant forms have been indiscriminately "lumped" into a single category, and other families where the most minute differences, probably reflecting a change in a single gene, have been dignified with a separate specific epithet.

Of course, the classic specific concept was based on the assumption of genetic discontinuity, which in some tropical groups may be the exception rather than the rule. The freedom with which fertile bigeneric hybrids are produced in the Orchidaceae, for example, would seem to

point in that direction. Quite obviously, those plant populations having the greatest degree of isolation, whether it be on islands, mountain peaks, or on the extreme limits of a group's geographic range, will show the greatest structural differences, and have the "best" genera and species, while those widely distributed in relatively homogeneous terrain will be genetically the most unstable, and present the greatest difficulties to the taxonomist. Variation, or the morphological reflection of genetic instability, may be the result of hybridization, or may be due to mutations, or to other less obvious factors, and may take endless forms, such as size, height, vegetative habit, soil preference, color, development or lack of fragrance, alkaloidal or other chemical content, tensile strength of bark fiber or dried latex. Obviously it is impossible to confine the taxonomic expression of such kaleidoscopic potentialities into a simple, unmodified binomial system. No very obvious solution presents itself, yet it must be apparent that taxonomy is faced with what in the final analysis must be an impossible task. Polynomials of impractical and unwieldy size would be required to convey any logical idea of the ultimate fractions in many of the commercially important families and genera. In spite of the difficulties, we may probably safely conclude that our binomials are with us to stay. We must rely on such monographic treatments as are available to us, realizing that the various workers may differ quite radically in their concepts of genera and species. In many groups there has been a strong tendency in recent years to recognize minor differences as "varieties" rather than as full-fledged species, and it appears likely that the trend may continue. In difficult families, such as *Euphorbiaceae* and *Lauraceae,* about the most that can be said for the names in our present list is that our specimens have been found to match those in various of the larger herbaria in the United States and Central America. In the cases where recent monographs have been available, or the specimens have been seen by specialists, the names can be received with considerably greater confidence.

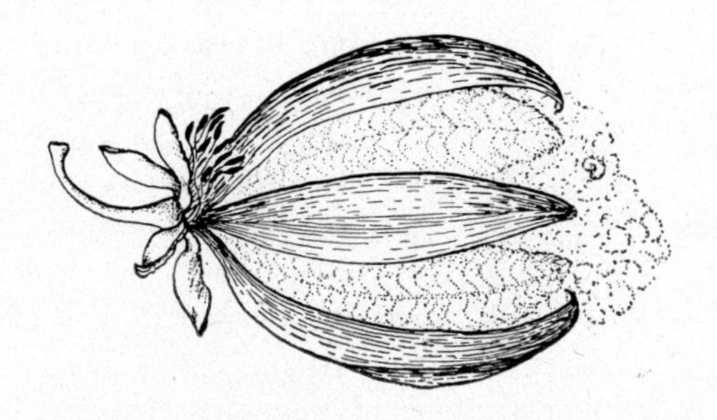

UTILIZATION LISTS

ALTHOUGH MUCH REMAINS TO BE DONE, it already appears certain that there are many potentially valuable hardwoods in the area that are not being used at present. Only high-priced species will bear the cost of transportation to distant markets, yet many might serve for local needs. Mr. N. E. Sanderson, manager of the Golfito Division, has pointed out that the United Fruit Company is its own best and only assured market for ties and lumber. Whether such material can best be grown, purchased from local contractors, or imported from Honduras or the United States becomes basically a problem in economics. It would be very easy to process a few test ties of some of the more common local timbers, using the simple and inexpensive method of treatment with penta-diesel already tried in the Chiriquí Division. Many of the local timbers would obviously be considerably improved by kiln drying, or in some cases if merely painted with creosote in exposed situations. An attempt has been made in the following lists to indicate some of the known possibilities of our timbers.

ECONOMIC SPECIES

General Heavy Construction

Andira inermis
Aspidosperma megalocarpon
Astronium graveolens
Caryocar costaricense
Dialium guianense
Gliricidia sepium
Hieronyma tectissima
Lonchocarpus latifolius
Minquartia guianensis
Mora oleifera
Sacoglottis excelsa
Swartzia panamensis
Sweetia panamensis
Tabebuia chrysantha
Vantanea Barbourii
Vitex Cooperi

UTILIZATION LISTS

Piling and Dock Fenders

Cassipourea podantha
Chlorophora tinctoria
Cordia alliodora
Dialium guianense
Hieronyma tectissima
Peltogyne purpurea
Pouteria heterodoxa
Rhizophora mangle
Sacoglottis excelsa

Mill Foundation Timbers

Caryocar costaricense
Chlorophora tinctoria
Gliricidia sepium
Minquartia guianensis
Sacoglottis excelsa
Vantanea Barbourii

House Posts

Calycophyllum candidissimum
Caryocar costaricense
Dialium guianense
Diphysa robinioides
Gliricidia sepium
Laguncularia racemosa
Minquartia guianensis
Rhizophora mangle
Tabebuia chrysantha
Vitex Cooperi

Beams

Aspidosperma megalocarpon
Astronium graveolens
Cordia alliodora

Rafters

Aspidosperma megalocarpon
Astronium graveolens
Cordia alliodora

Girders

Astronium graveolens
Cordia alliodora

Railroad Ties

Aspidosperma megalocarpon
Calophyllum braziliense var. *Rekoi*
Caryocar costaricense
Cassipourea podantha
Chimarrhis latifolia
Chlorophora tinctoria
Cordia alliodora
Dialium guianense
Diphysa robinioides
Gliricidia sepium
Hieronyma tectissima
Miconia argentea
Minquartia guianensis
Mora oleifera
Myroxylon balsamum var. *Perierae*
Pouteria heterodoxa
Rheedia edulis
Rhizophora mangle
Sacoglottis excelsa
Swartzia panamensis
Sweetia panamensis
Symphonia globulifera
Tabebuia chrysantha
Terminalia amazonia
Vantanea Barbourii
Vochysia hondurensis

UTILIZATION LISTS

Bridge Timbers

Chlorophora tinctoria
Dialium guianense
Pouteria heterodoxa
Sacoglottis excelsa
Sickingia Maxonii
Sweetia panamensis
Tabebuia chrysantha
Vantanea Barbourii

Bridge Boards

Anacardium excelsum
Calophyllum braziliense var. *Rekoi*
Carapa Slateri
Cordia alliodora
Hieronyma tectissima
Mora oleifera
Symphonia globulifera
Terminalia amazonia
Vitex Cooperi

Dragline Mats

Avicennia nitida
Chimarrhis latifolia
Pouteria heterodoxa
Terminalia lucida

Fence Posts

Anacardium excelsum
Calophyllum braziliense var. *Rekoi*
Carapa Slateri
Chimarrhis latifolia
Chlorophora tinctoria
Citharexylum viride
Curatella americana
Dialium guianense
Diphysa robinioides
Gliricidia sepium
Hieronyma tectissima
Laguncularia racemosa
Minquartia guianensis
Mora oleifera
Mouriria parvifolia
Pouteria heterodoxa
Rheedia edulis
Rhizophora mangle
Sickingia Maxonii
Swartzia panamensis
Tabebuia chrysantha
Terminalia amazonia
Vitex Cooperi

Living Fence Posts

Bursera simaruba
Diphysa robinioides
Erythrina costaricensis
Ficus Goldmanii
Gliricidia sepium
Spondias purpurea
Tabebuia chrysantha

General Carpentry and Interior Construction

Anacardium excelsum
Calophyllum braziliense var. *Rekoi*
Caryocar costaricense
Cassipourea podantha
Cedrela mexicana
Cedrela Tonduzii
Cordia alliodora
Couratari panamensis
Dialyanthera otoba
Guarea Hoffmanniana
Guarea longipetiola
Homalium eurypetalum

GENERAL CARPENTRY AND INTERIOR CONSTRUCTION — *Cont'd*

Hura crepitans
Nectandra latifolia
Nectandra salicifolia
Ocotea Ira
Ocotea veraguensis
Pithecolobium austrinum
Prioria copaifera
Pterocarpus officinalis
Rheedia edulis
Symphonia globulifera
Tabebuia pentaphylla
Terminalia amazonia
Terminalia lucida
Triplaris melaenodendron
Virola—all species
Vochysia—all species

SIDING

Calophyllum braziliense var. *Rekoi*
Carapa Slateri
Cedrela mexicana
Cedrela Tonduzii
Terminalia lucida
Vochysia—all species

FLOORING

Aspidosperma megalocarpon
Astronium graveolens
Calophyllum braziliense var. *Rekoi*
Chimarrhis latifolia
Chlorophora tinctoria
Cordia alliodora
Croton glabellus
Guarea longipetiola
Peltogyne purpurea
Pithecolobium austrinum
Platymiscium pinnatum
Terminalia amazonia
Terminalia lucida
Vitex Cooperi

PANELING AND INTERIOR TRIM

Cedrela mexicana
Cedrela Tonduzii
Cordia alliodora
Enterolobium cyclocarpum
Myroxylon balsamum var. *Pereirae*
Nectandra latifolia
Nectandra salicifolia
Ocotea veraguensis
Pithecolobium austrinum
Platymiscium pinnatum
Prioria copaifera
Protium copal var. *glabrum*
Tabebuia pentaphylla

WINDOWS AND DOORS

Calophyllum braziliense var. *Rekoi*
Cedrela mexicana
Peltogyne purpurea
Protium copal var. *glabrum*

FURNITURE AND CABINET WORK

Aspidosperma megalocarpon
Astronium graveolens
Calophyllum braziliense var. *Rekoi*
Cedrela fissilis
Cedrela mexicana
Cedrela Tonduzii
Chimarrhis latifolia
Chlorophora tinctoria
Cordia alliodora
Enterolobium cyclocarpum

Furniture and Cabinet Work — *Cont'd*

Guarea Hoffmanniana
Guarea longipetiola
Hippomane mancinella
Hura crepitans
Lafoensia punicifolia
Myroxylon balsamum var. *Pereirae*
Nectandra latifolia
Nectandra salicifolia
Ocotea veraguensis
Peltogyne purpurea
Pithecolobium austrinum
Platymiscium pinnatum
Protium copal var. *glabrum*
Tabebuia pentaphylla
Terminalia amazonia
Terminalia lucida
Tetragastris panamensis
Ximenia americana
Zanthoxylum procerum

Ornamental Boxes, Inlaid and Turned Work

Andira inermis
Aspidosperma megalocarpon
Astrocaryum Standleyanum
Astronium graveolens
Gliricidia sepium
Linociera panamensis
Peltogyne purpurea
Plumeria rubra forma *acutifolia*
Ximenia americana

Rough Boxes and Crates

Brosimum utile
Bursera simaruba
Ceiba pentandra
Dialyanthera otoba
Didymopanax Morototoni
Goethalsia meiantha
Hura crepitans
Prioria copaifera
Pterocarpus officinalis
Spondias mombin
Sterculia mexicana
Symphonia globulifera
Virola—all species
Vochysia—all species

Concrete Forms

Anacardium excelsum
Brosimum utile
Hura crepitans
Pterocarpus officinalis
Sterculia mexicana
Virola—all species
Vochysia—all species

Plywood Corestock

Carapa Slateri
Cedrela Tonduzii
Ceiba pentandra
Didymopanax Morototoni
Hura crepitans
Prioria copaifera
Protium copal var. *glabrum*
Sterculia mexicana
Symphonia globulifera
Terminalia amazonia
Terminalia lucida
Virola—all species
Vochysia—all species

UTILIZATION LISTS

Veneer

Astronium graveolens
Cedrela Tonduzii
Enterolobium cyclocarpum
Hippomane mancinella
Myroxylon balsamum var. *Pereirae*
Ocotea veraguensis
Platymiscium pinnatum

Boat Framing

Chlorophora tinctoria
Crescentia cujete
Myroxylon balsamum var. *Pereirae*
Rhizophora mangle
Terminalia amazonia

Boat Planking

Chlorophora tinctoria
Cordia alliodora
Terminalia amazonia

Boat Decking

Calophyllum braziliense var. *Rekoi*
Chlorophora tinctoria
Cordia alliodora
Terminalia amazonia
Vitex Cooperi

Interior Boat Work

Lafoensia punicifolia

Life Rafts

Ochroma lagopus

Oars

Cordia alliodora

Motor Car Bodies

Cordia alliodora

Wheelstock

Andira inermis
Calophyllum braziliense var. *Rekoi*
Swartzia panamensis

Ox Yokes

Crescentia cujete

Gun Stocks

Astronium graveolens
Enterolobium cyclocarpum
Hippomane mancinella

Canes and Fishing Rods

Astrocaryum Standleyanum
Guilielma utilis

UTILIZATION LISTS

Tanbark

Avicennia nitida
Conocarpus erecta
Hieronyma tectissima
Laguncularia racemosa
Psidium Guajava
Rhizophora mangle
Terminalia amazonia
Terminalia catappa

Paper Pulp

Ceiba pentandra
Didymopanax Morototoni
Goethalsia meiantha
Schizolobium parahybum
Symphonia globulifera

Axe and Tool Handles

Acacia farnesiana
Brosimum terrabanum
Calycophyllum candidissimum
Chione costaricensis
Chlorophora tinctoria
Crescentia cujete
Guazuma ulmifolia
Linociera panamensis
Rheedia edulis
Sickingia Maxonii
Swartzia panamensis
Talisia nervosa

Professional Instruments

Vitex Cooperi

Archery Bows

Astrocaryum Standleyanum
Calycophyllum candidissimum
Guilielma utilis
Tabebuia chrysantha

Slack Cooperage

Guazuma ulmifolia

Tight Cooperage

Symphonia globulifera

Barrel Hoops

Cordia alliodora

Combs

Calycophyllum candidissimum

Cordage

Apeiba Tibourbou
Cecropia obtusifolia
Cochlospermum vitifolium
Guazuma ulmifolia
Heliocarpus appendiculatus
Hibiscus tiliaceus
Muntingia calabura
Trema micrantha

Bark Cloth

Brosimum utile
Castilla fallax
Hibiscus tiliaceus

UTILIZATION LISTS

Kapok

Ceiba pentandra
Cochlospermum vitifolium
Ochroma lagopus

Cigar Boxes and Pencils

Cedrela fissilis

Edible Fruits

Acacia spadicigera
Acrocomia vinifera
Anacardium occidentale
Ardisia revoluta
Astrocaryum Standleyanum
Bactris balanoidea
Bellucia costaricensis
Brosimum alicastrum
Brosimum sapiifolium
Brosimum terrabanum
Byrsonima crassifolia
Byrsonima densa
Calocarpum borucanum
Carica papaya
Chrysophyllum mexicanum
Chrysophyllum panamense
Cocos nucifera
Diospyros ebenaster
Ficus lapathifolia
Guilielma utilis
Hedyosmum mexicanum
Herrania purpurea
Lacmellea panamensis
Mangifera indica
Muntingia calabura
Myrcia Oerstediana
Pachira aquatica
Persea americana
Posoqueria latifolia
Pourouma aspera
Pouteria neglecta
Pouteria triplarifolia
Pseudolmedia spuria
Psidium Guajava
Rheedia edulis
Rollinia Jimenezii
Simarouba glauca
Sloanea picapica
Spondias mombin
Spondias purpurea
Sterculia apetala
Symphonia globulifera
Terminalia Catappa
Theobroma angustifolium
Theobroma bicolor
Theobroma Cacao
Theobroma simiarum
Trophis racemosa
Ximenia americana

Palm Cabbage

Astrocaryum Standleyanum
Bactris, all species
Guilielma utilis
Hyospathe Lehmannii
Socratea durissima
Welfia Georgii

Palm Wine

Acrocomia vinifera

Palm Sugar

Cocos nucifera
Raphia taedigera

Edible Flowers

Gliricidia sepium

UTILIZATION LISTS

Edible Leaves

Spondias purpurea

Fodder for Stock

Brosimum alicastrum
Brosimum terrabanum
Pseudolmedia spuria
Trophis racemosa

Fruits Relished by Stock

Acrocomia vinifera
Brosimum alicastrum
Brosimum sapiifolium
Brosimum terrabanum
Guazuma ulmifolia

Flowers Eaten by Stock

Ceiba pentandra

Leaves Used for Thatch

Asterogyne Martiana
Cryosophila guagara
Geonoma congesta
Scheelea rostrata

Canes Used for House Walls

Geonoma congesta

Pack Saddles

Bauhinia ungulata
Crescentia cujete

Model Airplanes

Ochroma lagopus

Insulation for Refrigerators

Ochroma lagopus

Match Sticks

Bursera simaruba
Didymopanax Morototoni

Woods Which Will Burn When Green

Pterocarpus officinalis
Rhizophora mangle
Simarouba glauca
Sloanea laurifolia
Trichilia tuberculata

Latex Used for Candles and Torches

Symphonia globulifera

Seeds Rich in Fat and Oil

Acrocomia vinifera
Apeiba tibourbou
Corozo oleifera
Licania arborea
Ricinus communis
Scheelea rostrata
Virola—all species

UTILIZATION LISTS

Milk Substitutes

Brosimum, all species, but especially *B. utile*
Lacmellea panamensis

Sap Used to Clarify Sugar

Guazuma ulmifolia

Resins or Gums Used for Cement

Acacia farnesiana
Bursera simaruba

Latex Used for Caulking Boats

Castilla fallax
Symphonia globulifera

Dyes and Stains

Bixa orellana
Chlorophora tinctoria
Lafoensia punicifolia
Rhizophora mangle
Symphonia globulifera

Soap Substitutes

Sapindus saponaria

Woods Preferred for Charcoal

Byrsonima crassifolia
Conocarpus erecta
Curatella americana
Guazuma ulmifolia
Rhizophora mangle
Terminalia amazonia

Barbascos Used for Catching Fish

Diospyros ebenaster
Euphorbia cotinifolia
Hura crepitans
Randia armata
Sapindus saponaria

Perfumes and Incense

Acacia farnesiana
Myroxylon balsamum var. *Pereirae*
Protium copal var. *glabrum*

Baskets

Trema micrantha

Banana Prop Poles

Raphia taedigera

Leaves Used for Polishing

Curatella americana

MEDICINAL SPECIES

Cassia grandis
Cassia reticulata
Crescentia cujete
Malvaviscus arboreus
Myroxylon balsamum var. *Pereirae*
Quassia amara
Ricinus communis
Simaba cedron
Symphonia globulifera
Trichilia hirta
Vernonia patens

POISONOUS SPECIES

Poisonous Fruits or Seeds

Andira inermis
Crescentia cujete
Euphorbia cotinifolia
Hippomane mancinella
Hura crepitans
Jatropha curcas
Ricinus communis

Poisonous Bark

Andira inermis
Gliricidia sepium

Poisonous Leaves

Gliricidia sepium

Caustic or Poisonous Sap

Crataeva tapia
Euphorbia cotinifolia
Hippomane mancinella
Hura crepitans

Poisonous Sawdust

Enterolobium cyclocarpum
Hippomane mancinella

Trees Whose Pollen Produces a Rash or Hayfever-Like Allergy

Anacardium excelsum
Mangifera indica

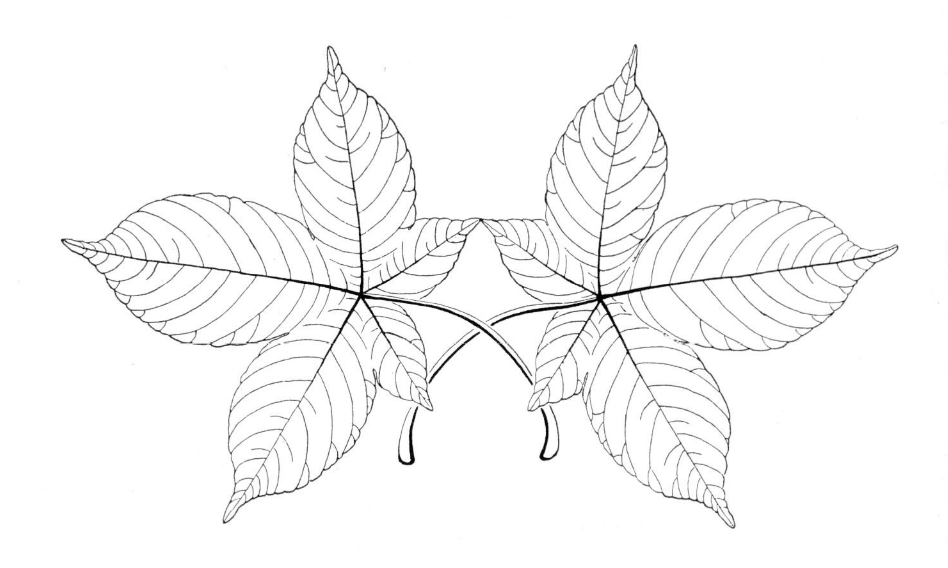

ALPHABETICAL INDEX TO THE FAMILIES, GENERA, SPECIES, AND COMMON NAMES

Abalche—See *Ximenia americana* (Guatemala).

ACACIA—KEY

1. Thorns large and conspicuous, hollow, usually inhabited by stinging ants.
   2. Thorns white or pale yellow .... *Acacia spadicigera*
   2. Thorns dark brown or black.
      3. Flowers in slender spikes .... *Acacia costaricensis*
      3. Flowers in globose heads .... *Acacia melanoceras*
1. Thorns slender, not hollow, not inhabited by stinging ants .... *Acacia farnesiana*

ACACIA COSTARICENSIS Schenck—Leguminosae; Mimosoideae—*Cornezuelo* (Costa Rica)—*Cuernito* (Panama)—*Carnezuelo* or *Cornezuelo* (Nicaragua)—*Cachito* or *Espino cachito* (Honduras).

Shrubs or small trees, to about 18 ft., with bipinnate leaves and paired, hollow bull's-horn thorns inhabited by stinging ants. The fragrant yellow flowers are produced in dense axillary fascicles of slender, elongate spikes, and are followed by curved, turgid, dehiscent pods. Known in our area only from the Terraba Valley.—Lagarto de Diquis, *Tonduz 4810.*

ACACIA FARNESIANA (L.) Willd.—Leguminosae; Mimosoideae—*Aromo* (Local, Costa Rica generally)—*Aromo* or *Cuernito* (Panama)—*Aromo, Espino blanco,* or *Clavito* (Nicaragua)—*Espino, Espino blanco, Cachito*

de aromo, or *Cornezuelo* (Honduras)—*Espino blanco, Espinal,* or *Subin* (Guatemala)—*Espino ruco* (Salvador). (Pl. 19)

Shrubs or small trees with slender, pale-tan or nearly white, needle-like spines, bipinnate leaves and bright-yellow, intensely fragrant flowers that are produced in small, globose heads and are followed by turgid, 2-valved fruits. The hard, heavy, fine-grained, reddish-brown or yellow wood is little used, but is suitable for good quality tool handles. The flowers are the basis of a perfume industry in the south of France, and are occasionally put among stored linens in Central America to give them a pleasant odor. The pods and bark are rich in tannin, and the mucilaginous fruits can be used to mend broken dishes. Known in our area only from the Terraba Valley.—Boruca, 1,500 ft., *Tonduz 3802*—Palmar, on rocky spoil banks, *Allen 5806 & 6699.*

ACACIA MELANOCERAS Beurl. — Leguminosae; Mimosoideae — *Cachito* (Panama).

Slender, erect, usually multiple-trunked trees, 20–35 ft. in height, occasional in pastures near Golfito. The leaves are bipinnate, and the fragrant yellow flowers are borne on globose heads, and usually produced in March. The trunks and lower branches have many short, woody shoots that are heavily armed with dark-brown, hollow, bull's-horn thorns, inhabited by stinging ants. The wood is not used. — Golfito, *Allen 5997.*

ACACIA SPADICIGERA Schlecht. & Cham.—Leguminosae; Mimosoideae—*Cornezuelo* (Costa Rica)—*Pico de gorrion, Pico de gurrion, Subin,* or *Subin blanco* (Guatemala).

Shrubs or small trees, to about 18 ft., with bipinnate leaves and yellow flowers produced in dense spikes. The short, dark-red, terete, indehiscent fruits are sometimes eaten locally, and are sold in the markets in Guatemala.—Known in our area only from Boruca, *Tonduz 4554.*

ACALYPHA DIVERSIFOLIA Jacq.—Euphorbiaceae—*Costilla de caballo* or *Costillo de danto* (Honduras)—*Cacucup, Ciiche,* or *Palo de sangre* (Guatemala).

Shrubs or small trees, to about 25 ft. in height. Leaves alternate, with slender petioles, the blades usually elliptic-lanceolate, acute or long-acuminate, with coarsely serrate margins. The minute, green or greenish-white flowers are produced in slender catkin-like spikes from the leaf axils. The wood is described as being yellowish brown, compact, and fine-textured. It has no local uses.—Forests near Palmar Norte, 100 ft., *Allen 5505 & 5882.*

ACANTHACEAE—One species, *Bravaisia integerrima.*

Aceituno—See *Hirtella americana* (Guatemala). Leaves simple.

Aceituno—See *Simarouba glauca* (Costa Rica and Salvador). Leaves pinnate.

Aceituno peludo—See *Hirtella americana* (Guatemala).

Achiote—See *Bixa Orellana* (Local and general).

Achiotillo—See *Vismia guianensis* (Local).

Achiotillo colorado—See *Hamelia patens* (Honduras).

Achote—See *Bixa Orellana* (Chiriquí and Honduras).

Achotillo—See Vismia guianensis (Costa Rica).

ACROCOMIA VINIFERA Oerst.—Palmaceae—*Coyol* (Local and general).

Single-trunked, rather stout, intensely spiny palms, with drooping, pinnate fronds, the spiny bases of which are usually long persistent. The large panicles of flowers, as yellow as ripening wheat, appear late in the dry season and are followed by elongate clusters of dark yellowish-green, smooth, globose fruits which average a little more than 1″ in diameter. The inner kernel has the flavor of coconut and is often eaten in Honduras. A refreshing cider-like wine is often made by fermenting the whitish sap obtained from deep rectangular incisions in the crown of the felled trunks. The trees are often left standing in pastures and the fruits are much relished by cattle. Locally common in dry, open situations, sometimes forming nearly pure stands near Rey Curre and Potrero Grande.

ACTINIDIACEAE—A single species, *Saurauia yasicae.*

AEGIPHILA—KEY

1. Inflorescences usually very slender, 3-flowered axillary cymes... *Aegiphila costaricensis*
1. Inflorescences many-flowered, terminal panicles... *Aegiphila martinicensis*

AEGIPHILA COSTARICENSIS Moldenke—Verbenaceae.

Shrubs or small trees, to about 20–30 ft. in height, with thin, opposite or often ternate leaves, the blades typically oblanceolate, with acute or acuminate apices. The relatively large white flowers are produced in December in very slender axillary, usually 3-flowered cymes, and are followed by ovoid fruits. The species is very different in appearance from others of the genus, and may possibly be referable elsewhere.—Esquinas Forest, *Skutch 5346.*

AEGIPHILA MARTINICENSIS Jacq.—Verbenaceae—*Juan de la verdad* or *Wild jasmine* (Panama).

Shrubs or small trees, to about 35 ft. in height, with opposite, lanceolate, acute or acuminate leaves and elongate terminal panicles of small,

pale-yellow flowers, which are produced in October and November. The plants, and particularly the short, truncate, cylindric fruits have much the appearance of *Citharexylum,* but are immediately separable by the paniculate rather than racemose inflorescence.—Pastures near Palmar Norte, 100 ft., *Allen 5346 & 6335.*

Agal—See *Myrcia Oerstediana* (Guatemala).

Aguacate—See *Persea americana* (Local and general).

Aguacatillo—A generic term, applied to many kinds of trees of the Lauraceae, in our area notably to *Nectandra salicifolia, Licaria Cervantesii, Nectandra perdubia,* and *Persea Skutchii,* which see.

Aguacaton—See *Ocotea Ira* (Local and Panama).

Aguja de arrea—See *Casearia aculeata* (Honduras).

Agujilla—See *Ladenbergia Brenesii* (Costa Rica).

Aji—See *Caryocar costaricense* (Costa Rica).

Ajillo—See *Caryocar costaricense* (Costa Rica).

Ajo—See *Caryocar costaricense* (Local).

Ajonocht—See *Enallagma latifolia* (Guatemala).

Alazano—See *Calycophyllum candidissimum* (Panama).

ALBIZZIA—KEY

1. Individual leaflets more than ¾″ in length .... *Albizzia longepedata*
1. Individual leaflets less than ½″ in length .... *Albizzia filicina*

ALBIZZIA FILICINA Standl. & L. Wms.—Leguminosae; Mimosoideae. (Pl. 21)

Slender, flat-topped, unarmed trees, 30–80 ft. in height, with conspicuously scaling bark and bipinnate leaves. The white, nearly globose heads of flowers are produced from early June until August, and are followed by flat, brown, linear pods. The trees often form conspicuously flat-topped groves along rocky, fast-flowing streams, such as the Río Piedras Blancas, and are among the most common and characteristic of our local species.—Esquinas pastures, 50 ft., *Allen 6249 & 6285.*

ALBIZZIA LONGEPEDATA (Pittier) Britt. & Rose ex Record—Leguminosae; Mimosoideae—*Cenicero macho* (Costa Rica)—*Frijolillo* or *Guachipili* (Honduras)—*Cadeno* (Guatemala).

Infrequent trees of moderate size, rather reminiscent of the common Saman on a somewhat smaller scale. The bipinnate leaves are usually deciduous during the dry season. The cream or pale-tan flowers are produced in axillary umbels in late December and early January, and are

followed by linear, densely pubescent fruits. The wood is described by Record as "dark brown, hard and heavy, medium-textured and probably fairly durable." It has no known local uses.—Hills above Palmar Norte, 1,000 ft., *Allen 5745.*

Albondiga—See *Alibertia edulis* (Guatemala).
Alcareto—See *Aspidosperma megalocarpon* (Panama).

ALCHORNEA GLANDULOSA Poepp. & Endl. var. PITTIERI Pax—Euphorbiaceae.

Shrubs or small trees, 15–35 ft. in height, with alternate, oblong or oblong-ovate, usually abruptly caudate-acuminate leaves which have coarsely sinuate-serrate margins. The minute flowers are produced in very slender axillary spikes, and are followed by small, 2- to 3-celled capsular fruits.—Agua Buena de Cañas Gordas, 3,500 ft., *Pittier 11101*—El Palmar de Boruca, *Tonduz 6757.*

Alcornoque—See *Mora oleifera* (Costa Rica and Panama). Leaves pinnate.
Alcornoque—See *Licania arborea* (Costa Rica). Leaves simple.
Alcornoque—See *Terminalia Catappa* (Nicoya). Leaves simple.
Alfaje—See *Trichilia tuberculata* (Panama, Chepo).
Alfajillo colorado—See *Trichilia tuberculata* (Chiriquí).
Alfeñique—See *Dialium guianense* (Local).
Algarrobo—See *Pithecolobium austrinum* (Local).
Algodon—See *Ochroma lagopus* (Salvador).
Algodoncillo—See *Hibiscus tiliaceus* (Panama).

ALIBERTIA EDULIS (L. Rich.) A. Rich.—Rubiaceae—*Madroño* or *Trompillo* (Costa Rica)—*Lagartillo, Madroño, Trompo,* or *Trompito* (Panama)—*Lirio* (Honduras)—*Torolillo* (Salvador)—*Albondiga, Guabillo, Guayabillo,* or *Guayaba de monte* (Guatemala).

Shrubs or small trees, to about 12–15 ft. in height, with opposite, dark-green, oblong-lanceolate, acuminate leaves, 2–8″ long, the blades barbate beneath in the axils of the nerves. The small white flowers are produced in sessile, terminal clusters, the tube relatively stout, about 1½ times as long as the lobes. The nearly globose, yellowish fruits average about 1″ in diameter, and are sometimes used by small boys for the making of tops.—Buenos Aires, *Pittier 4044*—Boruca, 1,500 ft., *Pittier 4141*—Sea shore, Santo Domingo de Osa, *Tonduz 10006*—Playa Blanca,

Golfo Dulce, *Valerio 439*—Floodplain of Río Sandalo, *Dodge & Goerger 10081*—Sea beaches near Santo Domingo de Osa, *Brenes 12188, 12226, & 12256.*

Almacigo—See *Bursera simaruba* (Costa Rica and Panama).
Almendro—See *Andira inermis* (Local and Guatemala). Leaves pinnate. Fruits globose.
Almendro—See *Lonchocarpus latifolius* (Guatemala). Fruits flat.
Almendro—See *Terminalia amazonia* (Honduras). Leaves simple.
Almendro—See *Terminalia Catappa* (Local and general). Leaves simple.
Almendro cimarron—See *Andira inermis* (Guatemala).
Almendro del rio—See *Andira inermis* (Salvador).
Almendro macho—See *Andira inermis* (Salvador).
Almendro montes—See *Andira inermis* (Salvador).
Almendro real—See *Andira inermis* (Salvador).
Amansa mujer—See *Prioria copaifera* (Panama).
Amapola—See *Malvaviscus arboreus* var. *penduliflorus* (Local and Guatemala).
Amapolilla—See *Malvaviscus arboreus* var. *penduliflorus* (Costa Rica).
Amarillo—See *Terminalia amazonia* (Local). Fruits winged.
Amarillo de fruta—See *Lafoensia punicifolia* (Panama). Fruits not winged.
Amarillo fruto—See *Lafoensia punicifolia* (Panama).
Amarillon—See *Terminalia amazonia* (Local).
Amarillo real—See *Terminalia amazonia* (Panama).
Amate—See *Ficus costaricana* (Salvador and Guatemala).
Amate—See *Ficus lapathifolia* (Guatemala).
Amate cusho—See *Fiscus lapathifolia* (Guatemala).

ANACARDIACEAE—Key

1. Leaves simple.
   2. Leaves narrowly lanceolate, acuminate at the apex .... *Mangifera*
   2. Leaves elliptic-oblong or oblanceolate, conspicuously rounded at the apex .... *Anacardium*
1. Leaves pinnate.
   2. Fruits yellow or red, plumlike .... *Spondias*
   2. Fruits not plumlike.
      3. Leaflets 11–15, with more or less serrate or dentate margins. Fruits with a conspicuous 5-parted calyx .... *Astronium*
      3. Leaflets 5–7, with entire margins. Fruits without a conspicuous 5-parted calyx .... *Mauria*

ANACARDIUM—Key

1. Large trees, to over 100 ft. in height. Leaves usually more than 10″ in length. A common forest tree, dominant over wide areas........................*Anacardium excelsum*
1. Small trees, seldom exceeding 20 ft. in height. Leaves usually less than 8″ long. Occasional near old house sites and in towns........................*Anacardium occidentale*

Anacardium excelsum (Bert. & Balb.) Skeels — Anacardiaceae — *Espavel* (Costa Rica)—*Espave* (Panama)—*Wild cashew* (Bocas del Toro and Limón). (Fig. 2)

Large trees, to over 100 ft. in height, very common throughout the area, often dominant on the floodplains, notable examples being the woodlands near Tinoco and Jalaca. The leaves are relatively large, obovate-oblong in outline, rounded at the apex and alternate on the stems. Many of the trees drop their leaves for a short period during late November and early December, giving a peculiar, wintry appearance to the forest. Bright-green flushes of new foliage appear in early January, and are soon followed by large, pale greenish-white terminal panicles of small white flowers. Individual flowers turn dark pink with age, and have a strong, delightful clovelike fragrance which permeates the forest. The fruits are kidney-shaped, and are subtended by a slender, strongly twisted receptacle. The nuts are reported to be edible when roasted, and the macerated bark is sometimes used in Panama as a barbasco for catching fish. The wood is used locally for bridge boards, fence posts, rough construction, and concrete forms. The Engineering Department in Golfito reports the wood to be susceptible to dry rot and termite attack, but preliminary tests in the Armuelles Division indicate that it can easily be treated with penta-diesel. Local sawmills complain that the wood tends to saw woolly, and is hard to finish smoothly. Suggested uses by the Yale School of Forestry are general carpentry and construction, wooden utensils, dugout canoes, interior trim and millwork, where strength is not an important factor. Buenos Aires, 750 ft., *Tonduz 4001 & 6671*—Forests near Palmar, *Allen 6033* & *Dorothy Allen 6679-A.*

Anacardium occidentale L.—Anacardiaceae—*Cashew* or *Marañon* (Local and general)—*Jocote marañon* (Honduras and Guatemala). (Pl. 14)

Small trees, often planted for their edible red or yellow fruits (hypocarps) and frequently persistent near old house sites and along trails. The terminal flushes of new leaves are red and conspicuous, and the small flowers are pleasantly fragrant.—Buenos Aires, 600 ft., *Tonduz 6669*—Santo Domingo de Osa, *Tonduz 10070.*

Fig. 2. ANACARDIUM EXCELSUM. 1 — Flowering branch. 2 — Individual flower. 3 — Individual fruit. 4 & 5 — Cross sections of flowers, showing sterile and fertile stamens.

Anatto—See *Bixa Orellana* (Bocas del Toro and Limón).

ANDIRA INERMIS (Wright) Urban—Leguminosae; Papilionatae—*Almendro* (Local)—*Cocu* or *Carne asado* (Costa Rica)—*Arenillo* or *Quira* (Chiriquí)—*Cocu* or *Pilon* (Panama)—*Cabbage bark* (Canal Zone, Bocas del Toro, and Limón)—*Chaperno* (Honduras)—*Almendro macho, Almendro del rio, Almendro montes,* or *Almendro real* (Salvador)—*Almendro, Almendro cimarron,* or *Guacamayo* (Guatemala). (Pl. 21)

Round-topped trees of varying size, mature specimens on forested hillsides usually about 80 ft. in height, with alternate, odd-pinnate leaves and large terminal panicles of lavender or deep-purple flowers which are followed by subglobose, 1-seeded fruits about 1½″ in diameter. These are reported to contain a poisonous or violently purgative aklaloid, which is also present in the bark. The wood is yellowish, or of varying shades of brown to almost black, and is hard, heavy, rather coarse-grained, tough, strong, and durable, being resistant to decay in contact with the ground and in water, and to insect attack. It is usually considered too hard to work, and is seldom used excepting for heavy construction. It has been utilized for piling, bridge timbers, railroad ties, logging cart wheels and spokes, billiard cue butts, canes, and miscellaneous articles of turnery. The genus is badly in need of revision, and it may be possible that two species are present in our area. Some of the smaller trees differ from the classic concept of *A. inermis* in their much larger panicles of darker flowers, rather peculiar calyx, and erratic flowering habit. After examining an extensive series of specimens it would appear however that the species is merely somewhat variable, particularly in the juvenile stage. The floral calyx may be relatively narrow and conspicuously lobed, or broad, shallow, and unlobed. The fruits may also vary considerably in size. The younger trees, at least in our area, have two flowering and fruiting seasons, which seem to be correlated with periods of drier weather. Flowers are usually produced in July or August during the "little dry season," known locally as the "*Canicula*" or "*Veranillo de San Juan*," and are followed by fruits in November or December. A second flowering often takes place in March or April, at the end of the longer dry season. Mature trees seem to bloom only at this latter time, possibly because their more extensive root system is less affected by minor differences in precipitation. The larger trees have smaller panicles of paler flowers, and are quite uniformly distributed on the hillsides in climax forest, where they are quite conspicuous and attractive, looking like arborescent lilacs when seen from the air.—Boruca, *Tonduz 6874*—Forests near Palmar Norte, *Allen 5220, 5609, & 6669.*

Anime—See *Protium neglectum* var. *sessiliflorum* (Panama).

ANISOMERIS RECORDII (Standl.) Standl.—Rubiaceae—*Clavo* (Guatemala).

Bushy trees, 20–30 ft. in height, with opposite, elliptic-lanceolate leaves on branchlets which are often spinose. The small white flowers have an elongate, slender tube, and are produced in June in few-flowered fascicles at the ends of the short lateral leafy spurs, and are followed by small drupaceous fruits. Locally common in lowland forest.—Jalaca, 100 ft., *Allen 5303*.

ANNONACEAE—KEY

1. Fruits more or less Annona-like, composed of many carpels which are fused into a compact, superficially undivided unit at maturity.
   2. Leaf petioles conspicuously thickened .... *Duguetia*
   2. Leaf petioles not thickened.
      3. Outer surface of petals with a longitudinal, winglike keel .... *Rollinia*
      3. Outer surface of petals without a keel .... *Annona*
1. Fruits not Annona-like, but composed of several distinct carpels, each with a slender pedicel.
   2. Outer petals imbricate in bud .... *Guatteria*
   2. Outer petals valvate in bud.
      3. Leaves narrowly lanceolate, acuminate, usually less than 4″ in length .... *Xylopia*
      3. Leaves more than 4″ in length (usually much more).
         4. Individual carpels of the fruits strongly asymmetrical, usually more than 2″ in length .... *Cymbopetalum*
         4. Individual carpels of the fruits symmetrical, about ½″ in length .... *Unonopsis*

ANNONA PITTIERI Donn. Sm.—Annonaceae.

Small trees, to about 25 ft. in height, with alternate, narrowly oblong, shortly acuminate, glabrous leaves 3–6½″ long and 1–1¾″ wide. The 3-petaled flowers are about 1¼″ long, and are followed by subconic fruits about 4–5″ in length.—Cañas Gordas, 3,400 ft., *Pittier 11108*.

Anona—See *Guatteria amplifolia* (Guatemala). Carpels of fruits free.

Anona—See *Rollinia Jimenezii* (Chiriquí and Guatemala). Carpels united.

Anonillo—See *Crataeva tapia* (Salvador). Leaves digitately compound.

Anonillo—See *Rollinia Jimenezii* (Local). Leaves simple.

Anum—See *Spondias purpurea* (Guatemala).

Añileto—See *Hamelia patens* (Costa Rica).

APEIBA—KEY

1. Leaf petioles conspicuously hirsute, the blades densely stellate-tomentose beneath, at least on the nerves. Fruits densely covered with long, flexible spines .... *Apeiba Tibourbou*

1. Leaf petioles and lower leaf surface essentially glabrous. Fruits covered with very short, hard, conical spines........ *Apeiba aspera*

APEIBA ASPERA Aubl.—Tiliaceae—*Peine de mico* or *Tapabotija* (Local) —*Monkey comb* (Canal Zone, Bocas del Toro, and Limón)—*Burillo* or *Tapabotija* (Nicaragua)—*Peine de mico* (Guatemala). (Pl. 31)

Medium-sized or large trees, to about 80 ft. in height, with alternate, ovate, acute leaves. The yellow flowers are followed by the curious sea urchin-like fruits which are black when mature and covered with short, hard spines. Very frequent in climax forests throughout the area, to about 3,000 ft. The soft white wood has no known uses.—Forested hills above Palmar Norte, 1,800 ft., *Allen 5922.*

APEIBA TIBOURBOU Aubl.—Tiliaceae—*Peine de mico* or *Majagua* (Local)—*Burio* (Costa Rica)—*Peinecillo* (Chiriquí)—*Peine de mico, Cortez,* or *Cortezo* (Panama)—*Monkey Comb* (Canal Zone, Bocas del Toro, and Limón)—*Burillo* (Nicaragua).

Small to medium-sized trees, 30–75 ft. in height, found throughout the area. The oblong-ovate, acute or acuminate leaves have minutely serrate margins, and are alternate on the branchlets. The pale-yellow flowers are produced from June to about December, and are followed in the early dry season by the dark-green or brown, sea urchin-like fruits which may readily be distinguished from those of the preceding species by the much longer, softer spines. The fresh seeds are rich in oil, and the bark contains a tough fiber which is sometimes used for making a coarse rope. The soft white wood is not used.—Diquis, 150 ft., *Pittier 12092*—Esquinas Forest, 200 ft., *Allen 5556 & 5803*—Forests near Palmar Norte, 100 ft., *Allen 5652*—Rocky beaches near the delta of the Río Esquinas, *Allen 5626.*

APOCYNACEAE—KEY

1. Flowers in elongate, pendulous cymes, which equal or exceed the leaves in length........ *Tabernaemontana*
1. Flowers not in elongate, pendulous cymes.
   2. Leaves conspicuously clustered at the ends of very thick branches. Flowers large, white, and showy........ *Plumeria*
   2. Leaves not conspicuously clustered at the ends of the branches. Flowers not white.
      3. Trunks armed with low, pyramidal spines. Milky latex sweet to the taste........ *Lacmellea*
      3. Trunks not armed with spines.
         4. Leaves opposite. Flowers more than 1″ in length, the lower portion contracted into a slender tube. Fruits always paired........ *Stemmadenia*
         4. Leaves alternate. Flowers less than ½″ long, the lower portion not contracted into a tube. Fruits never paired........ *Aspidosperma*

ARALIACEAE—Key

1. Lower leaf surface silky, pale tan or brownish in color. Flowers in open paniculate racemes............*Didymopanax*
1. Lower leaf surface green. Flowers in compact heads............*Oreopanax*

Arbol de sal—See *Avicennia nitida* (Salvador).

Arbol santo—See *Jatropha curcas* (Panama and Honduras).

ARDISIA—Key

1. Leaves very large, to about 2 ft. in length, conspicuously clustered at the ends of the branches............*Ardisia Cutteri*
1. Leaves usually less than 10″ long, not conspicuously clustered at the ends of the branches.
   2. Inflorescences less than 2″ long, the flowers subtended by many very large, conspicuous, persistent bracts............*Ardisia Dodgei*
   2. Inflorescences more than 2″ long, the flowers never subtended by conspicuous bracts.
      3. Leaves broadly elliptic, rounded or subacute at the apex............*Ardisia revoluta*
      3. Leaves lanceolate, acute or acuminate at the apex.
         4. Calyx lobes broadly rounded, with overlapping margins.
            5. Mature flower buds about ½″ long, the calyx lobes loosely spreading. Flowers dark pink............*Ardisia Cutteri*
            5. Mature flower buds less than ¼″ long, the calyx lobes tightly clasping. Flowers pale pinkish tan............*Ardisia Dunlapiana*
         4. Calyx lobes acute, the margins never overlapping. Flowers white............*Ardisia Standleyana*

Ardisia Cutteri Standl.—Myrsinaceae.

Slender trees, 12–25 ft. in height, with very large, oblanceolate leaves which are conspicuously clustered at the ends of the branches. The attractive, dark-pink flowers are produced in February and March in long pendulous panicles and are followed by blood-red, globose, fleshy fruits about ¾″ in diameter. An attractive species, fairly common in the Esquinas Forest, where it seems to be confined to clay ridges in dense shade.—Esquinas Forest, 200 ft., *Allen 5828, 6726, & 6749.*

Ardisia Dodgei Standl.—Myrsinaceae—*Guastomate* (Local).

Slender trees, 30–50 ft. in height, with alternate, elliptic-lanceolate, acute or shortly acuminate leaves having very short petioles, the blades usually 6–8″ long and 2–3″ wide. The pink, or sometimes nearly white flowers are produced in very short, terminal clusters, subtended by many conspicuous pinkish bracts. The fruit is a black berry. Frequent in climax forests.—Esquinas Forest, 200 ft., *Skutch 5310* & *Allen 6258.*

Ardisia Dunlapiana P. H. Allen, sp. nov.—Myrsinaceae.

Slender trees, 40–50 ft. in height, with alternate, essentially glabrous leaves, the lanceolate or elliptic-lanceolate blades 3–6″ long and 1–2″

wide, acute or shortly acuminate at the apex and narrowly wedge-shaped at the base so that the petiole is usually narrowly winged for almost its entire length. The small, pale pinkish-tan flowers are produced in April in a slender terminal panicle 3–4″ long, each flower having broadly rounded and closely imbricating calyx lobes.—Esquinas Forest, 200 ft., *Allen 5274.*

ARDISIA REVOLUTA HBK—Myrsinaceae—*Tucuico* (Costa Rica)—*Uvito, Margarita,* or *Fruta de pava* (Panama)—*Uva* (Honduras)—*Uva* or *Cerezo* (Salvador)—*Silacil, Sirasil, Cerecilla, Mora,* or *Morita* (Guatemala).

Usually shrubs, but sometimes small trees to about 18 ft. in height. The alternate, leathery, ovate or oblong-ovate leaves are rounded or subacute at the apex. The small, white or pinkish flowers are produced in panicled racemes, and are followed by small, black, juicy, edible fruits. Apparently known in our area only from the vicinity of Boruca.—*Tonduz 6854.*

ARDISIA STANDLEYANA P. H. Allen, sp. nov.—Myrsinaceae.

Small trees, about 20–25 ft. in height, with alternate, nearly sessile, glabrous leaves 7–10″ long and 2½–3″ wide, the lanceolate or elliptic-oblong blades acute or shortly acuminate at the apex and narrowly rounded at the base. The small white flowers have a deeply and conspicuously lobed calyx, and are produced in December in terminal branching panicles about 5″ in length.—Esquinas Forest, *Skutch 5384.*

Arenillo—See *Andira inermis* (Chiriquí).
Arito—See *Malvaviscus arboreus* (Salvador).
Aromo—See *Acacia farnesiana* (Costa Rica, Panama, and Nicaragua).
Arracheche—See *Mouriria parvifolia* (Panama).
Arrayan—See *Rheedia edulis* (Guatemala).

ASPIDOSPERMA MEGALOCARPON Muell. Arg.—Apocynaceae—*Alcareto* (Panama)—*Chaperno* (Honduras)—*My Lady* or *Malady* (British Honduras) *Chichica, Chichique, Fustan de vieja,* or *Milady blanco* (Guatemala). (Fig. 3)

Tall forest trees, averaging about 100 ft. in height and 18–24″ in trunk diameter, with alternate, leathery, oblong-ovate, acute leaves. The small, pale-yellow flowers are produced in July and early August in subterminal or axillary panicled cymes which about equal the leaves in length. The fruits are large, woody, flattened, obliquely and broadly obovate pods which contain the unique round, papery seeds which average about 3″ in diameter. The fruits have been collected in the Esquinas Forest in late May, but the curious seeds have been picked up at other times,

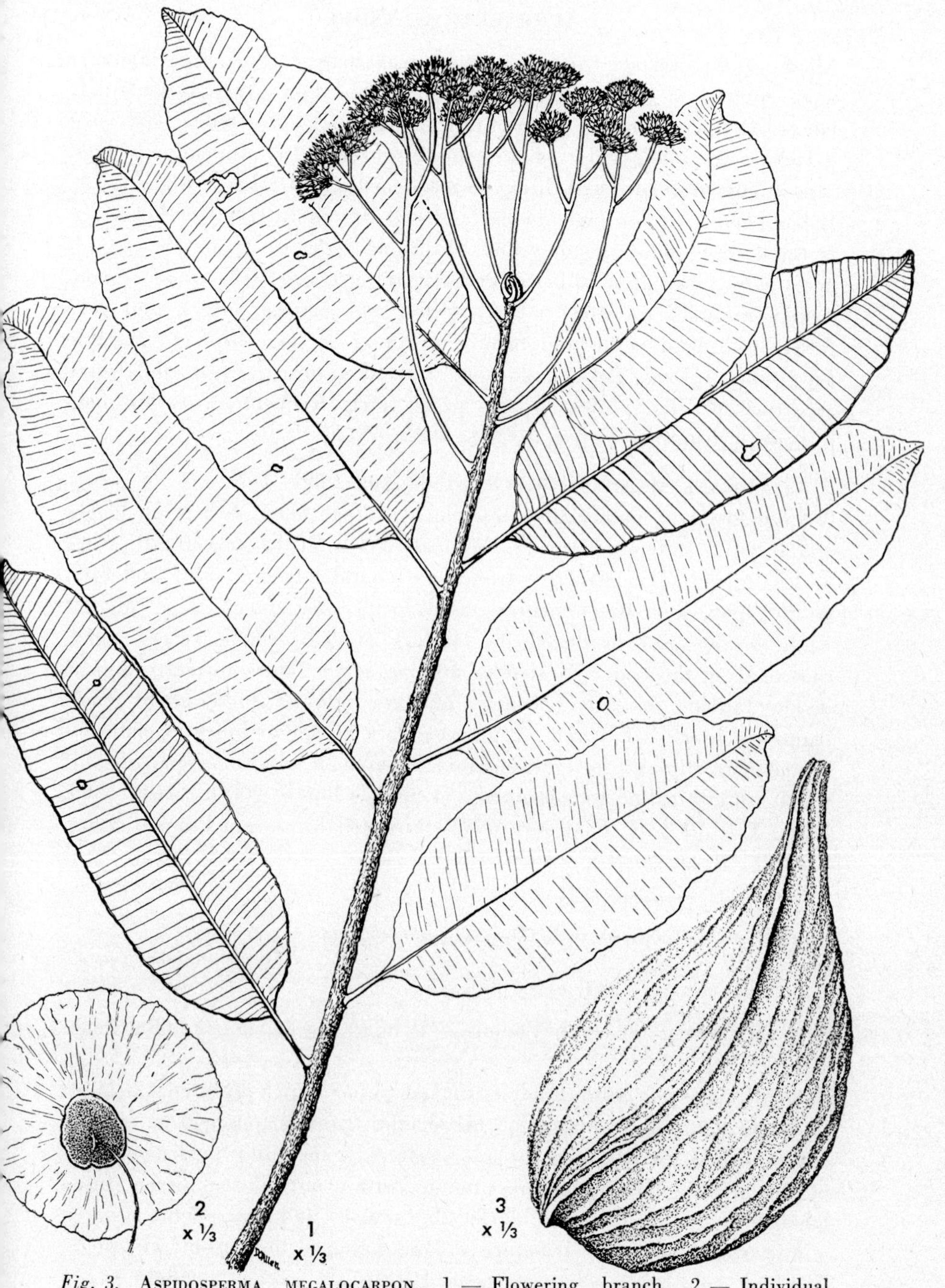

Fig. 3. ASPIDOSPERMA MEGALOCARPON. 1 — Flowering branch. 2 — Individual seed. 3 — Seed pod.

which would indicate that some fruits may mature from about January until August or September. The lower trunk shows very little if any latex, but it is abundant in the branches and immature fruits, and is of a rich, pale-orange color. The wood is of attractive appearance, light brown, fine-grained, hard, heavy, works easily, and takes a high polish. It has been used for second-class railroad ties in British Honduras, but is reported to have the defect of absorbing moisture after dry periods and tending thereafter to break under heavy pressure. The wood is moderately resistant to insect attack, decay, and marine borers, and very resistant to both white and brown rot fungi. It is considered suitable for beams, rafters, girders, flooring, furniture, turned work, and heavy construction. The species appears to be fairly frequent on well-drained hillsides.—Esquinas Forest, 200 ft., *Allen 5544.*

ASTEROGYNE MARTIANA Wendl.—Palmaceae. (Pl. 4)

Single-trunked, unarmed palms which average about 6–8 ft. in height and about 2″ in trunk diameter, the lower 2–4 ft. of the stem usually more or less repent. The attractive fronds are completely undivided and bifid at the apex, and are sometimes used locally for thatch. The fragrant white flowers are produced on clusters of 3–6 simple spikes which radiate from the end of a slender arching scape. They normally appear in late January or early February, and are followed in March by small fruits which are at first red, but become dark purple or black at maturity. Frequent in climax forests throughout the area. A handsome species, well worthy of cultivation for ornament.—Esquinas Forest, 200 ft., *Allen 5596, 5826, & 6752* & *H. E. Moore 6535*—Forested hills near Palmar Norte, 1,500 ft., *Moore 6530* & *Allen 6744.*

ASTROCARYUM—KEY

1. Fruits orange, unarmed, in conspicuous pendulous clusters .... *Astrocaryum Standleyanum*
1. Fruits not orange, densely armed with spines .... *Astrocaryum alatum*

ASTROCARYUM ALATUM Loomis — Palmaceae — *Coquillo* (Panama). (Pl. 8)

Small, intensely spiny, single-trunked palms which average 12–25 ft. in height. On close examination the slender trunk is found to be without spines, but is usually more or less covered by the spiny persistent bases of the fronds. The leaves are pinnate, with many of the pinnae fused together in broad confluent blocks, the midribs and bases being covered with a truly formidable armature of long, flattened lustrous-brown spines. The compact clusters of spiny, beaked fruits are erect or pendulous, and

are protected by a spiny spathe. They are usually in fruit in our area from October until late March. The prominently ringed trunks of mature specimens are unbelievably hard, and the black wood might be used for canes, fishing rods, or archery bows. A very common species, usually found in wet lowland forests, particularly near Tinoco and Jalaca.—Tinoco station, sea level, *Allen 6612.*

ASTROCARYUM STANDLEYANUM L. H. Bailey—Palmaceae—*Pejibaye* (Local)—*Black palm* (Canal Zone, Chiriquí, and Bocas del Toro)—*Chunga* or *Chonta* (Panama). (Pl. 6)

Tall, handsome palms, common to areas of climax forest, the individual specimens varying from about 40–65 ft. in height. The solitary black trunks are usually about 6–8″ in diameter, and are armed with broad bands of long, flat spines which are highly flammable. The fronds are pinnate and spinose, particularly near the base. The terminal bud, or "palmito," can be eaten. The fruits are produced from March to about June in long, pendulous, bright-orange, attractive clusters. Individual fruits are unarmed, and the rather scanty pulp surrounding the large seed is sweet and edible. The hard black wood is used for canes, fishing rods, ornamental boxes, inlaying, archery bows, and other similar purposes, and might be suitable for golf clubs.—Hills near Palmar Norte, 200–600 ft., *Allen 6662 & 6771.*

ASTRONIUM GRAVEOLENS Jacq.—Anacardiaceae—*Ronron* (Local)—*Zorro* (Panama)—*Ciruelillo, Frijolillo, Masicaran, Palo obero,* or *Ronron* (Honduras)—*Ciruelo, Culinzis, Jobillo, Palo obero, Ronron,* or *Quesillo* (Guatemala)—*Gonzalo alves* (U.S. trade name).

Tall forest trees, averaging about 100–120 ft. in height and 2–3 ft. in trunk diameter above the rather prominent thin buttresses. The sap and the crushed leaves have a strong, characteristic odor. The large, alternate, odd-pinnate leaves have 11–15 lanceolate, acute or acuminate leaflets with serrate margins, which average 3–4″ in length. The small flowers are produced in large axillary panicles, and are followed in February and March by the small, dry, 1-seeded fruits which are enclosed in the segments of the 5-parted, accrescent calyx. These expand at maturity and act as revolving parachutes for the dispersal of the seeds. The hard, heavy, fine-grained wood is variable in color from light to reddish brown, but can usually be recognized by the beautiful, bold, irregular blackish figure. The timber is reported to weather well, to be highly resistant to moisture (which however makes it glue rather poorly), to work and finish smoothly, and to take a beautiful polish. It is considered suitable for turned work

such as bowls and trays, gun stocks, furniture, beams, rafters, girders, flooring, heavy durable construction, and high-quality veneer and cabinet work. Considerable quantities have been exported to the United States, probably mostly for use in high-grade plywood and veneer.—Forested hills above Palmar Norte, 1,200 ft., *Allen 5998.*

Aticuej—See *Oreopanax xalapense* (Guatemala).
Atta—See *Bixa Orellana* (British Honduras).

AVICENNIA NITIDA Jacq.—Verbenaceae—*Mangle salado* or *Palo de sal* (Local)—*Culumate* (Costa Rica)—*White mangrove* or *Black mangrove* (Bocas del Toro and Chiriquí)—*Mangle* or *Mangle salado* (Panama)—*Palo de sal* (Nicaragua)—*Arbol de sal, Ishtaten, Istaten,* or *Palo de sal* (Salvador).

Common trees of the mangrove swamps, usually of medium size, but occasionally reaching 75 ft. in height and 24″ in trunk diameter. The lower trunk does not have stilt roots, but the species is characterized by the dense colonies of erect, finger-like breathing organs (pneumatophores) which project 2–3″ above the mud at low tide. The oblong or oblong-lanceolate, obtuse leaves are opposite on the stems, and the small fragrant white flowers are produced in March in axillary or terminal cymes, and are much frequented by bees at that season. The wood is dark brown to blackish, hard, heavy, coarse, with an uneven grain, very durable underground but perishable when exposed to the air. The bark contains about 12% tannin, and is occasionally exploited on a limited scale. The wood is difficult to work, and is seldom used excepting for fuel.—Tidal estuaries, Santo Domingo de Osa, *Tonduz 10060 & 10066*—Delta of the Río Esquinas, *Allen 5631.*

Avocado—See *Persea americana* (Local and general).
Azote caballo—See *Pithecolobium longifolium* (Local and Nicaragua).
Azulillo—See *Hamelia patens* (Costa Rica).

BACTRIS—KEY

1. Fronds undivided, bifid at the apex.
   2. Fronds more than 6 ft. in length. Plants confined to swampy forest at sea level........ *Bactris militaris*
   2. Fronds about 2 ft. long. Plants confined to forested ridges at 1,500–2,000 ft........ *Bactris* sp.—*Allen 6765*
1. Fronds pinnate, never bifid at the apex.
   2. Spathe conspicuously armed with needle-like spines.
      3. Spines conspicuously flattened, pale yellow in color, usually tipped with brown........ *Bactris divisicupula*

3. Spines not conspicuously flattened, dark brown or black in color........................ ....*Bactris balanoidea*
2. Spathe densely woolly, but never armed with spines.................*Bactris Baileyana*

BACTRIS BAILEYANA H. E. Moore—Palmaceae—*Hoja de duende* or *Huiscoyol* (Local).

Common spiny palms, with pinnate fronds and multiple trunks about 15 ft. in height. The short, broad spathes have a pale-brown, woolly covering, unique in our area in being unmixed with spines. The small, nearly globose fruits are red at maturity, and are usually found in October. A widespread and characteristic species of the forested ridges, to about 2,000 ft. in elevation.—Esquinas Forest, 200 ft., *Moore 6556* & *Allen 6606.*

BACTRIS BALANOIDEA (Oerst.) Wendl.—Palmaceae—*Huiscoyol* (Local). (Pl. 8)

Slender, intensely spiny palms, with multiple trunks and pinnate fronds, which form open colonies 10 to about 30 ft. in diameter, the interval between the canes being about 1–3 ft., depending upon the situation. The relatively large, dark-purple or brownish-purple fruits mature during March and April. They are frequently eaten locally, and have a pleasant, acidulous taste.—Palmar Norte, 100 ft., *Allen 6739*—Hills near Palmar, 200 ft., *Moore 6543 & Allen 6756.*

BACTRIS DIVISICUPULA Bailey—Palmaceae.

Slender palms, with 3 or 4 canes 12–18 ft. in height, terminating in 6–8 mature pinnate fronds, all parts of the plant conspicuously armed with white or pale-yellow, usually brown-tipped and flattened spines. Spathes tawny brown in color, also with many short, flat spines. Fruits about ½″ in diameter, the apex abruptly beaked; dark purple at maturity.—Forested hills above Palmar Norte, 1,200–1,500 ft., *Moore 6531* & *Allen 6746.*

BACTRIS MILITARIS H. E. Moore—Palmaceae. (Pl. 10)

A very striking, multiple-stemmed species, mature specimens of which average about 15 ft. in height. The large, arching fronds are completely undivided, a very unusual condition in this genus. The canes are slender, and armed with very long, black, needle-like spines. The small white flowers are produced from late April until early July and the attractive red fruits mature from about mid-August until November.—Very common in wet, lowland forests near Tinoco Station and Sierpe, *Allen 5276, 6264, & 6296.*

BACTRIS sp.—Palmaceae.

Spiny, stoloniferous palms, with 3–6 slender canes 6–10 ft. in height

and about ½″ in diameter, each with about 4–5 live, bifid, undivided fronds which may be either spiny or unarmed on the same plant.—Frequent in forested hills above Palmar Norte at 2,000–2,500 ft., *Allen 6765.*

Bala—See *Gliricidia sepium* (Panama).
Balo—See *Gliricidia sepium* (Panama).
Balsa—See *Heliocarpus appendiculatus* (Nicaragua). Fruits small, pink. Flowers small, yellow.
Balsa—See *Ochroma lagopus* (Local and general). Fruits elongate, individual flowers white, large.
Balsamo—See *Myroxylon balsamum* var. *Pereirae* (Local).
Balsam of Peru—See *Myroxylon balsamum* var. *Pereirae* (Costa Rica and U. S. Pharmacopoeia).
Balso—See *Ochroma lagopus* (Costa Rica).
Bambito—See *Ocotea veraguensis* (Panama).
Banak—See *Virola Koschnyi* (Nicaragua and British Honduras).
Banak—See *Virola sebifera* (Nicaragua).
Baraja—See *Cassia reticulata* (Honduras).
Barajo—See *Cassia reticulata* (Guatemala and Honduras).
Barba chele—See *Vochysia ferruginea* or *hondurensis* (Nicaragua).
Barenillo—See *Croton glabellus* (Honduras).
Barillo—See *Calophyllum braziliense* var. *Rekoi* (Salvador). Latex white.
Barillo—See *Symphonia globulifera* (Chiriquí, Guatemala, and Honduras). Latex yellow.
Barrabas—See *Euphorbia cotinifolia* (Costa Rica).
Barrigon—See *Bombax barrigon* (Panama).

BASILOXYLON EXCELSUM Standl. & L. Wms.—Sterculiaceae. (Fig. 4)

Tall forest trees, to about 150 ft. in height and 4–5 ft. in diameter above the slightly buttressed base. The broad, alternate leaves are nearly cordate in outline, with acute apices. The fruits are large, oblique, woody pods, produced in early March, the brown, winged seeds resembling those of mahogany (*Swietenia*). The wood is fairly hard, white, and fine-grained. Flowers have not been seen. Known only from the type collection.—Occasional in climax forest near Golfito, *Allen 5984.*

Bastard bay cedar—See *Trema micrantha* (British Honduras).
Bastard cedar—See *Guazuma ulmifolia* (Bocas del Toro).
Bastard timbersweet—See *Nectandra perdubia* (British Honduras).
Bateo—See *Carapa Slateri* (Panama).

Fig. 4. Basiloxylon excelsum. 1 — Leafy branch. 2 — Open pod, showing arrangement of seeds. Insert — Individual seed.

BATOCARPUS COSTARICENSE Standl. & L. Wms.—Moraceae—*Ojoche macho* (Local). (Pl. 24)

Tall, slender trees, to about 100 ft. in height, the lower trunk not prominently buttressed, apparently averaging about 18″ in diameter, at least in all the examples seen. The latex from the lower trunk is creamy white, while that from the branchlets and fruits is dull orange in color. The leaves are alternate, elliptic-lanceolate and abruptly acuminate at the apex. The flowers are dimorphic and axillary, the staminate form minute, on slender, catkin-like spikes and the pistillate nearly globose and solitary. The fleshy fruits are dark green, rather reminiscent of those of the common midwestern U. S. Osage orange (*Maclura*) on a smaller scale. The wood is white, hard, close-grained, and easy to work, but seems to have no local uses.—Forested hills above Palmar Norte, to about 1,500 ft., *Allen 5948 & 5971.*

BAUHINIA UNGULATA L.—Leguminosae; Caesalpinoideae—*Casco de venado* (Costa Rica)—*Pata de vaca* or *Pie de venado* (Guatemala)—*Pie de vaca* (Honduras)—*Pie de cabro* or *Pata de venado* (Salvador).

Usually a shrub, but occasionally a small tree to about 15 ft. in height. The leaves are alternate, the deeply bilobed blade resembling a cloven hoof in outline. The white or pinkish flowers are produced in elongate, many-flowered racemes, and are followed by brown, linear pods. The wood is sometimes used in Salvador for packsaddles because of its strength. The species is known in our area only from the vicinity of Buenos Aires.—*Tonduz 3803 & 6502.*

BELLUCIA COSTARICENSIS Cogn.—Melastomaceae—*Coronillo* or *Papaturro agrio* (Costa Rica)—*Manzana de montaña* (Guatemala).

Small trees usually less than 15 ft. in height, with broadly ovate, obtuse or abruptly acuminate, opposite, 4-ranked leaves. The relatively large white or pink flowers are produced singly or in small fascicles from the nodes or leaf axils. The fruits are large, juicy, and edible. The creamy-white wood is not used.—Terraba, *Tonduz 3916*—Buenos Aires, *Tonduz 4955.*

BELOTIA—KEY

1. Leaves narrowly linear-lanceolate, the upper surface glabrous, or with a very few, widely scattered, minute stellate hairs. Lower leaf surface densely and minutely stellate-tomentose, with the hairs of uniform size. Flowers pale pinkish lavender, appearing white when seen from a distance..*Belotia reticulata*
1. Leaves broadly ovate-lanceolate or elliptic-oblong, the upper surface stellate-tomentose, with scattering, much longer, conspicuous hairs (lens!). Lower surface densely stellate-tomentose, the hairs of very unequal size, some being much larger

than the others. Sepals deep pink and petals purple, the flowers appearing dark pink when seen from a distance........ *Belotia macrantha*

Belotia macrantha Sprague—Tiliaceae—*Plomo* (Local). (Pl. 23)
Small to medium-sized trees, which average about 20–40 ft. in height, but which may reach as much as 75 ft. under particularly favorable circumstances. The alternate, broadly ovate-lanceolate or elliptic-oblong leaves have 3 longitudinal nerves and minutely serrate margins. The attractive flowers have deep-pink sepals and purple petals, and are produced in profusion during January in axillary branching panicles, the color effect being dark pink when seen from a distance. They are followed in February by the typical, flattened, obcordate fruits. The species is very common throughout the area on the margins of forest, particularly near Golfito, where they form small groves of slender, straight-trunked specimens fronting many of the small lateral valleys near the airfield. The largest trees are to be found in the forested hills near the delta of the Río Esquinas. The soft white wood is not used locally.—Río Sandalo de Osa, *Dodge & Goerger 10077*—Forested hills above Palmar Norte, 1,800 ft., *Allen 5918*—Hills near Golfito, 200 ft., *Allen 6693.*

Belotia reticulata Sprague—Tiliaceae—*Capulin sabanero* (Nicaragua).
Slender trees, 15–65 ft. in height, with narrowly lanceolate, acuminate, 3-nerved leaves which are usually paler in color on the lower surface. The pale, pinkish-lavender, or nearly white flowers are produced in January in short axillary panicles, and are followed by 2-celled, flattened, obcordate fruits.—Hills near Golfito, *Allen 5750 & 6692.*

BIGNONIACEAE—Key

1. Leaves simple.
    2. Leaves produced in clusters of 3–7 from the nodes........ *Crescentia*
    2. Leaves solitary at each node........ *Enallagma*
1. Leaves compound.
    2. Leaflets 3. Flowers and fruits produced directly from the trunk........ *Parmentiera*
    2. Leaflets more than 3. Flowers and fruits not produced directly from the trunk.
        3. Leaves digitately compound.
            4. Leaflets 5–9. Flowers less than ¾″ long........ *Godmania*
            4. Leaflets 5. Flowers usually more than 2″ long........ *Tabebuia*
        3. Leaves pinnate.
            4. Leaves once pinnate. Flowers yellow........ *Tecoma*
            4. Leaves bipinnate. Flowers blue or purple........ *Jacaranda*

Billia colombiana Planch. & Lind.—Hippocastanaceae.
Large forest trees, to about 120 ft., with opposite, digitately 3-foliolate

leaves and terminal panicles of small white flowers with yellow centers, which turn red with age, the blooming season being February and March, during which time the fallen flowers are locally common along trails and in the forest. The wood is reddish brown, hard, tough, and strong. No local uses are known. The red-flowered species listed in *Flora of Costa Rica* under this name is actually *Billia Hippocastanum* Peyr., and seems to be confined to areas of greater elevation.—Common in lowland and hillside forests near Palmar Norte, to about 2,000 ft., *Allen 5982*.

Bimbayan—See *Vitex Cooperi* (Nicaragua).
Bitze—See *Inga punctata* (Guatemala).

BIXACEAE—One genus, *Bixa*.

BIXA—KEY

1. Fruits spiny .... *Bixa Orellana* forma *typica*
1. Fruits not spiny .... *Bixa Orellana* forma *leiocarpa*

BIXA ORELLANA L. forma TYPICA—Bixaceae—*Achiote* (Local)—*Achote* (Chiriquí)—*Anatto* (Bocas del Toro)—*Atta* or *Achiote* (British Honduras)—*Achote* (Honduras)—*Chaya, Xayau, Oox, Ox,* or *Achiote* (Guatemala).

Shrubs or small trees, to about 30 ft. in height, frequently cultivated and often persistent near old house sites. The alternate leaves are ovate, long-acuminate, and the large attractive white or pink flowers are produced in terminal panicles, followed by dark-red globose or ovoid capsules which are covered with spinelike bristles. The familiar "Achiote" or "Anatto" dye is obtained from the seeds, and is used locally for coloring rice, cheese, and butter. The soft white wood is not used.

BIXA ORELLANA L. forma LEIOCARPA (Kuntze) Standl. & Steyerm.—Bixaceae.

Identical with the typical form excepting for the fruits which are spineless.—Golfito, sea level, *Allen 5617*.

Black mangrove—See *Avicennia nitida* (Bocas del Toro and Chiriquí).
Black manu—See *Minquartia guianensis* (Local).
Black manwood—See *Minquartia guianensis* (Bocas del Toro and Limón).
Black palm—See *Astrocaryum Standleyanum* (Canal Zone, Chiriquí, and Bocas del Toro).
Black prickly yellow—See *Zanthoxylum procerum* (British Honduras).
Bloodwood—See *Pterocarpus officinalis* (Panama).

Boca de vieja—See *Posoqueria latifolia* (Panama).

Boca vieja—See *Posoqueria latifolia* (Panama).

Bocot—See *Cassia grandis* (Guatemala).

Bogamani—See *Virola,* various species (Chiriquí).

Bogamani verde—See *Dialyanthera otoba* (Chiriquí).

Bogum—See *Symphonia globulifera* (Bocas del Toro and Limón).

Bojon—See *Cordia alliodora* (Guatemala).

Bolador—See *Terminalia amazonia* (Honduras).

BOMBACACEAE—KEY

1. Leaves simple.
   2. Leaf blades with 3–5 longitudinal nerves.
      3. Flowers and fruits large, more than 4″ in length. Seeds imbedded in kapok-like fiber........ *Ochroma*
      3. Flowers and fruits small, usually less than 1″ in length........ *Hampea*
   2. Leaf blades with a single longitudinal nerve.
      3. Leaf petioles about equaling the blades in length. Huge trees, with tremendous plank buttresses........ *Huberodendron*
      3. Leaf petioles short, less than 1/10 the length of the blades. Trunks not conspicuously buttressed........ *Quararibea*
1. Leaves digitately compound.
   2. Flowers more than 10″ long. Fruits very large and woody, the fleshy seeds not surrounded by kapok-like fiber........ *Pachira*
   2. Flowers less than 6″ long. Seeds surrounded by kapok-like fiber.
      3. Flowers less than 2″ long. Large trees, with prominent buttresses........ *Ceiba*
      3. Flowers usually more than 4″ long........ *Bombax*

BOMBAX—KEY

1. Leaflets 7–9. Trunks usually somewhat swollen, with conspicuous vertical green stripes........ *Bombax barrigon*
1. Leaflets 5–7. Trunks not swollen or with vertical stripes........ *Bombax sessile*

BOMBAX BARRIGON (Seem.) Dcne.—Bombacaceae—*Ceibo barrigon* or *Ceibo pochote* (Local)—*Ceibo* (Costa Rica)—*Barrigon* (Panama).

Medium-sized, often flat-topped, unarmed trees, to 60–75 ft. in height, the lower trunk usually conspicuously swollen, with vertical green, gray, and yellowish stripes. The alternate, digitately 7–9 foliolate leaves are deciduous during the dry season, and the large white flowers, called "powder puffs" by the Canal Zone Americans and "motas" by the Panamanians, are produced from the ends of the bare branches from about January until March. They are followed by large woody capsules filled with gray, kapok-like wool which surrounds the seeds. The soft white wood is not used. Usually rather infrequent in the lowland forests, but notably represented in our area by large, handsome specimens conspicu-

ous from the sea in the region of Pavón Bay, where they form stately groves back of the fringing coconuts.—Forests near Palmar Norte, 250 ft., *Allen 6728.*

BOMBAX SESSILE (Benth.) Bakh.—Bombacaceae—*Ceibo* or *Ceibo nuno* (Panama).

Slender, unarmed trees, to about 60 ft. in height, with 5–6 parted, digitately compound leaves which are deciduous during the dry season. The pinkish, white, or pale-yellowish flowers are about 6″ long, and are produced from mid-December to about mid-February, and are followed by ovoid fruits from mid-February until April. The wood is apparently not used. The species is usually listed under the name of *Bombacopsis sessilis* (Benth.) Pittier in floras and handlists. Rather infrequent in our area.—Buenos Aires, 1,000 ft., *Pittier 3924*—Hills near Golfito, 500 ft., *Allen 5749.*

Bombon—See *Cochlospermum vitifolium* (Nicaragua and Honduras).

BORAGINACEAE—KEY

1. Style entire, or once bifid .... *Bourreria*
1. Style twice bifid .... *Cordia*

Botoncillo—See *Conocarpus erecta* (Salvador). Sap watery.

Botoncillo—See *Symphonia globulifera* (Costa Rica). Sticky yellow latex.

Boton de pega-pega—See *Vernonia patens* (Panama).

BOURRERIA LITORALIS Donn. Sm.—Boraginaceae—*Esquijoche* (Costa Rica).

Small trees, with alternate, ovate or obovate-elliptic leaves 3–5″ long and 2–2½″ wide and small white flowers produced in corymbose cymes which usually exceed the leaves in length. A poorly known species rather reminiscent of *Cordia gerascanthus* in superficial appearance.—Forests of the plain of Salinas, Golfo Dulce, *Pittier 2787.*

Brasil—See *Chlorophora tinctoria* (Costa Rica).

Brasil—See *Oreopanax xalapense* (Salvador).

BRAVAISIA INTEGERRIMA (Spreng.) Standl.—Acanthaceae—*Mangle de agua* (Local)—*Mangle blanco* or *Palo de agua* (Costa Rica, Panama, and Nicaragua). (Pl. 22)

Trees of varying size and habit, typical specimens having multiple trunks and many stilt roots, most examples seen being from about 30–50 ft. in height, but also represented in areas of climax forest by slender,

single-trunked trees with prominent stilt roots, which occasionally may reach as much as 65 ft. The leaves are opposite, with oblong-elliptic, acute or acuminate blades. The main flowering season is from January until mid-February, the tubular white flowers having lavender markings. They are produced in conspicuous terminal panicles. Mature fruits have been seen in mid-March, at the time when some of the trees in the drier locations were partially bare of leaves. The soft white wood has no known uses. One of the most common and characteristic trees of the region, found in pastures and forests from Palmar to Golfito.—Palmar Norte, 50 ft., *Allen 5714*.

Breadnut—See *Brosimum alicastrum* (British Honduras). Flowers in globose heads.

Breadnut—See *Brosimum terrabanum* (Bocas del Toro). Flowers in globose heads.

Breadnut—See *Trophis racemosa* (Bocas del Toro). Flowers in elongate spikes.

Bribri—See *Inga*, various species (Panama).

Bribri guavo—See *Inga spectabilis* (Panama).

BROSIMUM—KEY

1. Leaves thick and leathery, usually more than 3″ wide. Fruits about 1″ in diameter .... *Brosimum utile*
1. Leaves not notably thick and leathery, usually less than 3″ wide. Fruits less than 1″ in diameter.
   2. Leaves narrowly linear-lanceolate, long acuminate .... *Brosimum sapiifolium*
   2. Leaves oblong or elliptic-lanceolate, acute or shortly acuminate.
      3. Lateral veins of the lower leaf surface very prominently elevated, brownish beneath when dried .... *Brosimum terrabanum*
      3. Lateral veins of the lower leaf surface not prominently elevated, not brownish beneath when dried .... *Brosimum alicastrum*

BROSIUM ALICASTRUM Swartz—Moraceae—*Ojoche* (Local and general) —*Ujushte, Ujushte blanco, Masico, Ox, Ramon,* or *Ramon blanco* (Guatemala)—*Breadnut, Capomo,* or *Ramon* (British Honduras).

Trees 75 to about 100 ft. in height, with alternate, narrowly oblong-elliptic, abruptly acute, glabrous leaves 3½″–6½″ long and about 2″ wide. The globose fruits are about ¾″ in diameter, and are produced in March and April in the axils of the leaves. All parts of the tree have an abundant milky latex, which is agreeable to the taste. The leafy branches of this and other species are cut and fed to stock in British Honduras and Yucatan, and the fruits are also boiled and eaten.—Jalaca, sea level, *Allen 5207*.

BROSIMUM SAPIIFOLIUM Standl. & L. Wms.—Moraceae—*Morillo* or *Ojoche macho* (Local).

Tall, white-trunked, handsome trees, 90–120 ft. in height and 3–5 ft. in trunk diameter, with alternate, elliptic-lanceolate, abruptly long-acuminate leaves. The white flowers are produced in the leaf axils of the new foliage in December and January, and are followed in February by the small, greenish-yellow, nearly globose fruits, which are edible, having a pleasant flavor rather like that of avocado. The milky latex, as in most of the species, is sweet to the taste. Frequent in climax forests throughout the area.—Palmar Norte, 50 ft., *Allen 5877 & 6707.*

BROSIMUM TERRABANUM Pittier—Moraceae—*Ojoche* (Local)—*Breadnut* (Bocas del Toro)—*Masica* or *Ojoche* (Nicaragua)— *Pisma, Masica,* or *Masico* (Honduras)—*Masicaran* (British Honduras).

Tall forest trees, 90–120 ft. in height, with alternate, elliptic-lanceolate, acute leaves and minute white flowers borne on a small, axillary, globose receptacle. The flowering season is late January and early February. The latex is pale orange in color, and is of a sweetish taste. The fruits are edible, and are used in some parts of Nicaragua for making tortillas. The leaves are used as fodder for mules and cattle in forested areas of northern Central America. The trees are often left standing in pastures since the fallen fruits are relished by stock. The wood is pale brown to white, hard and strong, but is not durable. It is reported to burn green, and is principally used for fuel or charcoal. Fairly frequent throughout the area.—Esquinas Forest, 250 ft., *Allen 5809.*

BROSIMUM UTILE (HBK) Oken—Moraceae—*Palo de vaca* or *Vaco* (Local)—*Mastate Colorado* (Boruca)—*Lechero* or *Mastate* (Costa Rica) —*Palo de leche* (Panama). (Fig. 5 and Pls. 3 & 7)

Forest trees of noble aspect, with sheer gray trunks 100 ft. or more in height and up to 6 ft. in diameter above the somewhat buttressed base. The large, leathery leaves are alternate, with oblong-elliptic, shortly acuminate blades. They are renewed in November and December, the flushes of bright new foliage rendering the trees conspicuous from a distance against the darker greens of the hillsides, the forest floor at this season being deeply covered with the recently shed, dark-brown leaves. Young specimens, up to 10 or 15 ft. in height, have juvenile leaves which may often be as much as 30″ long and 9–10″ wide. The minute flowers are white, and are borne on a small, spherical, axillary receptacle from November until January, and are soon followed by the globose, 1-seeded fruits which average about 1″ in diameter. All parts of the plant produce an abundant, creamy latex, which is sweet and of pleasant flavor, particu-

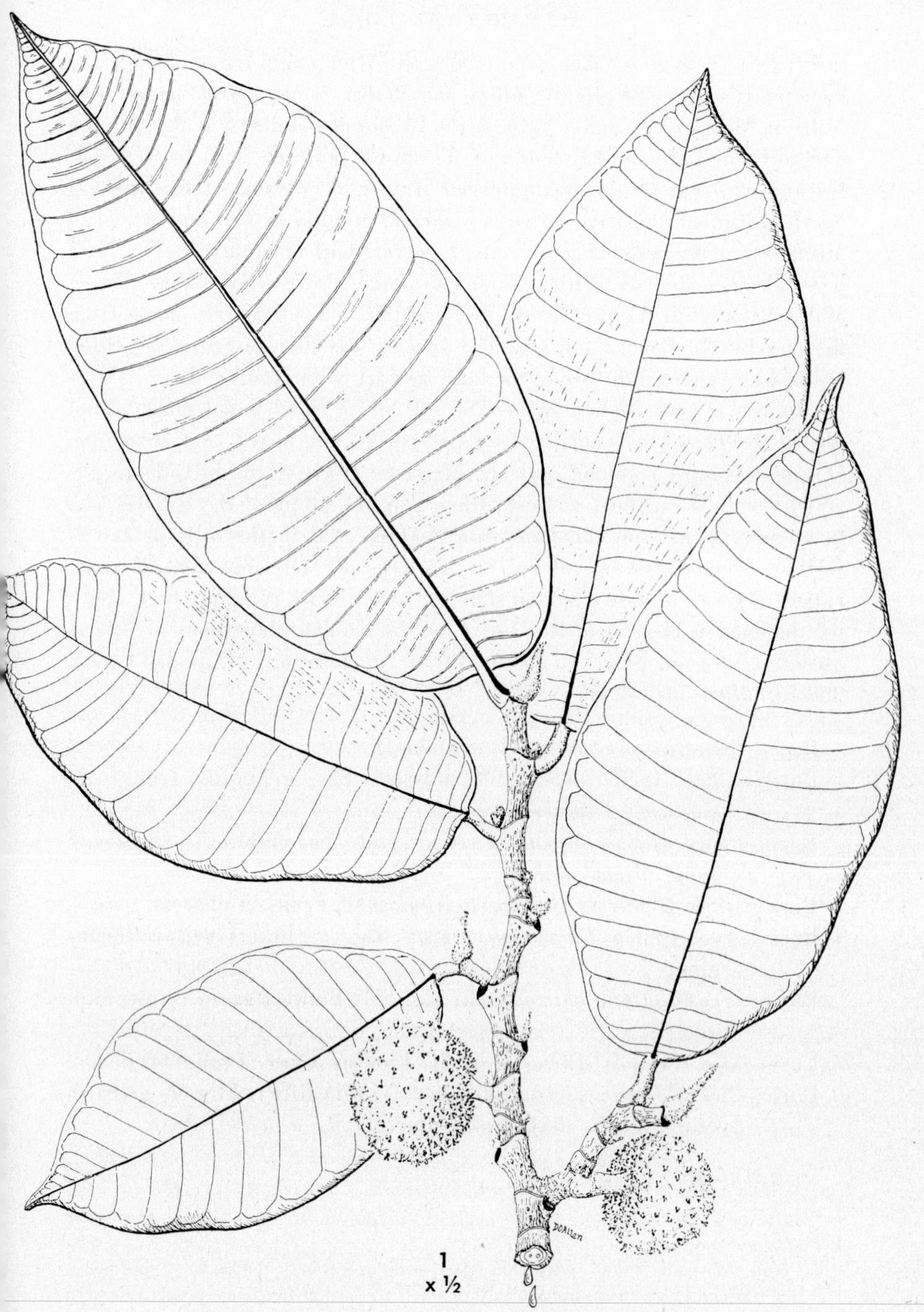

Fig. 5. Brosimum utile. Branch with mature fruits.

larly when it is first taken from the tree. After exposure to the air the taste tends to become slightly bitter and chalky (some detractors say like Milk of Magnesia!) and within about 24 hours solidifies in the tapping cuts to form a chicle-like substance. The fresh milk has been tried in coffee and can scarcely be distinguished from good cream, while chilled it can be whipped and flavored with sugar and vanilla extract and served to unsuspecting humans. Dogs or cats, however, will not touch it. The latex has also been used in British Guiana to make a sort of vegetable cheese; and a bark cloth, known locally as "mastate," was formerly made from the tree by the Boruca Indians. The species was first described by Humboldt, who reported the use of the latex as part of the routine diet of slaves on the north coast of Venezuela. The soft white wood is not durable, but has been utilized in Ecuador for concrete forms, boxes, and sheathing. One of the most common and characteristic trees of the hillside forests which occur in our part of Costa Rica from the gorge of the Río Terraba to its mouth, and on the Peninsula de Osa, and in the hills as far as Golfito in concentrations of one or two trees per acre, but often meet on ridges in great, almost unmixed stands, a triumphant company high above the thatched settlement of Rey Curre, or the banana plantations of Jalaca, ascending in some places to about 4,000 ft. in elevation.—Esquinas Forest, 250 ft., *Allen 5813 & 5833.*

Bucut—See *Cassia grandis* (Guatemala).

Burillo—See *Apeiba aspera* (Nicaragua). Sea urchin-like fruits with short, hard spines.

Burillo—See *Apeiba Tibourbou* (Nicaragua). Sea urchin-like fruits with long, pliant spines.

Burillo—See *Guatteria chiriquiensis* (Local). Fruits in clusters.

Burio—See *Apeiba Tibourbou* (Costa Rica). Flowers yellow. Fruits spiny.

Burio—See *Guatteria chiriquiensis* (Local). Flowers green. Fruits multiple.

Burio—See *Hampea Allenii* (Local). Flowers white. Fruits 3-celled.

Burio—See *Heliocarpus appendiculatus* (Costa Rica). Flowers greenish yellow. Fruits dark pink, spiny.

BURSERACEAE—Key

1. Plants epiphytic .... *Dacryodes*
1. Plants not epiphytic.
   2. Petals imbricate in bud. Bark thin, often peeling, coppery red in large specimens, but frequently greenish, particularly in small trees or living fence posts ...... *Bursera*

2. Petals valvate in bud. Bark not as above.
   3. Petals united .... *Tetragastris*
   3. Petals distinct .... *Protium*

BURSERA SIMARUBA (L.) Sarg.—Burseraceae—*Jiñocuavo* or *Indio desnudo* (Local)—*Caraña, Almacigo,* or *Jiñote* (Costa Rica)—*Almacigo* or *Carate* (Panama)—*Chinacuite, Chinacahuite, Chino, Copon, Gumbolimbo, Jenequite, Jicote, Jinicuite, Jiote, Indio desnudo, Mulato, Palo chino, Palo jiote, Palo mulato,* or *Torchwood* (Honduras)—*Cacah, Cajha, Chaca, Chacah, Chacah colorado, Chic-chica, Chinacahuite, Chino, Indio desnudo, Jiote, Palo chino, Palo jiote, Palo mulato, Solpiem,* or *Xacago-que* (Guatemala).

Common trees of varying size, often planted for living fence posts, but frequently seen in the forest, where they may reach 80 ft. in height and 24″ in trunk diameter. The larger specimens in particular may be readily recognized by the coppery-red bark that is constantly being renewed, peeling off in thin, irregular sheets. In smaller trees and especially living fence posts the bark may be greenish tan in color. The relatively large pinnate leaves are alternate and deciduous, the trees usually being bare from late December until about May. The small greenish or yellowish fragrant flowers are produced in axillary racemes and are followed by small, brown, 3-angled, dehiscent drupes, which last on the branches through most of the dry season. The trees often exude a brownish or reddish resin which has a strong turpentine-like odor and can be used for mending broken dishes. The white to light-brown wood is not durable, and is seldom used excepting for cheap packing boxes and match sticks.—Lagarto, *Tonduz 4536.*

Button bush—See *Conocarpus erecta* (British Honduras).
Button mangrove—See *Conocarpus erecta* (Canal Zone, Bocas del Toro, and Limón).
Buttonwood—See *Conocarpus erecta* (British Honduras).

BYRSONIMA—KEY

1. Leaves densely brown-tomentose on the lower surface .... *Byrsonima crassifolia*
1. Leaves not tomentose on the lower surface .... *Byrsonima densa*

BYRSONIMA CRASSIFOLIA (L.) HBK—Malpighiaceae—*Nance* (Local)—*Wild cherry* (Chiriquí)—*Nance* (Panama)—*Nancite* (Nicaragua)—*Nance, Nancito,* or *Crabo* (Honduras)—*Nance, Chi,* or *Tapal* (Guatemala). (Pl. 23)

Shrubs or trees, to about 40 ft. in height and 12″ in trunk diameter,

with opposite, rounded or oblong, acute or acuminate leaves and many-flowered racemes of yellow or orange flowers which are followed in April and May by small, nearly globose yellow fruits that are edible. The dark reddish- or pinkish-brown wood is rather hard, heavy, and strong, but is brittle, and is little used excepting for fuel and charcoal, and occasionally for the extraction of a red dye. The bark is reported to be used for tanning. The species is rather infrequent in pastures near Palmar Norte.—Boruca, 1,200 ft., *Tonduz 3935*—Buenos Aires, *Pittier 4035* & *Tonduz 6560*—Pastures near Palmar Norte, 100 ft., *Allen 6667*.

Byrsonima densa (Poir.) DC.—Malpighiaceae—*Nance* (Local).

Tall trees to about 100 ft. in height, with opposite, oblong-elliptic, acute leaves and terminal racemes of large yellow fruits that mature in August. A species of the Guianas, previously unreported from Central America.—Esquinas Forest, 200 ft., *Allen 5543*.

C'ab-choh—See *Oreopanax capitatus* (Guatemala).
Cabeza de negrito—See *Guazuma ulmifolia* (Panama).
Cabbage bark—See *Andira inermis* (Bocas del Toro and Canal Zone).
Cablote—See *Guazuma ulmifolia* (Honduras).
Caca de mico—See *Guazuma ulmifolia* (Salvador). Leaves alternate.
Caca de mico—See *Randia armata* (Salvador). Leaves opposite.
Caca de niño—See *Licania arborea* (Guatemala).
Cacagua—See *Gliricidia sepium* (Honduras).
Cacah—See *Bursera simaruba* (Guatemala).
Cacahuanance—See *Gliricidia sepium* (Salvador).
Cacaloguiste de flor quemada—See *Godmania aesculifolia* (Nicaragua).
Cacalojoche—See *Plumeria rubra* forma *acutifolia* (Costa Rica).
Cacao—See *Theobroma Cacao* (Local and general).
Cacao cimmaron—See *Herrania purpurea* (Panama). Leaves digitately compound. Flowers produced from the trunk.
Cacao cimmaron—See *Pachira aquatica* (Nicaragua). Leaves digitately compound. Flowers produced from the branches.
Cacao cimmaron—See *Theobroma angustifolium* (Chiriquí). Leaves simple. Sap not pink.
Cacao cimmaron—See *Virola guatemalensis* (Guatemala). Leaves simple. Sap pink.
Cacao de ardilla—See *Herrania purpurea* (Costa Rica).
Cacao de Costa Rica—See *Theobroma angustifolium* (Guatemala).
Cacao de mico—See *Theobroma angustifolium* (Local). Leaves simple. Fruits produced from branches.

Cacao de mico—See *Herrania purpurea* (Costa Rica). Leaves digitately compound.

Cacao de mico—See *Theobroma simiarum* (Local). Leaves simple. Fruits produced from trunk.

Cacao del monte—See *Herrania purpurea* (Local).

Cacao de playa—See *Pachira aquatica* (Nicaragua).

Cacao mani—See *Herrania purpurea* (Chiriquí).

Cacao pataste—See *Theobroma bicolor* (Costa Rica).

Cacao silvestre—See *Enallagma latifolia* (Costa Rica). Flowers greenish white.

Cacao silvestre—See *Theobroma angustifolium* (Costa Rica). Flowers orange.

Cacao volador—See *Virola guatemalensis* (Guatemala).

Cacau—See *Theobroma Cacao* (Guatemala).

Cache—See *Croton glabellus* (Guatemala).

Cachimbo—See *Crataeva tapia* (Honduras). Leaves digitately 3-foliolate.

Cachimbo—See *Platymiscium pinnatum* (Local). Leaves pinnate.

Cachito—See *Acacia costaricensis* (Honduras). Leaves bipinnate. Flowers in spikes.

Cachito—See *Acacia melanoceras* (Panama). Leaves bipinnate. Flowers in globose heads.

Cachito—See *Posoqueria latifolia* (Honduras). Leaves simple.

Cachito de aromo—See *Acacia farnesiana* (Honduras).

Cacho de venado—See *Mouriria parvifolia* (Guatemala).

Cacho venado—See *Mouriria parvifolia* (British Honduras).

Cacique—See *Diphysa robinioides* (Panama).

Caco—See *Theobroma Cacao* (Guatemala).

Cacucup—See *Acalypha diversifolia* (Guatemala).

Cadeno—See *Albizzia longepedata* (Guatemala).

Cagalera—See *Ximenia americana* (Honduras).

Cagalero—See *Randia armata* (Honduras). Leaves opposite.

Cagalero—See *Ximenia americana* (Honduras). Leaves alternate.

Caimito—See *Chrysophyllum mexicanum* (Local and general). Latex white.

Caimito—See *Chrysophyllum panamense* (Local). Latex white.

Caimito—See *Rheedia edulis* (Honduras). Latex yellow.

Caimito de montaña—See *Rheedia edulis* (Honduras).

Cainillo—See *Miconia argentea* (Panama).

Cainillo de cerro—See *Miconia argentea* (Panama).

Cajeto—See *Ochroma lagopus* (Guatemala).

Cajeton—See *Heliocarpus appendiculatus* (Guatemala).

Cajha—See *Bursera simaruba* (Guatemala).

Cakica-che—See *Casearia myriantha* (Guatemala).

Calabash—See *Crescentia cujete* (Local and general).

Calabasillo de la playa—See *Enallagma latifolia* (Local).

Calabazo—See *Crescentia cujete* (Panama).

Calan—See *Calycophyllum candidissimum* (Guatemala and Honduras).

CALLIANDRA GRANDIFOLIA P. H. Allen, sp. nov.—Leguminosae; Mimosoideae.

Small trees, about 18 ft. in height, with alternate, pinnate leaves, each composed of a single pair of leathery, glabrous leaflets 7½–9½″ long and 2½–4½″ wide, the elliptic-lanceolate, asymmetrical blades rather abruptly acuminate at the apex and with 3 very prominent longitudinal nerves. The bright-red flowers are produced in December in terminal or axillary, nearly globose heads of many slender stamens, the individual clusters being about 2½″ in diameter. A very remarkable species, apparently allied to *Calliandra arborea* Standl. and *Calliandra Brenesii* Standl., but differing in the much larger foliage and other characters. Known only from the type collection.—Esquinas Forest, between Golfo Dulce and Río Terraba, Prov. Puntarenas, 100 ft., *Skutch 5362* TYPE (Herb. Escuela Agrícola Panamericana).

CALOCARPUM BORUCANUM Standl. & L. Wms.—Sapotaceae—*Sapote* or *Zapote* (Local).

Large forest trees, 75–90 ft. in height and 2–3 ft. in trunk diameter, with alternate, slender-petiolate, glabrous leaves that are mostly clustered at the ends of the branchlets, the oblanceolate blades with acute apices and decurrent bases and averaging about 7 or 8″ in length. The relatively small white flowers are produced from the leaf axils in November and early December. The large pendulous fruits are elliptic in outline and about 5 or 6″ long and 3–3½″ in diameter, of a pale grayish-white color. The thick, dark-red flesh surrounding the 1 or 2 lustrous pale-brown fusiform seeds is edible.—Occasional in lowland forests near Palmar Norte, *Allen 5753 & 6636.*

CALOPHYLLUM BRAZILIENSE Camb. var. REKOI Standl.—Guttiferae—*Maria, Maria colorado,* or *Santa Maria* (Costa Rica)—*Maria, Santa Maria,* or *Palo de Maria* (Honduras)—*Mario, Barillo,* or *Varillo* (Salvador)—*Mario, Leche amarilla, Leche, Lech,* or *Santa Maria* (Guatemala).

Tall, straight-trunked forest trees, 50–120 ft. in height and 18–36″ in

trunk diameter, all parts of the plant having a sticky yellow latex. The shining leathery leaves are opposite, and are usually obovate or elliptic-oblong in outline and either obtuse or acute at the apex. The small white flowers have masses of bright-yellow stamens and are produced in our area in August and September in short axillary racemes. The fruit is a relatively large, yellowish, 1-seeded drupe, produced in September or early October. Specimens over 3 ft. in diameter are reported often to have defective hearts. The wood is pink to dull red in color, is easy to work, and is tough, strong, and fairly durable. It is suitable for canoes, general construction, bridge decking, ship masts and decks, wheel stock, sash and millwork, floors, siding, and fence posts. It has been used by the United Fruit Company for ties, interior construction, and furniture in the Armuelles Division.—Buenos Aires, 1,000 ft., *Tonduz 6713*—Santo Domingo de Osa, *Tonduz 10009*—Forested hills near the delta of the Río Esquinas, *Allen 5622*.

CALYCOPHYLLUM CANDIDISSIMUM (Vahl) DC.—Rubiaceae—*Conejo* (Local)—*Madroño, Harino,* or *Alazano* (Panama)—*Madroño, Salamo,* or *Surra* (Costa Rica)—*Calan, Madroño, Colorado,* or *Salamo* (Honduras)—*Calan, Canela, Chulub, Guayabillo, Madroño, Palo de peine, Salamo,* or *Uca* (Guatemala)—*Lancewood* or *Lemonwood* (U. S. trade names). (Pl. 31)

Medium-sized or rarely large forest trees, 40–75 ft. in height and 12 to about 30″ in trunk diameter, with reddish-brown, thin, scaling bark. Tne leaves are opposite, with broadly oval or ovate, abruptly acuminate blades and elongate, slender petioles. The trees are usually bare of leaves during the dry season. The small, fragrant white flowers are produced from about mid-November until early January, and are subtended by broad, creamy-white calyx lobes that render the trees conspicuous and attractive from a distance. The wood is yellowish or pale brown, with a beautiful close straight grain, and is hard, tough, elastic, and durable, but easy to work and finish smoothly. It has been used for house supports and farm tools in Cuba, and for firewood, charcoal, and the manufacture of fine-toothed combs in Honduras and Guatemala, as well as for archery bows and tool handles in the United States.—Fairly frequent in the forested hills above Palmar Norte at about 1,000–1,500 ft. elevation, *Allen 6309*.

Camaron—See *Mouriria parvifolia* (Salvador).
Camaroncillo—See *Hirtella triandra* (Panama).
Cambrillo—See *Zexmenia frutescens* (Guatemala).

Camche—See *Casearia arguta* (Guatemala).
Camfin—See *Trichilia tuberculata* (Local, Chiriquí, and Bocas del Toro).
Camibar—See *Prioria copaifera* (Costa Rica and Nicaragua).
Candalwood—See *Dracaena americana* (British Honduras).
Candelillo—See *Cassia spectabilis* (Costa Rica and Honduras). Flowers with 5 petals.
Candelillo—See *Tecoma stans* (Costa Rica). Flowers tubular.
Canela—See *Calycophyllum candidissimum* (Guatemala).
Canelilla—See *Ocotea veraguensis* (Costa Rica).
Canelito—See *Ocotea veraguensis* (Salvador).
Canelo—See *Nectandra salicifolia* (Local).
Canelo—See *Ocotea veraguensis* (Local and Salvador).
Canelo—See *Ocotea Williamsii* (Local).
Canilla de mula—See *Licania arborea* (Salvador). Branchlets not hollow.
Canilla de mula—See *Triplaris melaenodendron* (Salvador). Branchlets hollow.
Canilla de venado—See *Hamelia patens* (Nicaragua).
Canillito—See *Miconia Matthaei* (Costa Rica).
Canillo—See *Dialium guianense* (Honduras). Tall trees with pinnate leaves.
Canillo—See *Miconia argentea* (Panama). Small trees with simple leaves.
Canoh—See *Croton glabellus* (Guatemala).
Cantarillo—See *Cupania guatemalensis* (Costa Rica).
Canudo—See *Hamelia patens* (Guatemala).
Canum—See *Spondias purpurea* (Guatemala).
Canxan—See *Terminalia amazonia* (Guatemala).
Canxun—See *Terminalia amazonia* (Guatemala).
Caña de arco—See *Dracaena americana* (Guatemala).
Caña de danta—See *Geonoma congesta* (Local).
Caña fistula—See *Cassia grandis* (Panama). Leaves about 2″ long.
Caña fistula—See *Tachigalia versicolor* (Local). Leaflets about 6″ long.
Cañafistula—See *Cassia grandis* (Guatemala).
Capirote—See *Miconia argentea* (Nicaragua).
Capirote blanco—See *Miconia argentea* (Nicaragua).

CAPPARIDACEAE—Key

1. Leaves simple .... *Capparis*
1. Leaves digitately compound .... *Crataeva*

Capparis Sinclairii Benth.—Capparidaceae.

Trees, 18–60 ft. in height, with alternate, lanceolate, acuminate, glabrous leaves 4–5½″ long and about 1½″ wide, each leaf axil having a small but distinct top-shaped gland. The attractive white or greenish-white flowers are produced in March in short terminal 2–3-flowered clusters, and have rather thick petals about 1¼″ long and a great many showy white stamens. They are followed by slender, nearly terete, pale-brownish fruits which may be 4–5″ long.—Occasional in forests near Palmar Norte, 100 ft., *Allen 5510 & 6639.*

Capomo—See *Brosimum alicastrum* (British Honduras).

Capulin—See *Muntingia calabura* (Local and general). Flowers white.

Capulin—See *Trema micrantha* (Local and general). Flowers green.

Capulin blanco—See *Muntingia calabura* (Local and general).

Capulin blanco—See *Trema micrantha* (Local and general).

Capulincillo—See *Trema micrantha* (Salvador).

Capulin macho—See *Dicraspidia Donnell-Smithii* (Local). Large yellow flowers.

Capulin macho—See *Trema micrantha* (Chiriquí and Salvador). Minute green flowers.

Capulin montes—See *Trema micrantha* (Salvador).

Capulin negro—See *Trema micrantha* (Honduras).

Capulin sabanero—See *Belotia reticulata* (Nicaragua).

Capulin verde—See *Mouriria parvifolia* (Salvador).

Caracucha—See *Plumeria rubra* forma *acutifolia* (Panama).

Caracucha blanca—See *Plumeria rubra* forma *acutifolia* (Panama).

Caragra—See *Lippia oxyphyllaria* (Costa Rica).

Caragua—See *Cassia grandis* (Guatemala).

Carambano—See *Cassia grandis* (Nicaragua).

Caraña—See *Bursera simaruba* (Costa Rica). Leaves pinnate. Petals imbricate in bud.

Caraña—See *Carpotroche platyptera* (Costa Rica). Leaves simple.

Caraña—See *Protium* sp. (Boruca). Leaves pinnate. Petals valvate in bud.

Carao—See *Cassia grandis* (Local and general).

Carapa Slateri Standl.—Meliaceae—*Cedro bateo* or *Cedro macho* (Local)—*Bateo* or *Saba* (Panama). (Fig. 6 and Pl. 14)

Tall trees, to about 120 ft. in height and 3–4 ft. in trunk diameter, with very large, alternate, even-pinnate leaves. The small, greenish-white flowers are produced in slender, subterminal panicles at rather frequent intervals throughout the year, soon following the very showy blood-red

Fig. 6. Carapa Slateri. 1 — Flowering branch. 2 — Individual flower. 3 — Cross section of flower.

flushes of new foliage. The fruits are large, brown, nearly globose, woody pods which typically contain 8 large fleshy seeds. The wood is reddish brown, fairly hard and strong, but tends to saw woolly, is subject to transverse shear under stress, and is not durable in contact with the ground. The

species is most frequently found in relatively dense stands on low, swampy terrain, where strong buttresses are developed in old specimens. A few trees are also present on the ridges (usually unbuttressed) that are reported to have stronger wood than those from the swamps. The timber is used locally for general construction in labor camps, as bridge boards and fence posts. It is considered to be a second-grade lumber but better than Espavel *(Anacardium excelsum).* Considerable quantities have been exported to Germany in the past, presumably for use as plywood core-stock. One of our most common and characteristic species.—Esquinas Forest, *Allen 5202, 6608, & 6718.*

Carapato—See *Hirtella triandra* (Panama).

Carate—See *Bursera simaruba* (Panama).

Carbon—See *Guarea longipetiola* (Black River, Honduras). Leaves even-pinnate.

Carbon—See *Tetragastris panamensis* (British Honduras). Leaves odd-pinnate.

Carbon—See *Trichilia acutanthera* (Guatemala). Leaves odd-pinnate. Leaflets pubescent on the lower surface.

Carboncillo—See *Cupania guatemalensis* (Guatemala). Leaves pinnate, alternate. Flowers white.

Carboncillo—See *Swartzia panamensis* (Costa Rica). Leaves pinnate, alternate. Flowers yellow.

Carboncillo—See *Sweetia panamensis* (Costa Rica). Leaflets emarginate at the apex. Flowers white.

Carboncillo—See *Tecoma stans* (Costa Rica). Leaves pinnate, opposite. Flowers yellow.

Carboncillo—See *Terminalia amazonia* (Panama). Leaves simple.

Carboncillo—See *Trichilia acutanthera* (Guatemala). Leaves pinnate, alternate. Flowers greenish white, in axillary panicles.

Carbon colorado—See *Cupania macrophylla* (Guatemala).

Carbon de rio—See *Trichilia montana* (British Honduras).

Carbonero—See *Guarea Guara* (Boruca).

CARICACEAE—KEY

1. Leaves digitately compound .... *Jacaratia*
1. Leaves simple, but more or less deeply lobed .... *Carica*

CARICA—KEY

1. Fruits more than 4″ in diameter, edible, usually borne near the top of the tree near the crown of deeply lobed leaves .... *Carica papaya*
1. Fruits less than 3″ in diameter, not edible, usually borne low on the trunk. Leaves larger than those of *C. papaya,* darker green and less deeply lobed...... *Carica pennata*

CARICA PAPAYA L.—Caricaceae—*Papaya* (Local and general).

Small, slender trees, with broad, deeply lobed leaves, the very long, hollow petioles leaving a conspicuous scar on the trunk after they have fallen. The plants are commonly cultivated throughout the area for their edible, melon-like fruits, and are often persistent near old camp sites and along trails. The dried leaves are reported to be used in Honduras as a substitute for tobacco, and the dried latex forms the basis for various commercial preparations for tenderizing meat.

CARICA PENNATA Heilborn—Caricaceae—*Papayillo de venado* (Local) —*Papayo de monte* or *Papayo de montaña* (Guatemala).

Slender trees to about 15 ft. in height, with nearly circular, shallowly lobate leaves and small orange fruits which are produced low on the trunk, while those of the cultivated papaya are borne under the crown of foliage. The seeds have very strongly developed longitudinal, serrate ridges. Occasional in forests throughout the area.—Palmar Norte, 100 ft., *Allen 6274.*

Carne asada—See *Cupania guatemalensis* (Costa Rica). Flowers white.

Carne asado—See *Andira inermis* (Costa Rica). Flowers lavender or purple.

Carnezuelo—See *Acacia costaricensis* (Nicaragua).

CARPOTROCHE PLATYPTERA Pittier—Flacourtiaceae—*Caraña* (Costa Rica)—*Sucte* (Honduras)—*Sucte* or *Jaya* (Guatemala).

Small forest trees, 12–15 ft. in height, with alternate, oblong-oblanceolate leaves and white, unisexual flowers with attractive yellow stamens. The relatively large pistillate flowers are produced directly from the trunk and larger branches, in a manner reminiscent of Cacao, while the smaller staminate flowers are in small axillary racemes on the branches. The fruits are nearly globose, with 8 thin longitudinal wings. The wood is reported to be pinkish brown, moderately hard, fine-grained, and textured, but does not seem to be used, probably because of the small sizes obtainable.—Fairly frequent in the forested hills above Palmar Norte at about 2,000–2,500 ft., *Allen 5895 & 6757.*

CARYOCARACEAE—One genus, *Caryocar.*

CARYOCAR COSTARICENSE Donn. Sm.—Caryocaraceae—*Ajo* (Local)—*Aji* or *Ajillo* (Costa Rica)—*Manu* or *Plomillo* (Forests of Costa Rica). (Fig. 7)

Very large trees, 120–140 ft. in height and 4–6 ft. in diameter, very common in the hillside forests throughout the area. The opposite, digitately

Fig. 7. Caryocar costaricense. Flowering branch.

compound leaves are 3-parted, and the attractive, pale-yellow flowers are produced in erect, terminal racemes borne well above the foliage, the flowering season being in February. The fruit is a relatively large, green,

1- or 2-seeded drupe, which matures in early April. The white wood has a strong odor of spoiled vinegar when freshly cut. It has a reputation in some parts of the country of great durability in contact with the ground, felled trunks at Quepos having been reported to last up to 10 years in the banana plantations. The wood is moderately hard and heavy, and has been used for railroad crossties, house posts, framing, mill foundation timbers, and general heavy construction.—Esquinas Forest, 250 ft., *Allen 5840*—Forested hills above Palmar Norte, 1,500 ft., *Allen 5912*.

Cas—See *Psidium Friedrichsthalianum* (Costa Rica and general).

Cas acido—See *Psidium Friedrichsthalianum* (Costa Rica and general).

Casada—See *Psychotria chiapensis* (British Honduras).

Cascarilian—See *Croton glabellus* (Honduras).

Casco de venado—See *Bauhinia ungulata* (Costa Rica).

CASEARIA—KEY

1. Leaf axils of the new growth with conspicuous narrow stipules........ *Casearia arborea*
1. Leaf axils without conspicuous stipules.
    2. Leaves with very coarsely and densely serrate margins.................... *Casearia arguta*
    2. Leaves with minutely sinuate-serrate or entire margins.
        3. Flowers and fruits in pedunculate cymes, up to about 2″ in length. Tall, slender trees, to about 75 ft. in height.............................. *Casearia banquitana* var. *laevis*
        3. Flowers and fruits in sessile fascicles. Trees much less than 75 ft. in height.
            4. Flowers glabrous (lens!). Leaf margins entire.................... *Casearia sylvestris*
            4. Flowers pubescent (lens!). Leaf margins obscurely serrate or crenate.
                5. Leaf blades usually less than 3″ long, obtuse or narrowly rounded at the apex, the plants often armed with spines.............................. *Casearia aculeata*
                5. Leaf blades usually more than 4″ long, acute or acuminate at the apex, the plants never armed with spines.
                    6. Leaves oblanceolate, with minutely serrate margins...... *Casearia guianensis*
                    6. Leaves linear-lanceolate or elliptic-lanceolate, with finely sinuate-serrate margins.. *Casearia myriantha*

CASEARIA ACULEATA Jacq.—Flacourtiaceae—*Matacartago* (Costa Rica)—*Aguja de arrea, Escambron*, or *Guacuco* (Honduras)—*Limoncillo* or *Pullun* (Guatemala).

Shrubs or small trees, 12–20 ft. in height. The alternate, elliptic or obovate leaves are usually less than 3″ long, have obscurely crenate margins and are typically obtuse or narrowly rounded at the apex. The axillary fascicles of small greenish-white flowers appear in late March and early April. The short lateral branches are often armed with short, unbranched spines. The wood is described as being nearly white or yellowish, rather hard, straight-grained, fine-textured, tough, and more or less splintery. It has no local uses. A very common species found along trails

and in brushy places, particularly in the Terraba valley.—Palmar Norte, 100 ft., *Allen 5228 & 6006.*

CASEARIA ARBOREA (Rich.) Urban—Flacourtiaceae.

Slender trees, 35–60 ft. in height, with alternate, narrowly lanceolate, acuminate leaves 3½–4½″ long and 1–1¼″ wide, with minutely serrate margins and very short petioles, the leaf axils of the current growth with pairs of conspicuous linear stipules. The small white flowers are produced in March in short pedunculate cymes, and are followed by small red fruits. The branches and leaves have a drooping habit rather reminiscent of some of the species of *Xylopia.*—Fairly common on hillsides and cutover land near Palmar, *Allen 6769.*

CASEARIA ARGUTA HBK—Flacourtiaceae—*Huesillo, Palo Maria,* or *Paipute* (Costa Rica)—*Raspalengua* (Panama)—*Cuculmico* (Salvador) —*Camche, Guayabillo, Ixim-che, Manzanilla,* or *Raspa-lengua* (Guatemala). (Pl. 25)

Small, unarmed shrubs or trees, 10–20 ft. in height, with alternate, oblong-lanceolate, long-acuminate leaves which have conspicuously serrate margins. The trees flower several times at short intervals from October to February, the small, fragrant, greenish-white flowers being produced in dense axillary fascicles and are followed by dark-green, nearly globose fruits which are nearly ¾″ in diameter. Common along trails and in brushy places.—Santo Domingo de Osa, *Tonduz 9923 & 10065*—Trails near Palmar Norte, 100 ft., *Allen 5977, 6605,6678 & 6725.*

CASEARIA BANQUITANA var. LAEVIS (Standl.) I. M. Johnston—Flacourtiaceae. (Pl. 31)

In our area a slender forest tree, to about 75 ft. in height, with a short, spreading crown. The elliptic-lanceolate leaves are alternate, and the conspicuous reddish-brown fruits are produced in axillary panicles in late November and December. The plant is usually a shrub or small tree in other areas, and the present determination is open to some question. Fairly frequent in wet lowland forest from Lagarto Station to about Jalaca.—Esquinas Forest, sea level, *Allen 5713*—Coto Junction, sea level, *Allen 6657.*

CASEARIA GUIANENSIS (Aubl.) Urban—Flacourtiaceae—*Palo de la cruz* (Panama).

Shrubs or small trees, 6–20 ft. in height, with alternate, oblanceolate, acute or acuminate leaves about 5–6″ long and 1¾″ wide, which have minutely serrate margins and very narrowly wedge-shaped bases. The small white flowers are produced from March until early April in sessile axillary fascicles.—Santo Domingo de Golfo Dulce, *Tonduz 10092.*

CASEARIA MYRIANTHA Turcz.—Flacourtiaceae—*Mauro* (Panama)—*Cakica-che, Guayabillo, Mierda de gallina, Taixcaax,* or *Utaxcaax* (Guatemala).

Shrubs or small trees, to about 30 ft. in height, with shining, leathery, alternate leaves, the oblong or elliptic-oblong, acuminate blades entirely glabrous, and with inconspicuously serrate margins. The small greenish-white flowers are produced in dense axillary fascicles, and are followed by red or brown, pilose fruits.—Buenos Aires, 1,000 ft., *Tonduz 6665*—Santo Domingo de Osa, *Tonduz 9984* & *Donnell-Smith 9954.*

CASEARIA SYLVESTRIS Swartz—Flacourtiaceae—*Corta lengua* (Chiriquí)—*Comida de culebra* (Nicaragua)—*Sombra de armado* or *Sombra de conejo* (Honduras)—*Wild sage* (British Honduras)—*Coralillo* or *Sacmuba* (Guatemala).

Small, unarmed trees, to about 20 ft. in height, with lustrous, dark-green, lanceolate, acuminate leaves which have entire margins. The small, greenish-white flowers are produced in September in dense axillary fascicles. The hard, heavy yellow wood is reported to be used for construction in Guatemala. Fairly frequent in trailside thickets.—Boruca, 1,250 ft., *Tonduz 3945*—Santo Domingo de Osa, *Tonduz 10058*—Palmar Norte, *Allen 5667 & 6559.*

Cashew—See *Anacardium occidentale* (Local, Canal Zone, Bocas del Toro).

Caspiro—See *Inga punctata* (Guatemala).

CASSIA—KEY

1. Leaflets broadly rounded at the apex.
   2. Flowers yellow .... *Cassia reticulata*
   2. Flowers pink .... *Cassia grandis*
1. Leaflets acute or acuminate at the apex .... *Cassia spectabilis*

CASSIA GRANDIS L. f.—Leguminosae; Caesalpinoideae—*Carao* or *Saragundin* (Local)—*Sandalo* (Costa Rica)—*Caña fistula* (Panama)—*Stinking toe* (Canal Zone, Bocas del Toro, and Limón)—*Carao* or *Carambano* (Nicaragua)—*Horse cassia* (Honduras)—*Bocot, Bucut, Caragua, Carao, Cañafistula,* or *Mucut* (Guatemala). (Pl. 19)

Medium-sized or rarely large, spreading trees, averaging 40–60 ft. in height and 12–24″ in trunk diameter. The even-pinnate leaves are alternate, and the pink or old-rose flowers are produced in short axillary racemes from late February to early April, while the trees are nearly leafless. Well-developed specimens are very attractive when in bloom, and are conspicuous from a distance. The dark-brown or black, woody

pods are very large, and remain long on the branches. The pulp surrounding the seeds is edible, but is purgative. It is reported that an ointment made from the crushed leaves mixed with lard is used for treating mange in Guatemala. The coarse brown wood is usually more or less variegated with yellowish or purplish streaks and patches. It is reported to be firm and tough, heavy, easy to cut, but with a tendency to saw woolly. It is sometimes used in Central America for construction purposes in areas where it is abundant. The species has rather the aspect, at least near Palmar Norte, of being an introduced plant.—Santo Domingo de Osa, *Tonduz 10339*—Pastures near Palmar Norte, 100 ft., *Allen 5517*.

CASSIA RETICULATA Willd.—Leguminosae; Caesalpinoideae—*Saragundi* (Local)—*Laureño* (Panama)—*Wild senna* (Bocas del Toro)—*Soroncontil* (Nicaragua)—*Baraja* or *Barajo* (Honduras)—*Barajo* or *Sambran prieto* (Guatemala). (Pl. 13)

Shrubs or small, bushy trees, to about 18 ft., with large, alternate, even-pinnate leaves and showy terminal or axillary racemes of bright-yellow flowers, which last from late August to about February. The fruits are linear, flat, 2-valved, many-seeded pods. An infusion of the leaves is reported to be used locally to cure ringworm. Common throughout the area along streams, canals, in pastures and on cutover land. Occasionally planted for ornament in labor camps.—Boruca, 1,250 ft., *Tonduz 3834*—Santo Domingo de Osa, *Tonduz 10042*—Pastures near Palmar Norte, *Allen 5875*—Palmar Sur, sea level, *Allen 6685*.

CASSIA SPECTABILIS DC—Leguminosae; Caesalpinoideae—*Candelillo* (Costa Rica)—*Candelillo* or *Frijolillo* (Honduras)—*Pisabed* (British Honduras). (Pl. 19)

Small, spreading trees, to about 40 ft. in height, with alternate, pinnate leaves and terminal panicles of bright-yellow, fragrant flowers which may be produced from late August until late November or early December, and followed by clusters of long, slender, terete, indehiscent pods. Specimens growing in the Meseta Central tend to be taller than those in the lowlands. Common in our area along trails and in pastures, and occasionally planted in the townsites for ornament.—Palmar Norte, 100 ft., *Allen 5345 & 5610*.

CASSIPOUREA PODANTHA Standl.—Rhizophoraceae—*Goatwood, Huesito*, or Limoncillo (Panama)—*Waterwood* (British Honduras).

Medium-sized or tall forest trees, to about 90 ft. in height, with opposite, lanceolate, acuminate leaves 2½–3″ long and 1–1½″ wide and small white flowers which arc produced in short axillary fascicles in May. The

hard, durable wood has been used for dock piling in Bocas del Toro, and for railroad ties and house framing in British Honduras.—Esquinas Forest, 200 ft., *Allen 5536.*

Castaño—See *Sterculia apetala* (Honduras). Leaves palmately lobed.

Castaño—See *Sterculia mexicana* (Guatemala). Leaves digitately compound.

CASTILLA FALLAX O. F. Cook—Moraceae—*Caucho* (Local)—*Mastate blanco* (Boruca)—*Hule macho* or *Hule blanco* (Costa Rica)—*Ule* (Chiriquí)—*Rubber tree* (Bocas del Toro and Limón)—*Tuna* (Nicaragua).

Forest trees, to 50–80 ft. in height, notable for their narrow crowns and very large, alternate, elliptic-oblong leaves which are conspicuously drooping and 2-ranked on the branches, and are rough to the touch. The minute flowers are produced from October until early December on curious brown, scaly axillary receptacles, and are followed in February and March by the reddish-brown, attractive fruits. The smooth gray bark is reported to be soaked in water and beaten by the Sumo Indians of the Mosquito Coast of Nicaragua for the manufacture of a sort of bark cloth. The white, resinous latex is not elastic, and is not used except for caulking boats. The pale brownish-white wood is light and soft. A very common species, found throughout the area.—Santo Domingo de Osa, *Tonduz 10012*—Esquinas Forest, 200 ft., *Allen 5373*—Hills above Palmar Norte, 1,500 ft., *Allen 5954.*

Castor—See *Ricinus communis* (Costa Rica).

Castor bean—See *Ricinus communis* (General).

Catalox—See *Trophis racemosa* (Guatemala).

Catamericuche—See *Cochlospermum vitifolium* (Nicaragua).

Cativo—See *Cynometra hemitomophylla* (Costa Rica). Leaves with 1 pair of leaflets.

Cativo—See *Prioria copaifera* (Local and general). Leaves with 2 pairs of leaflets.

Caucho—See *Castilla fallax* (Local).

Caulote—See *Guazuma ulmifolia* (Honduras and Guatemala). Fruits circular in cross section.

Caulote—See *Luehea Seemannii* (Honduras). Fruits star-shaped in cross section.

Cecropia asperrima—See *Cecropia peltata.*

CECROPIA—Key

1. Leaves conspicuously whitish on the lower surface. Strands of the pistillate inflorescence about 4″ long.
   2. Spathes silvery white.......... *Cecropia peltata*
   2. Spathes red.......... *Cecropia Sandersoniana*
1. Leaves green on the lower surface. Strands of the pistillate inflorescence 9–12″ long.......... *Cecropia obtusifolia*

Cecropia obtusifolia Bertol.—Moraceae—*Guarumo* (Local and general).

Very common, slender, prominently ringed, green- or white-trunked weed trees, 20–60 ft. in height, with large, peltate, 8- to 15-lobed leaves which are green on the lower surface and clustered at the ends of the relatively few, candelabra-like branches. The young leaves and the petioles and veins of the undersurfaces of the mature foliage are often dark wine-red in color, as are also the floral spathes and bracts which envelop the terminal bud. The minute flowers are dioecious and are produced from August until November, the green pendulous strands of the pistillate inflorescences being 8–16″ in length. The hollow trunks are usually inhabited by colonies of stinging ants, as is usual in the genus. *Cecropias* are typical of the second-growth thickets and the margins of woodland, but large specimens are occasional in the hillside forests where they often develop conspicuous stilt roots. The larger trunks are sometimes split and used for water troughs. The bark contains a tough fiber which is sometimes used for making coarse rope. The genus is poorly understood, and early collections from the Golfo Dulce area are listed in *Primitiae Florae Costaricensis* as *Cecropia mexicana* Hemsl., but this concept has recently been reduced to synonomy under *C. obtusifolia* in the *Flora of Guatemala.* Our specimens differ from the closely allied *C. insignis* in the much longer fruiting strands, and also from the type description of *C. obtusifolia* in the green undersurface of the leaves. A series of specimens in the Herbarium of the Escuela Agrícola Panamericana from various localities shows a considerable degree of latitude in this character and it seems best to regard these local trees as a slightly aberrant form of the species. They are also quite closely allied to *Cecropia maxima* Snethlage, a rather poorly known species from Ecuador.—Cañas Gordas, 3,200 ft., *Pittier 11077*—Buenos Aires, 600 ft., *Tonduz 3999 & 6656*—Río Coto de Golfo Dulce, *Pittier 9996*—Margins of forest near Jalaca, 50 ft., *Allen 5302*—Forested hills above Palmar Norte, 1,800 ft., *Allen 5937*—Esquinas Station, 50 ft., *Allen 6545.*

CECROPIA PELTATA L.—Moraceae—*Guarumo* (Local and general). (Pl. 11)

Trees 20–50 ft. in height, with green or white, prominently ringed trunks. Similar to the preceding species, but differing in the mostly 9- to 10-lobed leaves which are white on the lower surface and in the much shorter, stouter grayish strands of the fruiting spikes. These are produced from August to about November, and average 3–4″ in length. Listed in the *Flora of Costa Rica* as *Cecropia asperrima* Pittier, but reduced to synonomy in the *Flora of Guatemala.*—Trailside thickets near Palmar Norte, 50 ft., *Allen 6543.*

CECROPIA SANDERSONIANA P. H. Allen, sp. nov.—Moraceae—*Guarumo* (Local and general). (Pl. 11)

Trees of varying size, mature, fruiting specimens 75–90 ft. in height and up to 36″ in trunk diameter above the buttressed base, the bark with many conspicuous corky lenticels. Young specimens up to 20 ft. or more in height with very large juvenile leaves are frequent along the railroad right of way and on other cutover land, their conspicuous black trunks and absence of prominent rings immediately separating them from other *Cecropias* in the area. Older trees show these same characters in the branching crown, the lower trunk in examples 50 ft. or more in height usually with stilt roots which later develop into buttresses. Mature leaves peltate, 8-lobed, 15–18″ in diameter, on long, terete, green, shallowly ribbed petioles, the minutely puberulent gray lower surface of the blade contrasting strongly with the dark-green, glossy, glabrous upper side. The juvenile leaves when they first unfold are a handsome pink with green veining, and emerge from a rich dragon's-blood-red spathaceous terminal bract which averages about 10″ in length and 2–2½″ in width. The pistillate inflorescences are produced in January and have clusters of 4–6 lemon-yellow terete strands 3–4″ long, which are at first enveloped in a coral-red spathe which is soon deciduous, both it and the strands slightly exceeding the subterete hispidulous peduncle in length. A common, handsome tree, found in areas of highest rainfall from Esquinas to Golfito, but with mature specimens apparently confined to the primary rain forest. The species is dedicated to Mr. N. E. Sanderson, Manager of the Golfito Division, who did so much to facilitate work in the region.—Hills near Golfito, 200 ft., *Allen 6688* TYPE (Herbarium Escuela Agrícola Panamericana).

Cedar—See *Cedrela mexicana* (General).

CEDRELA—Key

1. Capsules about 2″ long. Highland species, found above about 4,000 ft. .... *Cedrela Tonduzii*
1. Capsules 1½″ long or less. Lowland species.
   2. Lower surface of the leaflets densely pubescent. Capsules about 1⅛″ long .... *Cedrela fissilis*
   2. Lower surface of the leaflets glabrous. Capsules about 1½″ long .... *Cedrela mexicana*

CEDRELA FISSILIS Vell.—Meliaceae—*Cedro* (Local and Chiriquí)—*Cedro grenadino* or *Cedro real* (Chiriquí).

Fairly common trees, to about 75 ft. in height and 18–24″ in trunk diameter, with large, alternate, pinnate leaves which typically have 16–26 leaflets which are densely pubescent on the lower surface. The trees are usually deciduous during the dry season. The small white flowers are produced in May to July in terminal or subterminal panicles, and are followed in February and March by the brown, woody seed capsules which are about 1⅛″ in length. The wood is reddish brown, usually with a straight grain, is easy to work, takes a good polish, is very durable and resistant to insect attack. It is used for the manufacture of cigar boxes, turnery, furniture, and pencils.—Valley of the Río Terraba, *Tonduz 4740*—Vicinity of La Presa, 100 ft., *Allen 5280.*

CEDRELA MEXICANA Roem.—Meliaceae—*Cedro* or *Cedro Maria* (Local)—*Cedro amargo, Cedro blanco, Cedro dulce, Cedro colorado, Cobano,* or *Cedro cobano* (Costa Rica)—*Cedro cebolla* (Chiriquí)—*Cedar* (General)—*Spanish cedar* (General)—*Cedro real* (Salvador)—*Cedro, Cuche,* or *Yoxcha* (Guatemala). (Pl. 21)

Medium-sized or tall forest trees, 60–100 ft. in height and 2–4 ft. in trunk diameter, with large, pinnate, alternate leaves, which are deciduous during the dry season. The small, greenish-white, very fragrant flowers are produced in elongate, slender, terminal or subterminal panicles from June to early August, and are followed in the late dry season by brown, woody capsular fruits which contain the small, winged seeds. The wood has a characteristic garlic-like odor and bitter taste, particularly when freshly cut. Logs from the relatively dry parts of the Pacific coast have the reputation of having darker, denser, and stronger wood than those from the region of Limón. The wood is usually pinkish or reddish brown, relatively soft, light in weight, easy to work, durable, and resistant to insect attack. It is probably the most important single timber wood in the country, and was formerly exported to the United States in consider-

able quantities. It is used for doors and window frames, siding, interior trim, furniture, interiors of chests and closets, paneling, and general construction when available. Most of the readily accessible trees of large size have been cut in our area, but other stands may still exist on the higher slopes. Slender specimens to about 14–18″ in diameter are fairly frequent in pastures and along trails, particularly in the Terraba Valley, and along the railroad line between Progreso and Concepción in Chiriquí Province in Panama.—Palmar, *Allen 5585.*

CEDRELA TONDUZII C. DC.—Meliaceae—*Cedro* or *Cedro dulce* (Local) —*Cedro granadino* (Chiriquí).

Very large trees of the highland forests, to about 100 ft. in height and 4–5 ft. in trunk diameter, with large, alternate, pinnate leaves and brown woody capsules about 2″ in length. The attractive brown wood lacks the bitter taste typical of *C. mexicana* and is not resistant to insect attack, but is much used for paneling, furniture, and interior trim. Fairly frequent near Agua Buena at about 4,000 ft., and also in the Chiriquí highlands. Other suggested uses are for rotary veneering and plywood.

Cedrillo—See *Cupania guatemalensis* (Salvador). Trunk unarmed. Leaves even-pinnate.

Cedrillo—See *Trichilia acutanthera* (Local, Honduras, and Guatemala). Trunk unarmed. Leaves odd-pinnate.

Cedrillo—See *Virola Koschnyi* (Guatemala). Leaves simple.

Cedrillo—See *Zanthoxylum procerum* (Honduras). Trunk armed with prickles.

Cedro—See *Cedrela fissilis* (Local and Chiriquí). Capsules about 1⅛″ long.

Cedro—See *Cedrela mexicana* (Local and general). Capsules about 1½″ long.

Cedro—See *Cedrela Tonduzii* (Local). Capsules about 2″ long.

Cedro amargo—See *Cedrela mexicana* (Costa Rica).

Cedro bateo—See *Carapa Slateri* (Local).

Cedro blanco—See *Cedrela mexicana* (Costa Rica).

Cedro cebolla—See *Cedrela mexicana* (Chiriquí).

Cedro cobano—See *Cedrela mexicana* (Costa Rica).

Cedro colorado—See *Cedrela mexicana* (Costa Rica). Seeds winged.

Cedro colorado—See *Trichilia acutanthera* (Guatemala). Seeds not winged.

Cedro dulce—See *Cedrela mexicana* (Costa Rica). Capsules about 1½″ long.

Cedro dulce—See *Cedrela Tonduzii* (Local). Capsules about 2″ long.

Cedro espino—See *Trichilia acutanthera* (Honduras).

Cedro granadino—See *Cedrela Tonduzii* (Chiriquí). Highland species.

Cedro grenadino—See *Cedrela fissilis* (Chiriquí). Lowland species.

Cedro macho—See *Carapa Slateri* (Local). Fruits more than 4″ in diameter.

Cedro macho—See *Guarea Guara* (Panama). Fruits less than 1″ in diameter.

Cedro Maria—See *Cedrela mexicana* (Local).

Cedron—See *Simaba cedron* (Local).

Cedro real—See *Cedrela fissilis* (Chiriquí). Capsules about 1⅛″ long.

Cedro real—See *Cedrela mexicana* (Salvador). Capsules about 1½″ long.

Ceiba—See *Ceiba pentandra* (Local and general).

CEIBA PENTANDRA (L.) Gaertn.—Bombacaceae—*Ceiba* (Local and general)—*Cotton tree* (Bocas del Toro)—*Silk cotton tree* (General)—*Pochote* (Nicaragua)—*Ceiba, Inup, Nuo,* or *Mox* (Guatemala). (Fig. 8 and Frontispiece)

Giant forest trees with spreading crowns, often 120–140 ft. in height and 6–8 ft. in diameter above the tremendous buttresses. Young specimens usually have many large, stout, conical spines on the trunk, but in mature trees the bark is usually light gray, and may be relatively smooth or somewhat corrugated, and covered with distinct or obscure parallel lines of roughly circular, exfoliating lenticels which are somewhat prickly to the touch. The leaves are alternate and digitately compound, with 5–7 entire, acute or acuminate leaflets, the petioles of which are sometimes red in color. The trees are deciduous during the dry season, many individuals in our area dropping their leaves in late November, to be renewed from early December until late February, depending upon soil conditions and available moisture, while others stay green for a much longer time. The flushes of new leaves are sometimes a rich reddish brown and very conspicuous from a distance, but this is rather the exception, and it is not certain that trees which are showy in a given year are consistently so. The fragrant flowers are about 1½″ long, and are produced from the subterminal axils of the unarmed, brittle branchlets, usually in February. The color effect of the flowering trees when seen from a distance is greenish brown. The rather fleshy, green, cupular calyx is shallowly lobate, and the petals when fresh are pink, aging grayish lavender on the silky outer surfaces. The stamen column is prominently exserted,

Fig. 8. Ceiba pentandra. 1 — Flowering branch and leaf. 2 — Cross section of calyx, showing ovary and style.

with yellow, twisted anthers. When in flower, the trees are a favorite resort for parrots and the ground below is covered by the fallen blooms.

The fruit is a woody, elliptic-oblong capsule, about 3–4″ in length, produced in late March. They do not fall when mature, but open on the branches, freeing the fluffy pale-brown kapok and large brown seeds. Large specimens are often left standing in pastures for shade and the fallen flowers are reported to be eaten by stock. The light, elastic kapok fiber is used for filling pillows, life preservers, and mattresses and for insulation. The wood is grayish or pinkish white, and is soft, light, coarse-grained, easy to cut, tough, but not durable. It has sometimes been used for cheap packing boxes and match sticks. It is considered suitable for paper pulp and possibly plywood corestock. There are magnificent specimens along the edge of the Golfito golf course, near the Golfito dairy, and near the flying field in Palmar.—Palmar Sur, sea level, *Allen 5856*—Hills above Palmar Norte, 1,000 ft., *Allen 6248*—Hills near Golfito, 200 ft., *Allen 6690*.

Ceibillo—See *Zanthoxylum procerum* (Guatemala).
Ceibo—See *Bombax barrigon* (Costa Rica). Trunks swollen.
Ceibo—See *Bombax sessile* (Panama). Trunks not swollen.
Ceibo barrigon—See *Bombax barrigon* (Local).
Ceibo nuno—See *Bombax sessile* (Panama).
Ceibo pochote—See *Bombax barrigon* (Local).

CELASTRACEAE—One genus, *Maytenus*.

Cenicero macho—See *Albizzia longepedata* (Costa Rica).
Cenizo—See *Miconia argentea* (Honduras).
Cerbatana—See *Dracaena americana* (British Honduras and Guatemala).
Cerecilla—See *Ardisia revoluta* (Guatemala).
Cerezo—See *Ardisia revoluta* (Salvador).
Cerillo—See *Lacmellea panamensis* (Local). Trunk armed with spines.
Cerillo—See *Rheedia madruno* (Local). Trunk unarmed. Flowers white.
Cerillo—See *Symphonia globulifera* (Local and general). Trunk unarmed. Flowers red.
Cero—See *Rheedia edulis* (Chiriquí). Flowers whitish.
Cero—See *Symphonia globulifera* (Chiriquí). Flowers red.

CESPEDESIA MACROPHYLLA Seem. Ochnaceae—*Membrillo* (Panama)—*John crow wood* (Bocas del Toro).

Forest trees, 45–75 ft. in height and 15–30″ in trunk diameter, with very large, obovate-spatulate leaves with coarsely sinuate-serrate margins, which are clustered at the ends of the branches, in a manner somewhat

reminiscent of *Gustavia superba* or *Grias Fendleri* but on a much larger scale. The bright-yellow flowers are produced in showy terminal panicles from late September until early March. The reddish-brown, hard, heavy wood has a medium to coarse texture, is straight-grained, works well, and is considered to be fairly durable. The species is fairly frequent in the high hills between Palmar and Boruca at 1,000–4,000 ft., the reddish flushes of new leaves or yellow flowers often rendering the trees conspicuous from the air.

Chaca—See *Bursera simaruba* (Guatemala).
Chacah—See *Bursera simaruba* (Guatemala).
Chacah colorado—See *Bursera simaruba* (Guatemala).
Chac-ixcanan—See *Hamelia patens* (Guatemala).
Chajada amarilla—See *Pterocarpus officinalis* (Costa Rica).

CHAMAEDOREA—KEY

1. Plants dwarf, the trunks usually less than 1 ft. in height .... *Chamaedorea* sp.—*Allen 6742*
1. Plants of varying size, but with trunks at least 3 ft. in height.
   2. Trunks multiple, to about 35 ft. .... *Chamaedorea Woodsoniana*
   2. Trunks solitary.
      3. Scapes erect. Wet forests at sea level .... *Chamaedorea* sp.—*Allen 6262*
      3. Scapes pendulous. Forested ridges at 1,800–2,000 ft.
         4. Fruits obovoid, less than ½″ long .... *Chamaedorea* sp.—*Moore 6527*
         4. Fruits oblong, more than ½″ long .... *Chamaedorea Wendlandiana*

CHAMAEDOREA WENDLANDIANA (Oerst.) Hemsl.—Palmaceae.

Single-trunked, pinnate-leaved palms 10–20 ft. in height, with green, conspicuously ringed, unarmed canes, the lower portion often somewhat repent, with many adventitious roots. Inflorescences branching, pendulous, with a rather fleshy green rachis.—Forested hills above Palmar Norte, 2,000–2,500 ft., *Moore 6547* & *Allen 6761*.

CHAMAEDOREA WOODSONIANA Bailey—Palmaceae.

Slender, unarmed, pinnate-leaved palms with multiple trunks, 15–35 ft. in height. Scapes elongate, pendulous and branching, with a bright-orange rachis and black, globose fruits about ⅜″ in diameter.—Forested hills above Palmar Norte, 2,000–2,500 ft., *Moore 6549* & *Allen 6762*.

CHAMAEDOREA sp.—Palmaceae.

Dwarf, single-stemmed palms, 6–7 ft. in height, the trunk usually about 1 ft. or less in height, with 8–10 pinnate fronds. The plants bear a considerable superficial resemblance to those of *Neonicholsonia*, but may be immediately separated by the branching rather than spicate inflorescence.

—Locally common on steep forested ridges above Palmar Norte at about 1,800 ft., *Moore 6525* & *Allen 6742.*

CHAMAEDOREA sp.—Palmaceae.

Single-trunked palms, to about 8 ft. in height and 1″ in trunk diameter, the stem erect, green and ringed, bearing 5–6 spreading, pinnate fronds which typically have about 11 pairs of caudate-acuminate pinnae. Inflorescences about 3, either inter- or infra-foliaceous, the erect spadix having spreading, light-green rachillae and yellowish flowers.—Forest near Tinoco, sea level, *Moore 6533* & *Allen 6262.*

CHAMAEDOREA sp.—Palmaceae.

Slender, single-stemmed palms, to about 12 ft. in height, with a few dark-green, pinnate fronds. The spadix in young fruit is pendulous, with orange rachillae and green, obovoid fruits about 2/5″ long and 1/3″ in diameter.—Rather infrequent in the forested hills above Palmar Norte at about 1,800 ft. elevation, *Moore 6527.*

Chamah—See *Hamelia patens* (Guatemala).

Chancho blanco—See *Goethalsia meiantha* (Costa Rica.)

Chaparro—See *Curatella americana* (British Honduras and Guatemala).

Chapascuapul—See *Simarouba glauca* (Guatemala).

Chapel—See *Lonchocarpus guatemalensis* (Honduras).

Chapelno—See *Lonchocarpus minimiflorus* (Salvador).

Chapelno hediondo—See *Lonchocarpus guatemalensis* (Salvador).

Chapelno negro—See *Lonchocarpus minimiflorus* (Salvador).

Chaperno—See *Andira inermis* (Honduras). Leaves pinnate. Fruits globose.

Chaperno—See *Aspidosperma megalocarpon* (Honduras). Leaves simple.

Chaperno—See *Lonchocarpus guatemalensis* (Costa Rica). Leaves pinnate. Fruits flat, thickened along one edge.

Chaperno—See *Lonchocarpus minimiflorus* (Costa Rica, Salvador, and Guatemala). Leaves pinnate, leaflets less than 2″ long.

Chaperno—See *Lonchocarpus sericeus* var. *glabrescens* (Local and Bocas del Toro). Leaves pinnate. Sap turns blood red when exposed to the air.

Chaperno prieto—See *Lonchocarpus guatemalensis* (Salvador).

Chapparon—See *Rheedia edulis* (Salvador).

Chapulaltapa—See *Schizolobium parahybum* (Salvador).

Chapuno—See *Lonchocarpus minimiflorus* (Salvador).

Chate—See *Dialium guianense* (Guatemala).

Chaya—See *Bixa Orellana* (Guatemala).
Cheja—See *Pterocarpus Hayesii* (Guatemala).
Cherry—See *Pseudolmedia spuria* (British Honduras).
Chi—See *Byrsonima crassifolia* (Guatemala).
Chicharillo—See *Mouriria parvifolia* (Guatemala).
Chicharron—See *Guazuma ulmifolia* (Salvador). Inflorescences axillary.
Chicharron—See *Hirtella triandra* (Panama). Inflorescences terminal.
Chic-chica—See *Bursera simaruba* (Guatemala).
Chichica—See *Aspidosperma megalocarpon* (Guatemala).
Chichicaste—See *Myriocarpa longipes* (Guatemala). Inflorescences threadlike, pendulous.
Chichicaste—See *Urera alceifolia* (Guatemala). Inflorescences compact. Leaf bases wedge-shaped.
Chichicaste—See *Urera caracasana* (Guatemala). Inflorescences compact. Leaf bases broadly rounded.
Chichicaste colorado—See *Myriocarpa longipes* (Salvador).
Chichicaste de hormiga—See *Urera caracasana* (Guatemala).
Chichicaste de montaña—See *Urera alceifolia* (Guatemala).
Chichicaste manso—See *Myriocarpa longipes* (Guatemala).
Chichicastillo—See *Myriocarpa longipes* (Honduras).
Chichicaston—See *Urera caracasana* (Guatemala).
Chichipate—See *Swartzia panamensis* (Salvador and Honduras). Leaf tips entire.
Chichipate—See *Sweetia panamensis* (Salvador, Honduras, and Guatemala). Leaf tips emarginate.
Chichipin—See *Hamelia patens* (Guatemala).
Chichipince—See *Hamelia patens* (Salvador).
Chichipinte—See *Hamelia patens* (Salvador).
Chichique—See *Aspidosperma megalocarpon* (Guatemala).
Chilaca—See *Myroxylon balsamum* var. *Pereirae* (Local).
Chilamate—See *Ficus Tonduzii* (Costa Rica).
Chilca—See *Tecoma stans* (Nicaragua).
Chile—See *Phyllanthus acuminatus* (Local).
Chilil—See *Oreopanax xalapense* (Guatemala).
Chilil mazorco—See *Oreopanax xalapense* (Guatemala).
Chilillo—See *Phyllanthus acuminatus* (Costa Rica).
Chilujushte—See *Trophis racemosa* (Salvador).

CHIMARRHIS LATIFOLIA Standl.—Rubiaceae—*Yema de huevo* (Local)—*Jagua amarilla* (Chiriquí). (Fig. 9)

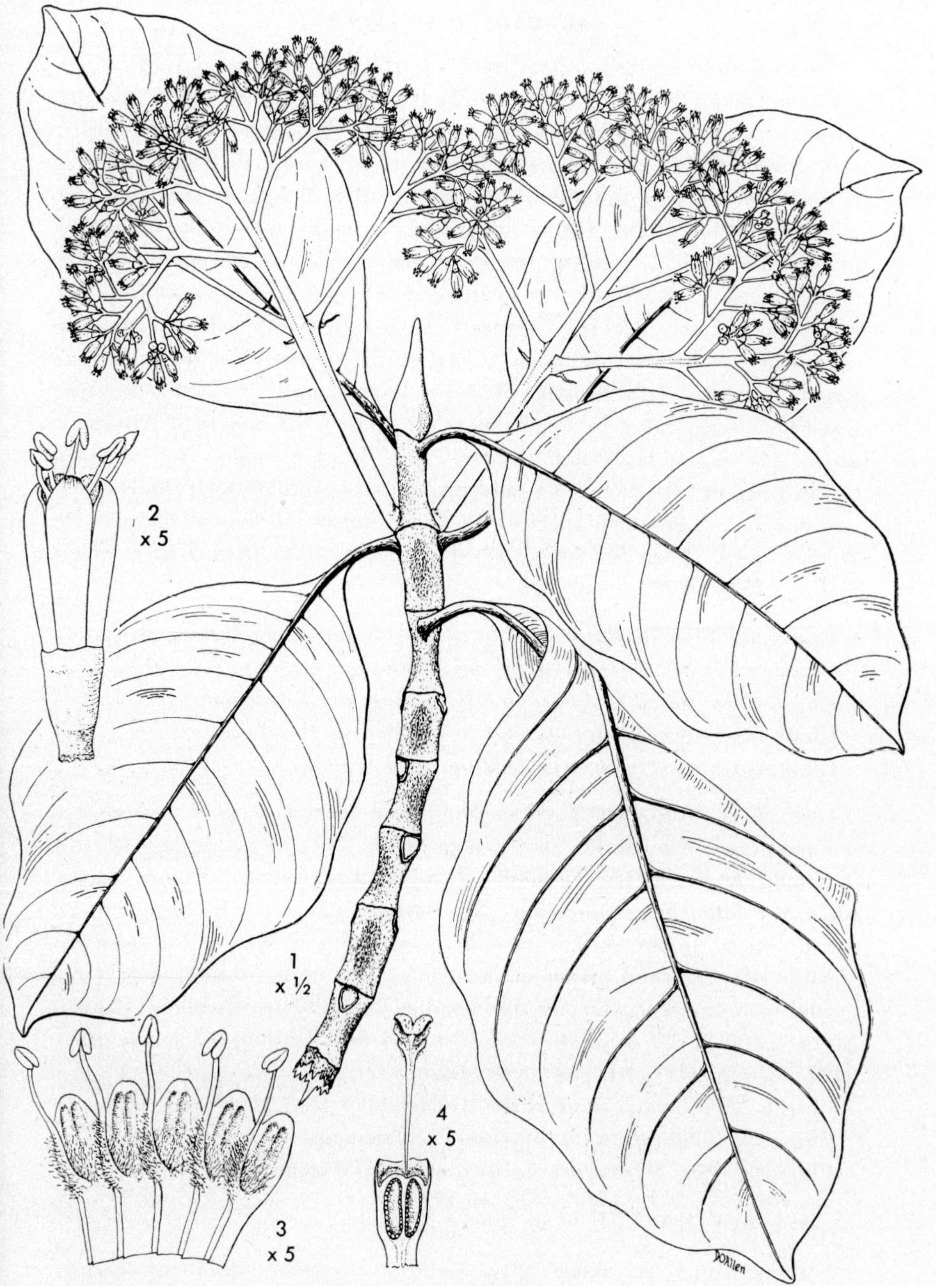

Fig. 9. CHIMARRHIS LATIFOLIA. 1 — Flowering branch. 2 — Individual flower. 3 — Corolla opened to show attachment of stamens. 4 — Cross section of calyx, showing style.

Medium-sized or large forest trees, 50–100 ft. in height and 18–30″ in diameter above the rather prominently developed buttresses. The broadly elliptic, abruptly acute leaves are opposite on the stems, and the small, very fragrant white flowers are produced in dense axillary cymes in October and November. Individual specimens often flower several times at intervals of 4–6 weeks, but it is unusual for many trees to be in bloom at once. The fruits are small, dry, bivalvate capsules, which mature during the dry season. The freshly cut heartwood is bright orange in color and is fine-grained, fairly heavy, of a rather waxy appearance when finished. It is reported to be durable when buried in the ground, but not when exposed to the air. It is considered to be the best local wood for dragline mats, and is also used to some extent for fence posts and fuel. A closely allied species has been used by the United Fruit Company for railroad ties in Boca del Toro, and it would probably be suitable for cabinet work and flooring. A common species, found in forests throughout the area.—Palmar Norte, *Allen 5613 & 5647*—Esquinas Forest, *Allen 5762*—Jalaca Station, *Allen 6302*.

Chinacahuite—See *Bursera simaruba* (Honduras and Guatemala).
Chinacuite—See *Bursera simaruba* (Honduras).
Chino—See *Bursera simaruba* (Honduras and Guatemala).
Chintonrol—See *Posoqueria latifolia* (British Honduras).
Chintorol—See *Posoqueria latifolia* (Guatemala).

CHIONE COSTARICENSE Standl.—Rubiaceae—*Fruta de pava* (Local).
Forest trees of moderate size, averaging 40–75 ft. in height and 12–18″ in trunk diameter, with oblong-elliptic, obtuse, rather leathery leaves which are opposite on the stems. The small, white, very fragrant flowers are produced in terminal cymes in February and March, followed by clusters of bright-red, conspicuous fruits. The hard, white wood is reported to be used locally for axe handles.—Forested hills above Palmar Norte, 1.800 ft., *Allen 5930*—Forested hills near Golfito, sea level, *Allen 5990*—Esquinas Forest, sea level, *Allen 5321*.
Chirca—See *Croton xalapensis* (Guatemala).
Chirimoya—See *Rollinia Jimenezii* (Guatemala).
Chirraca—See *Myroxylon balsamum* var. *Pereirae* (Local).

CHLORANTHACEAE—One genus, *Hedyosmum*.

CHLOROPHORA TINCTORIA (L.) Gaud.—Moraceae—*Mora* or *Morillo* (Local)—*Brasil* (Costa Rica)—*Macano* (Panama)—*Fustic* (Bocas del Toro)—*Mora amarilla* (U. S. trade name). (Fig. 10)

Fig. 10. Chlorophora tinctoria. 1 — Branch with staminate (male) inflorescences. 2 — Branch with pistillate (female) inflorescence.

Forest trees 40–90 ft. in height and 18–36″ in trunk diameter, with alternate, ovate, acuminate, often deciduous leaves which have serrate or entire margins. Branchlets, particularly in young specimens often have more or less deeply lobed foliage and are frequently spinose. The dioecious flowers are produced in July and August, those of the staminate form being in greenish-white axillary catkins, while the fleshy-green pistillate inflorescences are nearly globose. These soon mature, and contain the numerous small black seeds. The fresh heartwood is bright yellow, becoming reddish or brownish on exposure to air. It is hard, strong, and heavy, with a somewhat interwoven grain, and does not crack, check, or warp, works fairly easily, finishes smoothly, takes a high polish, and is resistant to insect attack and to decay in contact with the ground. It is considered by the Engineering Department of the United Fruit Company to be its best all-purpose construction timber. It is used locally for railroad ties (which are reported to last 10–12 years), bridge timbers, piles, and fence posts. It is also suitable for interior trim, cart wheels, mine timbers, decking, planking, and framing in boat construction, and for flooring, furniture, and tool handles. A common species generally distributed throughout the area.—Palmar Norte, sea level, *Allen 5601.*

Cho—See *Cochlospermum vitifolium* (Guatemala).
Chocolatillo—See *Herrania purpurea* (Panama).
Chocomico—See *Ximenia americana* (Guatemala and Honduras).
Chonta—See *Astrocaryum Standleyanum* (Panama). Trunks spiny.
Chonta—See *Socratea durissima* (Boruca). Trunks with stilt roots.
Chonta negra—See *Iriartea gigantea* (Boruca).
Choonte—See *Zanthoxylum procerum* (Guatemala).
Chorequillo—See *Godmania aesculifolia* (Guatemala).

CHRYSOPHYLLUM—KEY

1. Leaves conspicuously reddish brown on the lower surface .... *Chrysophyllum mexicanum*
1. Leaves somewhat paler on the lower surface, but never reddish brown .... *Chrysophyllum panamense*

CHRYSOPHYLLUM MEXICANUM Brand. ex Standl.—Sapotaceae—*Caimito* (Local and general).

Forest trees, 40–75 ft. in height with elliptic or elliptic-oblong, acute or acuminate leaves which are conspicuously reddish brown on the lower surface. All parts of the plant have a white, sticky latex. Known in our area only from sterile material, so that determination must be regarded as provisional.—Primary forest near Coto Junction, 50 ft., *Allen 6656.*

CHRYSOPHYLLUM PANAMENSE Pittier—Sapotaceae—*Caimito* (Local). (Pl. 24)

Medium-sized forest trees, 60–75 ft. in height, with alternate, elliptic, acute or shortly acuminate leaves and small pale-green flowers which are produced in short axillary fascicles during February and March. The fruits are dark purple when ripe and seem to be consistently 1-seeded and about 1½″ in diameter. They are reported to be edible when fully mature, the green fruits, as well as all other parts of the plant being supplied with a sticky white latex. The species is fairly frequent in the hillside forests throughout the area.—*Allen 5953 & 6008.*

Chuchupate—See *Guarea longipetiola* (Chiriquí).
Chuh—See *Oreopanax capitatus* (Guatemala).
Chulub—See *Calycophyllum candidissimum* (Guatemala).
Chulujushte—See *Trophis racemosa* (Salvador).
Chumico—See *Curatella americana* (Local and Panama).
Chumico de palo—See *Curatella americana* (Local and Panama).
Chunga—See *Astrocaryum Standleyanum* (Panama).
Churrusco—See *Trema micrantha* (Salvador).
Chutras—See *Protium sessiliflorum* (Chiriquí).
Cierito—See *Mouriria parvifolia* (Chiriquí).
Ciiche—See *Acalypha diversifolia* (Guatemala).
Cincho—See *Lonchocarpus guatemalensis* (Salvador). Fruits conspicuously thickened along one margin.
Cincho—See *Lonchocarpus latifolius* (Honduras). Margins of fruits not thickened.
Cirri amarillo—See *Mauria glauca* (Costa Rica).
Ciruela—See *Spondias purpurea* (Nicaragua and Honduras).
Ciruela de jobo—See *Spondias mombin* (Nicaragua).
Ciruela de monte—See *Spondias mombin* (Honduras).
Ciruelillo—See *Astronium graveolens* (Honduras). Large trees.
Ciruelillo—See *Phyllanthus acuminatus* (British Honduras). Small trees.
Ciruelo—See *Astronium graveolens* (Guatemala). Margins of leaflets serrate.
Ciruelo—See *Spondias purpurea* (Costa Rica and Honduras). Leaflet margins not serrate.

CITHAREXYLUM VIRIDE Moldenke—Verbenaceae—*Corrimiento* (Chiriquí).

Shrubs or small trees, to about 20 ft. in height, with opposite, oblong-

elliptic, acute or acuminate leaves and mostly terminal racemes of small, bright-orange fruits. Fairly frequent in old pastures and in brushy places. The wood of a closely allied species has been found to be very satisfactory for fence posts in Puerto Rico.—Puerto Jiménez, *Brenes 12332*—Pastures near Puerto Cortés, *Allen 5595*.

CLARISIA MEXICANA (Liebm.) Lanjouw—Moraceae.

Forest trees, 75–100 ft. in height and 24–36″ in trunk diameter, with alternate, elliptic-oblong, short-acuminate leaves and dioecious flowers which are produced in the leaf axils in February and March. The staminate flowers are borne in small, bracteate panicles or racemes, while the small, green, ovoid pistillate flowers are usually in pairs, each with two prominently exserted orange stigmas at the apex. The fruits are described as being globose or ovoid, apparently with a single seed. The wood is bright yellow when freshly cut, becoming darker on exposure, and is of variable and uneven texture. It tends to saw woolly, is difficult to finish smoothly, is not durable in contact with the ground, and in general is of dubious commercial possibilities.—Forested hills above Palmar Norte, 1,500 ft., *Allen 5955*.

Clavito—See *Acacia farnesiana* (Nicaragua). Plants armed with spines.
Clavito—See *Hamelia patens* (Guatemala). Plants not spiny.
Clavo—See *Anisomeris Recordii* (Guatemala).

CLETHRACEAE—A single genus, *Clethra*.

CLETHRA LANATA Mart. & Gal.—Clethraceae—*Nance macho* or *Nance* (Costa Rica).

Shrubs or small trees, to about 35 ft. in height and 12–36″ in trunk diameter, with alternate, obovate, acute, leathery leaves and dense terminal racemes of small white flowers which are produced in March and April. A species usually confined to rather dry highland ridges or savannas. The wood is apparently not used locally.—Boruca, 1,300 ft., *Tonduz 6869*—Savannas of Cañas Gordas, 3,300 ft., *Pittier 11115*.

Cobano—See *Cedrela mexicana* (Costa Rica).

COCCOLOBA—KEY

1. Leaves very large, usually exceeding 12″ in length.................*Coccoloba Standleyana*
1. Leaves usually less than 8″ long.
   2. Inflorescences very long and slender, much exceeding the leaves in length.............. ....*Coccoloba acuminata*
   2. Inflorescences about equaling the leaves in length or shorter.

3. Secondary nerves of the lower leaf surface inconspicuous, about 6 pairs; trees commonly found along sea beaches.......... *Coccoloba padiformis*
3. Secondary nerves of the lower leaf surface very prominent, about 8 pairs; trees on forested slopes from about 1,000 to 3,000 ft.......... *Coccoloba roseiflora*

COCCOLOBA ACUMINATA HBK—Polygonaceae—*Papaturrillo* (Nicaragua).

Shrubs or small trees 12–18 ft. in height, with alternate, lanceolate or elliptic-lanceolate, acute or acuminate leaves, the basal portion of the very short petioles enveloping the stems. Inflorescences elongate, slender, usually more or less pendulous spikes, which conspicuously exceed the leaves in length. The calyxes are at first red, becoming fleshy, translucent, and white at maturity. A very common species, found everywhere along trails in the Terraba Valley.—Palmar Norte, Cañablancal trail, 100 ft., *Allen 5226.*

COCCOLOBA PADIFORMIS Meiss.—Polygonaceae.

A shrub tree or tree of unknown size, the alternate, leathery leaves elliptic or oblong in outline and often quite variable in size, ranging from 4–8″ in length and 1¼–3¼″ in width.—Santo Domingo de Osa, *Tonduz 9875 & 9951.*

COCCOLOBA ROSEIFLORA Standl. & L. Wms.—Polygonaceae.

Forest trees, 40–90 ft. in height, with alternate, lanceolate, acute leaves and slender axillary racemes of small, dark-pink flowers which have conspicuous white stamens, the flowering season being in February. The flushes of new leaves which appear at the same season are dark pink, and render the trees conspicuous from a distance. The trees form small, nearly pure stands at the heads of valleys on the summit of the high forested hills above Palmar Norte, at about 2,500 ft. elevation, but also occur in the mixed stands down to about 1,000 or 1,500 ft. The fruits have not been seen. The hard reddish or pinkish wood appears to be of good quality, but is not used locally.—Forested hills above Palmar Norte, *Allen 5944 & 5964.*

COCCOLOBA STANDLEYANA P. H. Allen, sp. nov.—Polygonaceae—*Papaturo* (Local).

Slender trees, about 40–60 ft. in height, with very large, alternate, broadly elliptic or obovate, acute or shortly acuminate, essentially glabrous leaves. The blades are 14–20″ long and 8–12″ wide, with shallowly or deeply cordate bases and densely hispidulous veins and petioles. A very striking species, probably related to *C. hirsuta* Standl. and *C. belizensis* Standl. but amply distinct in the much smaller ocreae and other characters, which is dedicated to Dr. Paul C. Standley, Dean of Central

American Botanists. The species is thus far known only from sterile specimens collected in the Esquinas Forest at about 200 ft. elevation, where it is a rather infrequent element in the climax forest.—*Allen 6645* TYPE (Herbarium Escuela Agrícola Panamericana).

COCHLOSPERMACEAE—One genus, *Cochlospermum.*

Cochlospermum vitifolium (Willd.) Spreng.—Cochlospermaceae—*Poroporo* (Local)—*Poroporo* (Panama)—*Bombon, Poroporo,* or *Catamericuche* (Nicaragua)—*Bombon* or *Jicarillo* (Honduras)—*Wild cotton* or *Pochote* (British Honduras)—*Cho, Pochote, Pomp, Pumpo, Pumpumjuche, Pumpunjuche, Tecomasuchil, Tecomasuche, Tecomajuche, Tecomatillo,* or *Tsuyuy* (Guatemala). (Pl. 12)

Small trees of pastures and cutover land, which vary from 6 to about 40 ft. in height and with a maximum trunk diameter of about 10″. The alternate, long-stalked, 5-lobed, palmately compound leaves are deciduous during the dry season. The very large and attractive golden-yellow flowers are produced in short erect terminal racemes from December until April and are followed by 5-valved ovoid, pendulous capsules which contain a grayish fiber which resembles kapok and can be used for filling pillows and similar purposes. The trees when not in bloom are of rather ordinary appearance, but the flowers are exceptionally beautiful and are produced when the plants are relatively small so as to be suitable for garden use. They are occasionally cultivated for ornament in Florida, and a double-flowered form is known. The soft light wood is not used, but the bark supplies a fiber which is sometimes utilized for cordage. Common in our area in the Terraba Valley.—Buenos Aires, 750 ft., *Tonduz 3931 & 6556*—Boruca, 1,500 ft., *Tonduz 4788*—Pastures near Palmar Norte, 100 ft., *Allen 5781.*

Coco—See *Cocos nucifera* (Local and general).
Cocobolito—See *Psychotria chiapensis* (Chiriquí).
Coconut—See *Cocos nucifera* (Local and general).

Cocos nucifera L.—Palmaceae—*Coco, Cocotero, Coconut,* or *Pipa* (Local and general). (Pl. 9)

Coconut palms are the dominant element along sandy beaches in the entire Golfo Dulce region, forming small picturesque groves or, particularly from the delta of the Río Coto to Banco Point, stretching out in a thin line for miles in front of the darker, broad-leaved vegetation. The trees have every appearance of being wild and are universally believed to be so by the local inhabitants, since they regenerate spon-

taneously without the aid of man, often far from any present habitation. In view of the recent revival of controversy as to the origin of the coconut it is perhaps of some interest that Burica Point was particularly mentioned by Oviedo as early as 1526 as having large and thriving stands, evidently much resembling those which are found today.

Cocotero—See *Cocos nucifera* (Local and general).

Cocu—See *Andira inermis* (Costa Rica and Panama).

Cohetillo—See *Oreopanax capitatus* (Guatemala).

Cojon—See *Stemmadenia Donnell-Smithii* (Salvador and Guatemala).

Cojon de burro—See *Stemmadenia Donnell-Smithii* (Costa Rica and Honduras).

Cojon de caballo—See *Stemmadenia Donnell-Smithii* (Costa Rica and Guatemala).

Cojon de mico—See *Stemmadenia Donnell-Smithii* (Honduras).

Cojon de puerco—See *Stemmadenia Donnell-Smithii* (Salvador and Guatemala).

Cojoton—See *Stemmadenia Donnell-Smithii* (British Honduras).

Cola de pava—See *Cupania guatemalensis* (Guatemala). Inflorescences terminal.

Cola de pava—See *Guarea longipetiola* (Local). Inflorescences lateral.

Cola de pavo—See *Trichilia acutanthera* (Guatemala and Salvador). Leaves odd-pinnate.

Coligallo—See *Trichilia acutanthera* (Honduras and Guatemala).

Coloradillo—See *Hamelia patens* (Honduras).

Coloradito—See *Heisteria longipes* (Local).

Colorado—See *Calycophyllum candidissimum* (Honduras).

COMBRETACEAE—KEY

1. Leaves opposite........ *Laguncularia*
1. Leaves alternate.
   2. Flowers and fruits in dense, conelike heads........ *Conocarpus*
   2. Flowers and fruits in slender spikes or racemes........ *Terminalia*

Comenegro—See *Dialium guianense* (Nicaragua). Flowers inconspicuous.

Comenegro—See *Swartzia panamensis* (Local). Flowers yellow.

Comida de culebra—See *Casearia sylvestris* (Nicaragua).

Comida de mono—See *Protium neglectum* var. *sessiliflorum* (Chiriquí).

COMPOSITAE—KEY

1. Leaves opposite. Flowers yellow........ *Zexmenia*

1. Leaves alternate. Flowers white or lavender.
   2. Flowers white.
      3. Leaves glabrous, smooth to the touch .... *Oliganthes*
      3. Leaves puberulent, especially on the lower surface, rough to the touch .. *Vernonia*
   2. Flowers lavender .... *Tessaria*

COMPSONEURA SPRUCEI (A. DC.) Warb.—Myristicaceae—*Sangre* (Honduras and Guatemala). (Pl. 31)

Slender trees, 40–60 ft. in height, with alternate, oblong, acute or acuminate leaves and minute flowers in axillary racemes, followed by rather large, yellow, nutmeg-like fruits, the chocolate-brown, mottled seeds covered by a showy, undivided, blood-red aril which is exposed when the fruits are mature in March. The common name is in reference to the sap, which is red. The wood is yellowish brown, fairly hard but light in weight, straight-grained, and even-textured, but is not durable and has no local uses.—Forested hills above Palmar Norte, 250 to about 1,500 ft., *Allen 5942 & 6009.*

Conacaste—See *Enterolobium cyclocarpum* (Guatemala).

CONDAMINEA CORYMBOSA (R. & P.) DC.—Rubiaceae.

A shrub or small tree, to about 20 ft. in height, with very large, opposite, obovate, acute leaves (to about 30″ long and 12–14″ wide) and large terminal panicles of flowers, the corolla tube described as being creamy white below and red above, followed in February by obovate, shallowly bisulcate capsular fruits.—Fairly frequent along trails in the forested hills above Palmar Norte at 1,800–2,500 ft., *Allen 5910.*

Conejo—See *Calycophyllum candidissimum* (Local).
Conejo colorado—See *Trichilia acutanthera* (Chiriquí).

CONOCARPUS ERECTA L.—Combretaceae—*Mangle marequita* (Local)—*Mangle negro, Marequito,* or *Mariquito* (Costa Rica)—*Zaragoza, Mangle piñuelo,* or *Mangle torcido* (Panama)—*Button mangrove* (Canal Zone, Bocas del Toro, and Limón)—*Botoncillo* (Salvador)—*Buttonwood* or *Buttonbush* (British Honduras). (Pl. 20)

Shrubs or small, usually bushy trees, to about 35 ft. in height, common to the margins of mangrove swamps and sea beaches. The alternate, lanceolate leaves are obtuse or acute at the apex and have 2 elliptic glands at the base of the blade. The small white flowers are produced in August and September in terminal or axillary nearly globose racemes, followed by clusters of spherical, conelike fruits which are long persistent on the branches. The wood is yellowish to olive brown, fine-grained,

moderately hard, heavy, tough, strong, and durable, but is little used excepting for fuel and charcoal, largely because of the difficulty in finding large or straight pieces. The bark contains 16% tannin, and might be worth exploiting in conjunction with that of *Rhizophora* and *Avicennia.*—Santo Domingo de Osa, *Tonduz 9949*—Sea beaches near the delta of the Río Esquinas, *Allen 5638*—Puerto Jiménez, *Brenes 12248.*

Contamal—See *Guazuma ulmifolia* (Guatemala).

Copal—See *Protium costaricense* (Costa Rica). Leaves pinnate. Flowers pedicellate.

Copal—See *Protium neglectum* var. *sessiliflorum* (British Honduras). Leaves pinnate. Flowers sessile.

Copal—See *Stemmadenia Donnell-Smithii* (Guatemala). Leaves simple.

Copal—See *Tetragastris panamensis* (British Honduras). Leaves pinnate. Fruits top-shaped.

Copalchi—See *Croton glabellus* (Local and Panama).

Copal macho—See *Protium neglectum* var. *sessiliflorum* (British Honduras).

Copete—See *Tecoma stans* (Panama).

Copon—See *Bursera simaruba* (Honduras).

Copte—See *Schizolobium parahybum* (Guatemala).

Coquillo—See *Astrocaryum alatum* (Panama). Spiny palm.

Coquillo—See *Jatropha curcas* (Costa Rica). Unarmed shrub.

Coquito—See *Corozo oleifera* (Nicoya). Palm.

Coquito—See *Jatropha curcas* (Costa Rica). Shrub.

Coral—See *Hamelia patens* (Honduras).

Coralillo—See *Casearia sylvestris* (Guatemala). Leaves simple, alternate.

Coralillo—See *Hamelia patens* (General). Leaves simple, in whorls of 4 at each node.

Coralillo—See *Picramnia latifolia* (Local). Leaves pinnate.

Coratu—See *Enterolobium cyclocarpum* (Panama).

Corcho—See *Ochroma lagopus* (Guatemala).

CORDIA—KEY

1. Forks of the young branches usually with hollow, gall-like swellings, inhabited by ants.
   2. Flowers about ¾″ in diameter .... *Cordia gerascanthus*
   2. Flowers less than ½″ in diameter .... *Cordia alliodora*
1. Forks of the young branches without hollow, gall-like swellings.
   2. Slender forest trees, seldom exceeding 15 ft. in height, the leaves usually more than 1 ft. long .... *Cordia protracta*

2. Trees more than 15 ft. in height, the leaves less than 1 ft. long.
   3. Leaves abruptly caudate-acuminate at the apex........................*Cordia lasiocalyx*
   3. Leaves acute at the apex..*Cordia collococca*

CORDIA ALLIODORA (Ruiz & Pavon) Cham.—Boraginaceae—*Laurel* (Local)—*Laurel negro* or *Onion cordia* (Costa Rica)—*Laurel blanco* (Panama, Honduras, and U. S. trade)—*Laurel, Laurel negro,* or *Laurel macho* (Nicaragua)—*Laurel, Bojon,* or *Laurel blanco* (Guatemala). (Pl. 23)

Medium-sized trees 25 to about 80 ft. in height and 12–36″ in trunk diameter, usually with relatively narrow crowns of alternate, elliptic-oblong, acute or acuminate leaves, which, with the bark, have a garlic- or onion-like odor when crushed. The forks of the young branches usually have hollow, gall-like swellings, which are inhabited by small colonies of stinging ants. The small, white, intensely fragrant flowers are produced from about mid-January to early March in large, showy terminal panicles, rendering the trees attractive and conspicuous from a distance. The wood is described by Record as being of two types: (1) grayish or yellowish, and (2) variegated brown, suggesting walnut, the differences being probably the result of growing conditions. The wood is pleasantly scented when freshly cut, seasons readily, with very little warping or checking, is fairly light, straight-grained, strong, durable, easy to work, finishes smoothly, and is highly resistant to insect attack. It is considered suitable for bridge decking, piling, railroad ties, beams, rafters, girders, flooring, paneling, interior trim, furniture, ships decking and planking, oars, motorcar bodies, and general construction. The smaller branches are sometimes used for making barrel hoops. One of the best, and most generally useful local timber species, and one that lends itself particularly well to reforestation projects, since the growth rate is rapid, and remarkably uniform on cleared land. Fairly frequent in the Terraba Valley, and to be expected in dryer situations such as Buenos Aires, Boruca, and the Peninsula de Osa.—Pastures near Palmar Norte, 200 ft., *Allen 5879.*

CORDIA COLLOCOCCA L.—Boraginaceae.

Trees, 15 to about 60 ft. in height, with alternate, usually elliptic-lanceolate, rather leathery, acute leaves 5–9″ long and 2–3½″ wide. The white flowers are produced in broadly spreading subterminal cymes, and are followed in late August and September by bright-red, juicy fruits about ¼″ in diameter.—Pastures near Palmar Norte, 50 ft., *Allen 6558.*

CORDIA GERASCANTHUS L.—Boraginaceae—*Laurel negro* (Honduras).

A large tree, similar to *C. alliodora* in superficial appearance, but with

much larger flowers.—Golfo Dulce, without definite locality, *Manuel Valerio 573*—Buenos Aires, 1,000 ft., *Tonduz 6701.*

CORDIA LASIOCALYX Pittier—Boraginaceae.

Small, spreading trees, to about 40 ft. in height, with alternate, broadly oblong-elliptic, shortly acuminate leaves and pseudoaxillary diffuse cymes of small white flowers. Fairly common along small streams in pastures near Palmar Norte, flowering in late December and early January, the scarlet fruits being produced about a month later.—Palmar Norte, *Allen 5736.*

CORDIA PROTRACTA I. M. Johnston—Boraginaceae.

Slender forest trees, about 12 ft. in height, with very large, oblong-lanceolate, long-acuminate leaves which average about 14–18″ in length. The small white flowers appear in December, and the ovoid, abruptly pointed fruits in January, both on an elongate, pendulous cyme.—Esquinas Forest, 100 ft., *Allen 5772 & 5827* & *Skutch 5303.*

Cordoncillo—A generic term, applied to many species of *Piper*, notably to *P. reticulatum* (Local).

Cornezuelo—See *Acacia costaricensis* (Costa Rica and Nicaragua). Thorns brown or black, hollow.

Cornezuelo—See *Acacia farnesiana* (Honduras). Thorns not hollow.

Cornezuelo—See *Acacia spadicigera* (Costa Rica). Thorns white or pale yellow, hollow.

CORNUTIA GRANDIFOLIA (Schlecht. & Cham.) Schauer—Verbenaceae—*Palo de danto* (Local)—*Pavilla* (Costa Rica)—*Murcielago* (Chiriquí)—*Cuatro caras* or *Palo cuadrado* (Panama)—*Cucaracho* or *Zopilote* (Honduras)— *Flor lila* or *Hoja de zope* (Guatemala).

Shrubs or small, bushy trees, to about 12–25 ft. in height, with opposite, broadly elliptic-lanceolate or nearly ovate, acuminate leaves, plants in our area usually having leaves with coarsely serrate margins. The rather attractive violet or purple flowers are produced in slender terminal panicles which usually exceed the leaves in length. Fairly frequent in brushy places along trails and on cutover land throughout the area.—Palmar Norte, 100 ft., *Allen 5581.*

Corocito—See *Corozo oleifera* (Panama).

Coronillo—See *Bellucia costaricensis* (Costa Rica).

Corotu—See *Enterolobium cyclocarpum* (Panama).

Corozo—See *Corozo oleifera* (Local and Panama).

Corozo colorado—See *Corozo oleifera* (Panama).

CoROZO OLEIFERA (HBK) L. H. Bailey—Palmaceae—*Corozo* (Local and general)—*Tuskra (?)* (Boruca)—*Coquito* or *Palmiche* (Nicoya)—*Corozo colorado* or *Corocito* (Panama). (Pl. 8)

Common palms, with massive, often more or less prostrate trunks and very large pinnate fronds, the basal portions of which are armed with short, stout spines which represent modified pinnae. The yellow or reddish-orange fruits average about 1″ in length, and are very densely crowded on the large, sessile, axillary panicles. The fleshy pericarp is utilized in some places, notably in the Perlas Islands of Panama Bay, for the extraction of an edible oil. The species is related to the African Oil Palm, and is superficially similar to it in appearance. Frequent in wet pastures and swampy forest throughout the area.—Pastures near Palmar, *Allen 6768.*

Corrimiento—See *Citharexylum viride* (Chiriquí).
Corta lengua—See *Casearia sylvestris* (Chiriquí).
Cortes—See *Godmania aesculifolia* (Guatemala). Flowers greenish yellow.
Cortes—See *Tabebuia chrysantha* (Local). Flowers bright yellow.
Cortes amarillo—See *Tabebuia chrysantha* (Local).
Cortes blanco—See *Godmania aesculifolia* (Salvador).
Cortes negro—See *Guarea longipetiola* (Local).
Cortez—See *Apeiba Tibourbou* (Panama).
Corteza—See *Tabebuia chrysantha* (Costa Rica).
Corteza amarilla—See *Tabebuia chrysantha* (Costa Rica).
Corteza de chivo—See *Godmania aesculifolia* (Costa Rica).
Cortezo—See *Apeiba Tibourbou* (Panama).
Costilla de caballo—See *Acalypha diversifolia* (Honduras).
Costilla de danto—See *Acalypha diversifolia* (Honduras). Leaves simple.
Costilla de danto—See *Didymopanax Morototoni* (Nicaragua). Leaves digitately compound.
Cotonron—See *Luehea Seemannii* (Honduras and Guatemala).
Cotton tree—See *Ceiba pentandra* (Bocas del Toro). Leaves digitately compound.
Cotton tree—See *Ochroma lagopus* (Panama). Leaves simple.

COURATARI PANAMENSIS Standl.—Lecythidaceae. (Fig. 11)

Tall forest trees, 90–120 ft. in height and 3–4 ft. in diameter above the prominently developed buttresses. The leaves are alternate, elliptic-obovate in outline, and obtuse or shortly acute at the apex and are usually

Fig. 11. Couratari panamensis. 1 — Branch with leaves. 2 — Seed. 3 — Seed pod.

deciduous when the trees are in flower in late September. The relatively small flowers are about 3/4″ in diameter, and are produced in fairly large terminal or axillary panicles, the attractive pink or old-rose color rendering the trees conspicuous from afar. The flowers have a faint, but rather unpleasant odor. The fruits are elongate, tubular, box-like pods, 3 or 4″ long and about 1½″ in diameter, with a triangular, detractable plug, and many thin, winged seeds. The reddish-brown wood is hard and heavy and of attractive appearance, and might be suitable for interior trim and carpentry. Occasional in the forested hills between the Río Esquinas and Golfito, the trees remaining in bloom from mid- or late September to about mid-October.—Esquinas Forest, 150 ft., *Allen 5686 & 6347.*

COUSSAPOA PARVICEPS Standl.—Moraceae.

Trees, often epiphytic, sometimes as much as 30–75 ft. in height, with alternate, elliptic-lanceolate, abruptly acute leaves which average about 4″ in length. The dioecious flowers are produced in the leaf axils in late February and early March, the minute greenish-tan staminate form being in tiny globose heads on a slender, diffuse panicle.—Cañas Gordas, *Pittier 11166*—Forested hills above Palmar Norte, 1,500 ft., *Allen 5949.*

Cowee—See *Pterocarpus officinalis* (Honduras).
Coyokiche—See *Nectandra latifolia* (Guatemala).
Coyol—See *Acrocomia vinifera* (Local and general).
Coyote—See *Platymiscium pinnatum* (Nicaragua).
Crabo—See *Byrsonima crassifolia* (Honduras).

CRATAEVA TAPIA L.—Capparidaceae—*Muñeco* (Costa Rica)—*Estrella* (Chiriquí)—*Palo de guaco* (Panama)—*Manzana de playa* (Nicaragua)—*Cachimbo* or *Tortugo* (Honduras)—*Anonillo* or *Granadillo macho* (Salvador)—*Matasanillo, Granadillo,* or *Tortugo* (Guatemala).

Small or medium-sized trees, 20–60 ft. in height and 10–18″ in trunk diameter, with alternate, digitately 3-parted, long-petiolate leaves and conspicuous terminal racemes of greenish-white flowers, which are produced in February and March, followed by globose or ovoid green fruits about 1 or 2″ in diameter. The bark has a disagreeable odor, and the sap will blister the skin. The wood is white or yellowish, and has a garlic-like odor when freshly cut. It is easy to work, but is coarse-textured, brittle, and not durable.—Santo Domingo de Osa, *Tonduz 10036 & 10037*—Forests near Tinoco Station, sea level, *Allen 5471.*

CRESCENTIA CUJETE L.—Bignoniaceae—*Calabash* (Local and general) —*Jicaro* (elongate form, local)—*Guacal* (round form, local)—*Calabazo, Totumo,* or *Totumbo* (Panama)—*Guacal* (Nicaragua)—*Morro* (Honduras)—*Wild Calabash, Jicara,* or *Guiro* (British Honduras)—*Guacal, Jicaro de cuchara,* or *Cutuco* (Salvador)—*Jicaro, Xigal,* or *Hom* (Guatemala).

Small, spreading trees, to about 35 ft. in height, with firm, oblanceolate or spatulate leaves which are produced in sessile, 3–12 foliolate fascicles. The large, greenish-white or greenish-brown flowers are borne directly on the trunk and larger branches, and are followed by smooth, green, gourdlike fruits which are much used for domestic utensils. The pulp of the fruit is purgative, according to Standley, and has been accused of causing abortion in cattle that eat the fallen fruits. Frequently planted, and occasionally persistent as old orchid-laden specimens near former house sites and along trails. The tough wood is sometimes used for saddle trees, ox yokes, tool handles, vehicle parts, and boat building.

Crespon—See *Urera caracasana* (Costa Rica).
Criollo—See *Minquartia guianensis* (Chiriquí).
Cristobal—See *Platymiscium pinnatum* (Local and general).

CROTON—KEY

1. Inflorescences equaling or exceeding the leaves in length. Sap dries blood red.
   2. Leaves usually densely stellate-tomentose, particularly on the lower surface (lens!) .... *Croton xalapensis*
   2. Leaves usually sparsely stellate-tomentose on the lower surface... *Croton panamensis*
1. Inflorescences much shorter than the leaves. Sap not red.......... *Croton glabellus*

CROTON GLABELLUS L.—Euphorbiaceae—*Copalchi* (Local and Panama) —*Quizarra copalchi* (Costa Rica)—*Barenillo, Cascarilian,* or *Lian* (Honduras)—*Cache, Canoh, Fruta de chancha, Paujil,* or *Perescuch* (Guatemala).

Small or medium-sized forest trees, to about 50 ft. in height and 10–12″ in trunk diameter, with alternate, oblong or oblong-elliptic, acuminate leaves and short axillary racemes of small white flowers which are produced from about October until April, and followed by small, nearly globose, tuberculate green fruits, most trees having both flowers and fruits at any given time. The brownish or yellowish wood is moderately hard, dense, fine-grained, and durable. It is reported to be used for house construction, particularly flooring. A very common species, found throughout our area.—Boruca, 1,300 ft., *Tonduz 3877 & 4651*—Santo Domingo de Osa, *Tonduz 9971* & *Pittier 7159*—Forests near Jalaca, *Allen*

5213—Pastures near Palmar Norte, *Allen 5716*—Esquinas Forest, 250 ft., *Skutch 5387* & *Allen 5816 & 6630.*

CROTON PANAMENSIS Muell. Arg.—Euphorbiaceae—*Targua* or *Targua colorado* (Costa Rica). (Pl. 18)

Shrubs or small trees, to about 25 ft. in height, with alternate, long-petiolate leaves, the large, ovate, acuminate blades rounded or deeply cordate at the base. The small, greenish-white flowers are produced in long, slender, erect, terminal spikes from September to November and are followed by small, 3-celled fruits. The trees often form small, almost pure stands on the margins of forest or on cutover land. The sap turns red on exposure to the air. The soft white wood is not used.—Boruca, *Tonduz 4784*—Golfito, *Allen 6627.*

CROTON XALAPENSIS HBK—Euphorbiaceae—*Terre, Targua,* or *Targua blanco* (Costa Rica)—*Pela nariz* (Honduras)—*Drago, Chirca,* or *Llora sangre* (Guatemala).

Shrubs or small trees to about 18 ft. in height, with alternate, ovate or oblong-ovate, acute or acuminate leaves, which are paler in color and usually densely stellate-tomentose on the lower surface. The flowers are produced in slender terminal racemes which equal or exceed the leaves in length. All parts of the plant have a watery sap which turns red on exposure to the air. Locally common, particularly near Buenos Aires and Boruca, often forming thickets on cutover land.—*Tonduz 4650.*

Crucetilla—See *Randia armata* (Honduras).
Crucillo—See *Randia armata* (Costa Rica).
Crucito—See *Randia armata* (Guatemala).

CRYOSOPHILA GUAGARA Allen—Palmaceae—*Guagara* (Local). (Pl. 6)

Slender, single-trunked, fan-leaved palms, 12–20 ft. in height, the trunks armed throughout with extensive branching root spines. Mature plants with 12–15 live fronds, the flabellate blades averaging about 5 ft. in diameter, the dark, glossy-green upper surface contrasting strongly with the silvery-white lower side. The fronds are strongly bifid, with a deep central cleft which divides the blade to within about 1″ of the short, broadly triangular hastula. The inflorescences are elongate and pendulous, and are covered throughout their length with 25 to about 40 broad, papery, yellowish bracts. The flowers are produced on short, branching panicles, which are spirally arranged on the main axis of the inflorescence. Fruiting clusters are commonly seen without bracts, which tend to fall before the globose, waxy fruits mature. A very striking species, immediately separable from all others in the genus by the

elongate, pendulous inflorescences which are covered to the apex by the conspicuous bracts. Common in the lowland forests throughout the area, the fanlike fronds being much used for thatch.—Forests near the Tinoco Station, sea level, *Allen 6602.*

Cuajada—See *Vitex Cooperi* (Local).
Cuajada negro—See *Minquartia guianensis* (Local and Chiriquí).
Cuajiniquil—See *Inga* spp.—A generic term, of rather wide application.
Cuajiniquil blanco—See *Inga punctata* (Guatemala).
Cuatchi—See *Dialium guianense* (Guatemala).
Cuatro caras—See *Cornutia grandifolia* (Panama).
Cucaracho—See *Cornutia grandifolia* (Honduras).
Cuche—See *Cedrela mexicana* (Guatemala).
Cuculat—See *Theobroma Cacao* (Guatemala).
Cuculmico—See *Casearia arguta* (Salvador).
Cucte—See *Schizolobium parahybum* (Guatemala).
Cucunango—See *Vernonia patens* (Honduras).
Cuernito—See *Acacia costaricensis* (Panama). Thorns hollow.
Cuernito—See *Acacia farnesiana* (Panama). Thorns not hollow.
Cuerno de venado—See *Mouriria parvifolia* (Guatemala).
Cuetillo—See *Hamelia patens* (Guatemala).
Cuje—See *Inga multijuga* (Guatemala).
Cujincuil—See *Inga laurina* (Salvador).
Cukil—See *Dracaena americana* (Guatemala).
Culinzis—See *Astronium graveolens* (Guatemala).
Culumate—See *Avicennia nitida* (Costa Rica).
Cumpap—See *Plumeria rubra* forma *acutifolia* (Guatemala).

CUPANIA—KEY

1. Leaflets completely glabrous on the lower surface. Fruits glabrous .... *Cupania macrophylla*
1. Leaflets puberulent or pilose, at least on the veins of the lower surface. Fruits densely pilose.
   2. Leaflets about 5. Lower surface with a conspicuous, elevated, reticulate venation .... *Cupania largifolia*
   2. Leaflets 6–10. Lower surface densely and uniformly pilose, the venation not elevated or conspicuous .... *Cupania guatemalensis*

CUPANIA GUATEMALENSIS (Turcz.) Radlk.—Sapindaceae—*Cantarillo, Carne asada, Huesillo, Pozolillo,* or *Manteco* (Costa Rica)—*Cola de pava* or *Carboncillo* (Guatemala)—*Cedrillo, Miacaguite,* or *Huesito* (Salvador).

Shrubs or small trees, to about 30 ft. in height, with alternate, pinnate leaves, the 6–10 narrowly lanceolate leaflets densely and uniformly pilose on the lower surface, the margins variously dentate or entire. The small white flowers are produced in terminal panicles and are followed by 3-celled, capsular, pubescent fruits.—*Tonduz 6651* & *Biolley 10629.*

CUPANIA LARGIFOLIA Radlk.—Sapindaceae.

Small to medium-sized trees, 20 to 50 ft. in height, with rather large, alternate, pinnate leaves, the leaflets typically 5 in number. The small white flowers are produced in July and August in terminal panicles, and are soon followed by puberulent, brown, 3-angulate capsules.—Rather frequent in pastures and on the margins of forest near Palmar Norte, *Allen 5602.*

CUPANIA MACROPHYLLA A. Rich.—Sapindaceae—*Carbon colorado* (Guatemala).

Small or medium-sized forest trees, 18–60 ft. in height, with alternate, pinnate leaves, the 2–4 large pinnae completely glabrous on both surfaces. The very small white flowers are borne in terminal paniculate racemes, and are followed in July and August by the glabrous, 3-celled, capsular fruits.—Forested hills near Palmar Norte, 200 ft., *Allen 6286.*

CURATELLA AMERICANA L.—Dilleniaceae—*Chumico* (Boruca)—*Chumico de palo, Hoja chigue, Raspa guacal,* or *Yahal* (Costa Rica)—*Sandpaper tree* (Canal Zone)—*Chumico* or *Chumico de palo* (Panama)—*Malcajaco* or *Lengua de vaca* (Salvador)—*Chaparro* or *Saha* (Guatemala)—*Chaparro* or *Yaja* (British Honduras).

Small, gnarled trees, seldom exceeding 20 ft. in height, typical of areas of dry, open grassland where they often form local, open stands of considerable extent. The stiff leaves have a rough, sandpapery feel, and are alternate on the stems, the blades being oval or elliptic in outline, usually with distinctly undulate margins. The small, greenish-white flowers are produced from January to April in short axillary panicles. The leaves are sometimes used for scouring and polishing kitchen utensils, and the rather hard, heavy, reddish-brown wood is sparingly utilized in some localities for fence posts, fuel, and charcoal. Common on grassy ridges near Boruca and Potrero Grande to about 2,000 ft.—Buenos Aires, *Tonduz 6571*—Boruca, 1,250 ft., *Tonduz 6847.*

Curtidor—See *Hieronyma alchorneoides* (Honduras).

Cutuco—See *Crescentia cujete* (Salvador).

Cuxamate—See *Ficus costaricana* (Guatemala).

Cuyapo—See *Lafoensia punicifolia* (Salvador).

Cuyus—See *Sapindus saponaria* (Nicaragua).

CYMBOPETALUM COSTARICENSE (Donn. Sm.) Fries—Annonaceae—*Huevo de toro* (Honduras).

Shrubs or small trees 15–30 ft. in height, with alternate leaves, the oblong or obovate blades 6–11″ long and 1¾–3½″ wide. The pendulous clusters of green flowers are produced from the leaf axils, and are followed by 8–10 stipitate carpels of the fruits. A species closely allied to *Cymbopetalum penduliflorum*.—Cañas Gordas, 3,200 ft., *Pittier 11106.*

CYNOMETRA HEMITOMOPHYLLA (Donn. Sm.) Britt. & Rose—Leguminosae; Caesalpinoideae—*Guapinol negro* (Local)—*Cativo* (Costa Rica).

Medium-sized or tall forest trees, 50–100 ft. in height and 1–3 ft. in trunk diameter above the rather prominently developed buttresses, the bark typically with many small corky lenticels. The leaves are alternate and pinnate, but the pinnae consist of a single pair of obliquely lanceolate, acuminate leaflets, the flaccid flushes of new foliage rather superficially reminiscent of those of *Brownea.* The small, pale-brown flowers are produced from late August to early November, in short, axillary fascicles, and are followed in April and May by the hard, brown, woody, 1-seeded indehiscent fruits. Fairly frequent in forests and along small streams throughout the area. The hard, heavy wood is apparently not used.—Santo Domingo de Osa, *Tonduz 9972*—Pastures near Palmar Norte, *Allen 5264*—Forested hills near Palmar Norte, 500 ft., *Allen 5649.*

CYPHOMANDRA COSTARICENSIS Donn. Sm. — Solanaceae — *Zopilote* (Local)—*Pepinillo* (Costa Rica).

Small trees, to about 18 ft., with broadly spreading, typically flat-topped crowns. The broad, thin, alternate leaves vary from ovate to oblong-elliptic in outline, with broadly rounded, deeply cordate, or strongly asymmetrical bases and acute or shortly acuminate tips. The green flowers have rich purple stamens, and are borne on pendulous, scorpioid cymes, and are followed by mottled, green, fleshy, egg-shaped fruits. Very common along the railroad right of way, and on cutover land throughout the area.—Esquinas, *Skutch 5276* & *Allen 5241*—Tinoco Station, *Allen 6617.*

DACRYODES EPIPHYTICA Standl. & L. Wms.—Burseraceae.

Epiphytic trees, 20–40 ft. in height, always found growing in the tops of giant specimens of the floodplain forests, particularly "Espavel" *(Anacardium excelsum).* The species is characterized by thin, coppery-red bark, which is almost identical in superficial appearance with that

of *Bursera simaruba*. Large specimens are often quite conspicuous in the woodlands when the sun strikes through the pendant fringes of loose, red, translucent bark, producing a burst of brilliant color. The leaves are alternate and pinnate, usually with 5 broadly ovate, shortly acute leaflets. The very small, fragrant white flowers are produced in February in fascicles of short, slender, subterminal panicles, and are followed in July by small, 3-angled fruits. Unique in being the only epiphytic member known in the Burseraceae. Locally common, especially near the Tinoco Station, Sierpe, Jalaca, and in the Esquinas Forest.—*Allen 5884, 5966, & 6271.*

Damajao colorado—See *Heliocarpus appendiculatus* (Honduras).
Dantisco de montaña—See *Mauria glauca* (Central America).
Danto—See *Roupala complicata* (Costa Rica).
Danto hediendo—See *Roupala complicata* (Costa Rica).
Dialium divaricatum—See *Dialium guianense.*

DIALIUM GUIANENSE (Aubl.) Sandwith—Leguminosae; Caesalpinoideae—*Alfeñique* (Local)—*Tamarindo* (Costa Rica)—*Fria* or *Tamarindo* (Panama)—*Sangrillo negro* (Chiriquí)—*Monkey apple* (Bocas del Toro)—*Comenegro* or *Tamarindo montero* (Nicaragua)—*Canillo, Paleta, Paleto, Paleto negro, Tamarindo,* or *Tamarindo prieto* (Honduras)—*Paleta, Cuatchi, Chate, Palo de Lacandon, Tamarindo, Tamarindo prieto,* or *Uapake* (Guatemala).

Tall forest trees, to about 120 ft. in height and 2–4 ft. in trunk diameter above the moderately developed buttresses. Leaves alternate and pinnate, with 5–7 lanceolate, acuminate leaflets. The small, inconspicuous olive-green flowers are produced from late September to early December in minutely tomentulose, greenish-brown, spreading terminal panicles, followed by dark greenish-brown, obovoid or nearly globose, 1-seeded fruits. There is some indication that the trees may flower and fruit twice during the year, since mature fruits have been found in January and again in August. When in sterile condition the trees bear a considerable superficial resemblance to *Pterocarpus officinalis* (Sangrillo), which explains the common name of "Sangrillo negro" used in Chiriquí. The trees are found only on high, well-drained land. The wood is pale brown when freshly cut, becoming darker upon exposure. It is hard, heavy, tough, strong, and highly durable; has an interwoven grain which makes it very difficult to saw, but it can be finished smoothly. It is a first quality construction timber much used by the Engineering Department of the United Fruit Company in the Armuelles Division,

who compare it to teak in general usefulness. It is suitable for bridge timbers, railroad ties, house and fence posts, piling, and heavy construction. The logs will not float, and the wood is considered too hard and heavy for furniture. Frequent in hillside forests throughout our area.—Esquinas Forest, 250 ft., *Allen 5810*—Forested hills near Golfito, *Allen 6299, 6318, & 6695.*

DIALYANTHERA OTOBA (H. & B.) Warb.—Myristicaceae—*Sebo* (Costa Rica)—*Bogamani verde, Roble,* or *Saba* (Chiriquí)—*White cedar* (Bocas del Toro).

Tall forest trees, to about 90 ft. in height and 36″ in trunk diameter, with alternate, elliptic, rather leathery, obtuse or acute leaves and short axillary racemes of small yellow flowers, which are followed in September by globose fruits about 1″ in diameter. The single seed is covered by a red, laciniate aril. The pinkish-brown wood is light and soft, straight-grained, and easy to work, but is not durable. It is considered suitable for boxes and interior construction. Fairly frequent in hillside forests throughout the area.—Esquinas Forest, 250 ft., *Allen 5367 & 5591.*

DICRASPIDIA DONNELL-SMITHII Standl.—Tiliaceae—*Capulin macho* (Local). (Pl. 34)

Small trees, 10 to about 30 ft. in height, common to stream banks and second-growth scrub throughout the region. The alternate, oblong-lanceolate, acuminate leaves have coarsely serrate margins, and are dark green above and conspicuously whitish on the lower surface. The species may readily be distinguished from all others in the area by the unique round foliaceous stipules which cover the bases of the leaves. The large, solitary, handsome bright-yellow flowers are produced from the upper leaf axils during the dry season, from about December to April, and are followed by the truncate-globose, many-seeded pendulous fruits which retain the segments of the calyx at the apex.—Corredor de Golfo Dulce, *Pittier 11172*—Río Piedras Blancas, *Allen 5836*—Second-growth scrub along railroad near Kilometer 42, *Allen 5434*—Banks of small streams near Golfito, *Allen 5993.*

DIDYMOPANAX MOROTOTONI (Aubl.) Dcne. & Planch.—Araliaceae—*Probado* (Local)—*Pava, Pavo,* or *Pavilla* (Costa Rica)—*Pava, Mangabe,* or *Gargoran* (Panama)—*Costilla de danto* (Nicaragua). (Pl. 12)

Tall, slender trees with grayish-white trunks and broad, spreading crowns, common to areas of old second growth and the margins of forest. They often form nearly pure stands which vary in height, accord-

ing to age, old specimens sometimes being as much as 75 ft. in height and up to about 40″ in trunk diameter. The very large, long-petioled, digitately compound leaves are clustered at the ends of the branches, the 7–10 elliptic or oblong, acuminate leaflets laxly pendulous, the undersurfaces usually pale brown in color, contrasting strongly with the dark-green upper surfaces when turned by the wind, often producing the illusion of a crown of flowers when seen from a distance. The small white flowers are produced from November to about February in large axillary panicles, and are followed in March and April by small, dark greenish-blue, laterally compressed fruits. The wood is white or grayish or brownish in color and also variable in density from light and soft to moderately hard and heavy. It is straight-grained, somewhat brittle and not durable. It has been used for match sticks, boxes, and crates and might be suitable for paper pulp or plywood corestock.—Buenos Aires, *Tonduz 3980*—Hills near Palmar Sur, *Allen 6004.*

DILLENIACEAE—One genus, *Curatella.*

Dimorphandra megistosperma—See *Mora oleifera.*

DIOSPYROS EBENASTER Retz.—Ebenaceae—*Zapote de mico* (Salvador) —*Zapote negro* or *Matasano de mico* (Guatemala). (Pl. 24)

Forest trees, to about 80 ft. in height and up to 30″ in diameter, mature specimens being characterized by their black bark and deeply furrowed trunks. The dark-green lanceolate, acute or acuminate leaves are alternate on the branchlets and average about 6″ in length. The small, whitish, fragrant flowers are borne in the leaf axils and are followed in February and March by the dark-green, depressed-globose fruits which are about 3″ in diameter, and are filled with a sweet, but unattractive inky-black pulp which surrounds the large brown seeds. The fruits can be used as a fresh fruit or for preserves, or can be fermented to produce a brandy-like liquor. The green fruits have been reported to be utilized in the Philippines as a barbasco for catching fish. The hard yellowish wood has a few streaks of ebony-black near the center of old trunks, usually in areas of injury. The species is related to the common persimmon of the southern United States and to the true ebonies of the Old World tropics. Locally frequent in hillside forests, but no use appears to be made of the wood.—Forested hills above Palmar Norte, 1,500 ft., *Allen 5908.*

DIPHYSA ROBINIOIDES Benth.—Leguminosae; Papilionatae—*Guachipelin* (Local)—*Cacique, Guachepelin,* or *Macano* (Panama)—*Guachipilin* or

Quebracho de cerro (Honduras)—*Guachipilin, Palo amarillo,* or *Much* (Guatemala). (Pl. 17)

Small or medium-sized trees, to about 45 ft. in height and 10–16″ in trunk diameter, with alternate, pinnate, deciduous leaves and axillary racemes of attractive bright-yellow flowers, which may be produced from December until February. They are followed by inflated, papery, indehiscent pods, which average about 3″ in length. Viable seeds are hard to find in our area, since most of the pods are usually wormy. The trunks, even of mature specimens, are seldom straight, undoubtedly partly because of the adverse conditions under which they usually grow. The hard, heavy, yellow or olive-brown wood is highly durable in contact with the ground and in water, and is often used for house posts, fence posts, and occasionally railroad ties. The principal limiting factor is the difficulty in securing adequate supplies of the larger sizes. Supply seldom equals demand and plantations might prove profitable in some of the drier, better drained parts of the country, particularly since planted trees should tend to be straighter. Small scale experimental plantings at Esquinas were unsuccessful, probably because of the excessively high rainfall. The species is frequently planted for living fence posts and may possibly be native near Boruca and Buenos Aires.—Boruca, *Tonduz 4763* —Palmar Norte, *Allen 6321.*

Dipterodendron costaricense Radlk.—Sapindaceae—*Iguano* or *Loro* (Costa Rica)—*Harino* or *Jarino* (Panama). (Pl. 18)

Medium-sized, or occasionally large trees, sometimes reaching 85 ft. in height, with relatively large, alternate, bipinnate, fernlike leaves, which are deciduous during the late dry season. The small white flowers are produced in axillary panicles, and are followed, usually in February and March, by dark-red, 2- or 3-celled capsular fruits. The pinkish or brownish wood is rather hard, heavy, and strong, but has an irregular grain. It is reported to be easy to work, but is not used locally. The species is fairly common throughout the area in pastures and forest from sea level to about 1,800 ft.—Palmar Norte, *Allen 5738.*

Disciplina—See *Chamaedorea* sp. (Boruca).

Dogwood—See *Lonchocarpus latifolius* (Bocas del Toro). Leaves completely glabrous.

Dogwood—See *Lonchocarpus sericeus* var. *glabrescens* (Bocas del Toro). Leaves minutely puberulent on the lower surface.

Dos caras—See *Miconia argentea* (Panama). Leaves whitish on the lower surface.

Dos caras—See *Miconia impetiolaris* (Panama). Leaves not whitish on the lower surface.

DRACAENA AMERICANA Donn. Sm.—Liliaceae—*Isote* or *Izote* (Honduras)—*Candalwood* or *Cerbatana* (British Honduras)—*Caña de arco, Cerbatana, Cukil, Halal, Ilcaax,* or *Izote de montaña* (Guatemala). (Pl. 22)

Small trees, 20 to about 40 ft. in height, with short thick trunks and stout arching branches which terminate in dense clusters of glossy, linear, acuminate leaves. The small white flowers are produced in large terminal panicles from January until April and are followed in August and September by the fleshy, orange, nearly globose, large-seeded fruits. The whitish or pale-brownish wood is described as being light, firm, easy to cut, but is coarse-textured, difficult to finish, and poorly resistant to decay. It has no known uses. The species is a very striking one, and unique in our local forests, bearing a considerable superficial resemblance to *Yucca elephantipes* ("Itabo") so common to gardens and hedgerows in the Meseta Central. The genus *Dracaena* is otherwise restricted in its distribution to the Old World tropics. Frequent in pastures and forest, particularly near Palmar Norte up to about 2,500 ft. in elevation.—Tinoco Station, sea level, *Allen 5497*—Palmar Norte, sea level, *Allen 5508, 5600, & 5801.*

Drago—See *Croton xalapensis* (Guatemala). Leaves paler in color on the lower surface.

Drago—See *Virola Koschnyi* (Guatemala). Leaves brownish on the lower surface.

DUGUETIA PANAMENSIS Standl.—Annonaceae. (Pl. 14)

Slender trees, 20–40 ft. in height, with large, alternate, linear or oblong, abruptly acuminate leaves and dark-yellow supra-axillary flowers which are produced in February and March, followed in August and September by the nearly globose, many-carpeled fruits which are at first green, but become yellow at maturity. Fairly frequent in hillside forests from sea level to about 1,500 ft. in elevation.—Hills near Palmar Norte, *Allen 5958, 5999, & 6552.*

DUSSIA—KEY

1. Floral calyx divided nearly to the base. Flowers pinkish lavender, inconspicuous from a distance .... *Dussia macrophyllata*
1. Floral calyx tubular, with short, distinct marginal teeth. Flowers pink, conspicuous from a distance .... *Dussia mexicana*

DUSSIA MACROPHYLLATA (D. Sm.) Harms—Leguminosae; Papilionatae. (Pl. 15)

Tall forest trees, 90 to about 125 ft. in height, with a broad, spreading crown. The trunk, in mature specimens may reach 3 ft. or slightly more in diameter, above the strongly developed buttresses. The sap is blood red, like that of *Pterocarpus officinalis.* The pinnate leaves are alternate on the branches, with 7–9 large, oblong or oblanceolate leaflets, which are rounded or obscurely acute at the apex. The inconspicuous flowers are produced in January in pseudoterminal racemes, individual flowers having a dark-brown, deeply cleft calyx; the expanded limb of the corolla is pale pink, with darker markings, the general effect being more or less pinkish lavender. The large, fleshy yellowish fruits are pendulous, and conspicuous from a distance. They contain from 1–3 rather succulent scarlet seeds. The pale-yellow sapwood is nearly identical in color with the heartwood, which appears to be firm, and of good texture, but is not used locally. The species is occasional in lowland climax forests.—Coto Junction, 50 ft., *Allen 6655*—Hills near Golfito, 100 ft., *Allen 6703.*

DUSSIA MEXICANA (Standl.) Harms—Leguminosae; Papilionatae. (Pl. 15)

Tall forest trees, 125–150 ft. in height and about 4 ft. in diameter above the strongly developed buttresses. The large pinnate leaves are alternate on the branchlets, and are deciduous from late February until mid-March. During this season the handsome racemes of pink flowers are produced, rendering the trees very conspicuous and attractive. The axillary and terminal buds are small and completely glabrous, as are the flushes of new growth and leaf petioles. There is no difference in appearance between the heartwood and sapwood, both being relatively soft and white. Frequent in the forested hills near Golfito, where it is very showy during the flowering season.—*Allen 5988.*

Ear tree—See *Enterolobium cyclocarpum* (Canal Zone and Honduras).

EBENACEAE—One genus, *Diospyros.*

Elaphrium simaruba—See *Bursera simaruba.*

Elequeme—See *Erythrina costaricensis* (Costa Rica).

ENALLAGMA LATIFOLIA (Mill.) Small—Bignoniaceae—*Calabasillo de la playa* (Local)—*Cacao silvestre* (Costa Rica)—*Wild calabash* (Bocas del Toro)—*Wild calabash* or *Morito de rio* (British Honduras)—*Ajonocht* (Guatemala). (Pl. 16)

Small trees, to about 30 ft. in height, with alternate, broadly elliptic or obovate, acute leaves and tubular, greenish-white flowers, which have an elongate, deeply cleft, green calyx. The fruits resemble a small green

calabash. Occasional along sea beaches and near mangrove swamps in the Golfo Dulce area.—Vicinity of the delta of the Río Esquinas, Golfo Dulce, *Allen 5639.*

Encino—See *Licania arborea* (Guatemala). Individual fruits not enclosed in a basal cup.

Encino—See *Quercus* sp. (Costa Rica). Individual fruits in a basal cup.

ENTEROLOBIUM CYCLOCARPUM (Jacq.) Griseb.—Leguminosae; Mimosoideae—*Guanacaste* (Local and general in northern Central America)—*Coratu* or *Corotu* (Panama)—*Genisero* or *Jarina* (Costa Rica)—*Ear tree* (Canal Zone)—*Tuburus* (Sumo, Nicaragua)—*Conacaste, Guanacaste,* or *Pit* (Guatemala). (Fig. 12)

Very large trees, with broad, spreading crowns of rather brittle branches, mature specimens often as much as 85 ft. in height and up to 6–8 ft. in trunk diameter above the occasionally well-developed buttresses. The bipinnate leaves are alternate, and are often deciduous during the dry season. The small white flowers are produced in globose heads in early December and are followed in March and April by the unique dark-brown, coiled, ear-shaped, indehiscent seed pods. In spite of their relatively large size, clear lengths of more than 40 ft. are unusual, since the branching habit tends to be low. The brown or reddish-brown wood somewhat resembles walnut and has been used in the United States for paneling and interior trim. It has a reputation for being very durable in water, but not when exposed to the air, and is not termite resistant, but works well, finishes smoothly, and takes a good polish. It is considered suitable for dugouts, gunstocks, furniture, veneer, paneling, and general inside carpentry. The sawdust is very disagreeable to the workmen in sawmills, and is reported to be poisonous to fish and cattle if dumped into streams. The species is not common in our area, which may be explained in part by the high rainfall, and by the curious circumstance that most of the fruits are parasitized by a gall insect, so that large specimens may not mature more than a half dozen normal pods in a given season. In areas where the trees fruit freely the pods are often relished by cattle, and the trees deliberately left standing in pastures for this reason and for their shade.—Diquis (Terraba) Valley, *Pittier 12217*—Pastures near Palmar Norte, *Allen 5717 & 6001.*

ERYTHRINA COSTARICENSIS Micheli—Leguminosae; Papilionatae—*Poro, Poro espinas,* or *Poro trinidad* (Local)—*Poro colorado* or *Elequeme* (Costa Rica).

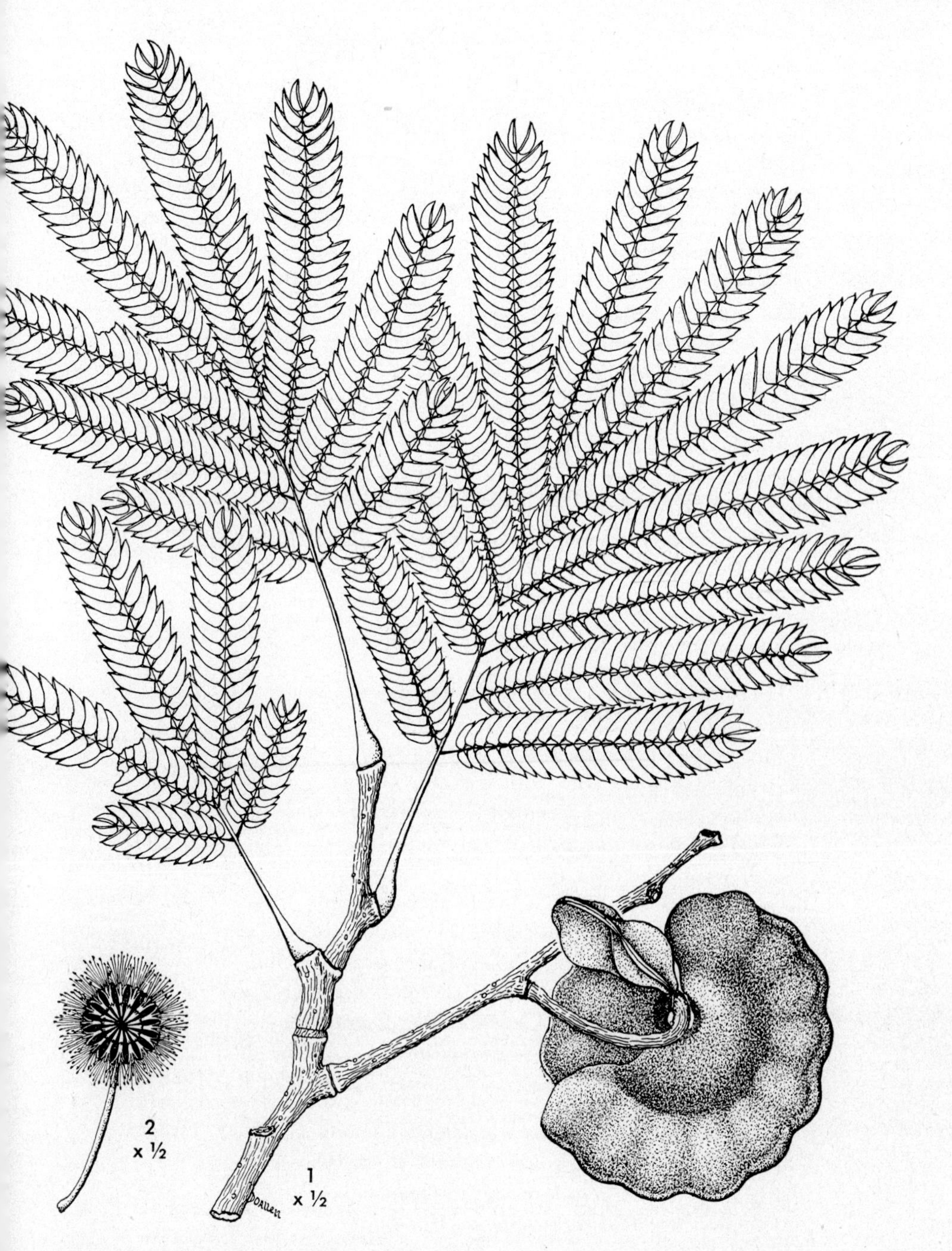

Fig. 12. Enterolobium cyclocarpum. 1 — Fruiting branch. 2 — Cross section of inflorescence.

Small to medium-sized trees, 15 to about 45 ft. in height, with somewhat thorny branches and alternate, 3-foliolate, pinnate leaves, which are deciduous during the dry season. The showy, blood-red flowers are produced in terminal racemes from November through February, and are soon followed by the pendulous beanlike pods, which are deeply constricted between the seeds. Occasional in our area as living fence posts, or persistent near old house sites, or as apparently spontaneous specimens in the forest. The soft wood is not used.—Boruca, 1,500 ft., *Tonduz 4801*—Santo Domingo de Osa, *Tonduz 10050*—Margins of forest near Palmar Norte, 100 ft., *Allen 5792*.

ERYTHROXYLON LUCIDUM HBK—Erythroxylaceae.

Shrubs or small trees, to about 30 ft. in height, with alternate, short-petiolate, lanceolate or elliptic-lanceolate, acute or acuminate, rather leathery leaves. The very small white flowers are produced in dense, sessile axillary fascicles and are followed by small, drupaceous fruits.—Santo Domingo de Golfo Dulce, *Tonduz 10092*—Boruca, *Pittier 11100 & 11950*.

Escambron—See *Casearia aculeata* (Honduras).

ESCHWEILERA CALYCULATA Pittier—Lecythidaceae—*Mata cansada* or *Ollito* (Chiriquí). (Pl. 14)

Slender, narrow-crowned trees, to about 75 ft. in height and 12–18″ in trunk diameter, with rather large, alternate, elliptic-oblong, shortly acuminate leaves and terminal or axillary racemes of attractive, fleshy, pale-yellow flowers, which are produced in late March and April. The round, woody boxlike fruits have a detachable lid and are filled with large brown seeds which considerably resemble those of the *Sapucaya* nut, but are bitter and inedible. They mature in our area in August and September. The wood of some of the related South American species contains silica and has a reputation for being highly resistant to marine borers. Nothing is as yet known of the potentialities of our local species in this respect, and it does not seem to be used either here or in Panama. Occasional in hillside forests.—Esquinas Forest, 250 ft., *Allen 6298*.

Espave—See *Anacardium excelsum* (Panama and U. S. trade).
Espavel—See *Anacardium excelsum* (Costa Rica).
Espinal—See *Acacia farnesiana* (Guatemala).
Espino—See *Acacia farnesiana* (Honduras).
Espino blanco—See *Acacia farnesiana* (Nicaragua, Honduras, and Guatemala).

Espino cachito—See *Acacia costaricensis* (Honduras).

Espino ruco—See *Acacia farnesiana* (Salvador).

Espinudo—See *Lacmellea panamensis* (Local).

Esquijoche—See *Bourreria littoralis* (Costa Rica).

Estrella—See *Crataeva tapia* (Chiriquí). Leaves digitately compound.

Estrella—See *Myriocarpa longipes* (Local and Boruca). Leaves entire.

EUGENIA—Key

1. Leaves mostly about 5″ long. Fruits about ¾″ in diameter.
   2. Stems, petioles, and veins of the lower leaf surface densely hirtellous........ *Eugenia* sp.—*Allen 6553*
   2. Stems, petioles and veins of the lower leaf surface glabrous......*Eugenia palmarum*
1. Leaves mostly about 2½″ long. Fruits about ¼″ in diameter........ *Eugenia* sp.—*Allen 5979*

Eugenia palmarum Standl. & L. Wms.—Myrtaceae.

Slender forest trees, 25–40 ft. in height and 6–10″ in trunk diameter, with opposite, lanceolate, acuminate, leathery leaves which average about 5″ in length. The dark-purple globose fruits are sweet to the taste, and are produced in December. The trunks are smooth, with nearly pure-white, or pale-brown mottled bark, rather reminiscent of that of the common guava. Common, and locally very striking, due to the unusual bark.—Hillside forests above Palmar Norte, 1,000–1,500 ft., *Allen 6331.*

Eugenia sp.—Myrtaceae.

Slender trees, about 30 ft. in height, with rather small, opposite, lanceolate, acuminate leaves and small white fruits, which are produced in short axillary or pseudoterminal racemes in late February and early March.—Forests near Palmar Norte, 100 ft., *Allen 5979.*

Eugenia sp.—Myrtaceae.

Small trees, about 20 ft. in height with opposite, elliptic-lanceolate, acuminate leaves about 4½–5½″ long, the stems, petioles and veins of the lower leaf surface densely hirtellous, the lateral nerves of the blade elevated and conspicuous. The nearly globose fruits are about ¾″ in diameter, and are solitary in the leaf axils. Closely allied to *E. oreinoma* Berg, and possibly identical.—Forested hills near Palmar Norte, 1,000 ft., *Allen 6553.*

EUPHORBIACEAE—Key

1. Leaves peltate, deeply palmately lobed........*Ricinus*
1. Leaves not peltate.
   2. Individual leaflets usually less than 1½″ in length, the lateral branches having the superficial appearance of pinnate fronds........*Phyllanthus*

2. Individual leaflets more than 2″ long, the lateral branches never appearing as pinnate fronds.
   3. Leaves usually shallowly 3– 5-lobed. Small trees, usually found near house sites, or as living fence posts...*Jatropha*
   3. Leaves not lobed.
      4. Trunks intensely spiny. Fruits many-seeded, pumpkin-like in shape.........*Hura*
      4. Trunks not spiny. Fruits 1– 3-seeded, not pumpkin-like in shape.
         5. Leaf blades nearly as broad as long, broadly rounded at the apex...*Euphorbia*
         5. Leaf blades usually longer than broad, not rounded at the apex.
            6. Fruits black, less than ¼″ in diameter. Sap of lower trunk black..........................*Hieronyma*
            6. Fruits not black. Sap of trunk not black.
               7. Fruits green, crab-apple-like in superficial appearance but poisonous. Species confined to sea beaches..........................*Hippomane*
               7. Fruits not crab-apple-like. Trees not found on sea beaches.
                  8. Trunk and branches with abundant milky latex. Leaf petioles with a single pair of conspicuous glands near the base of the blade...*Sapium*
                  8. Plants without milky latex. Glands sometimes present, but never as a single pair.
                     9. Flowers in branching panicles..........................*Alchornea*
                     9. Flowers in spikes or racemes.
                        10. Staminate flowers without petals..........................*Acalypha*
                        10. Staminate flowers with petals..........................*Croton*

EUPHORBIA COTINIFOLIA L.—Euphorbiaceae—*Barrabas* (Costa Rica)—*Hierba mala* (Guatemala).

Shrubs or small trees, to about 15–18 ft. in height, with opposite or ternate leaves, the slender, often pink petioles about equaling the broadly ovate or sometimes nearly circular blades in length. The small, white flowers are produced in terminal cymes and are followed by 3-celled capsular fruits, the seeds being violently purgative. All parts of the plant have a caustic milky sap which will blister the skin, and can be used as a barbasco for catching fish. The species is usually found at about 3,000–5,000 ft., and is very commonly used for hedges and living fence posts in Guatemala.—*Tonduz 4501.*

EUTERPE PANAMENSIS Burret—Palmaceae.

Slender, unarmed, single-trunked palms 30–65 ft. in height, with attractive crowns of pinnate fronds. Common on the crests of forested hills between Palmar and Boruca at 2,500–4,000 ft. The inflorescences are borne directly below the crownshaft, the deep-purple or black, globose fruits averaging about ⅜″ in diameter.—Hills above Palmar Norte, 2,500 ft., *Moore 6553* & *Allen 6766.*

FAGACEAE—One genus, *Quercus.*

FARAMEA—Key

1. Leaves with 3 longitudinal nerves, the outer pair marginal. Blades conspicuously narrowed at the base, with a short but distinct petiole about ¼″ long........................ ....Faramea suerrensis
1. Leaves with a single longitudinal nerve. Blades rounded at the base, almost completely sessile, the petioles ⅛″ or less in length........................*Faramea sessifolia*

FARAMEA SESSIFOLIA P. H. Allen, sp. nov.—Rubiaceae.

Small, handsome trees, to about 15 ft. in height, with opposite, essentially sessile, glabrous leaves, the oblong or oblong-lanceolate, 1-nerved blades 7–9″ long and 1½–3″ wide, the apex abruptly and often falcately acuminate, the base rounded or subcordate in outline. The small, attractive blue flowers are produced in May and June in slender, long-pedunculate terminal panicles. The species differs from all the other blue-flowered members of the genus in the very distinctive sessile foliage.—Occasional on clay ridges in the Esquinas Forest, 200 ft., *Allen 5539* TYPE (Herbarium Escuela Agrícola Panamericana).

FARAMEA SUERRENSIS Donn. Sm.—Rubiaceae.

Shrubs or small trees to about 10 ft. in height, with grayish-green, opposite, linear, or narrowly oblanceolate leaves which are long-acuminate at the apex. The small, dark-blue flowers are produced in terminal cymes and are followed in late January and early February by the blue, obovoid fruits. Rather infrequent in hillside forests, but notable because of the blue color of the flowers which, excepting for the preceding species is almost completely absent in woody material from this level of the forest.—Esquinas Forest, 250 ft., *Allen 5802*.

FICUS—Key

1. Leaves very large, usually more than 10″ long. Fruits solitary.
   2. Leaf blades with about 10 pairs of lateral nerves........................*Ficus Tonduzii*
   2. Leaf blades with 20–25 pairs of lateral nerves........................*Ficus Werckleana*
1. Leaves usually less than 10″ long. Fruits in pairs.
   2. Leaf petioles and veins of the lower surface with a conspicuous reddish-brown velvety covering........................*Ficus Bullenei*
   2. Leaf petioles and veins not reddish brown or velvety.
      3. Leaf blades mostly broadest above the middle.
         4. Fruits sessile, about ⅜″ in diameter........................*Ficus costaricana*
         4. Fruits with a distinct peduncle: about ¼″ in diameter........................*Ficus Oerstediana*
      3. Leaf blades broadest at about the middle.
         4. Leaf blades glabrous beneath. Low, spreading trees, commonly grown for living fence posts........................*Ficus Goldmanii*
         4. Leaf blades pilose or puberulent beneath, sometimes becoming glabrate with age. Forest trees, never grown for living fence posts........................*Ficus lapathifolia*

FICUS BULLENEI I. M. Johnston—Moraceae.

Trees 30–75 ft. in height with milky latex. Leaves alternate, oblong or sometimes nearly obovate, rounded or subcordate at the base and shortly acute at the apex, the blades 5½″–9″ long and 3½–4½″ wide. The upper surface is rough to the touch, at least in dried specimens and the nerves of the lower surface, petioles, and branchlets have a dense and conspicuous reddish-brown tomentum. Mature specimens in pastures near Palmar Norte and in the Esquinas Forest are often as much as 30″ in trunk diameter and seem to be completely independent, yet younger plants have rather the aspect of a strangler epiphyte.—Pastures near Palmar Norte, 100 ft., *Allen 6638.*

FICUS COSTARICANA (Liebm.) Miq.—Moraceae—*Higueron* or *Higueron colorado* (Costa Rica)—*Higo* (Panama)—*Higuero* or *Higuillo* (Honduras) —*Amate, Cuxamate, Higo,* or *Matapalo* (Guatemala)—*Amate* (Salvador).

Forest trees of varying size, frequently starting as epiphytes, mature specimens sometimes reaching 100 ft. in height, often with a more or less cylindrical, hollow center representing the space occupied by the original host tree. Leaves alternate, oblong or oblanceolate, obtuse or acute at the apex, the small, nearly globose fruits being produced in the leaf axils in June.—Forests near Jalaca, 100 ft., *Allen 5309.*

FICUS GOLDMANII Standl.—Moraceae.

Spreading, usually flat-topped trees 12–25 ft. in height with alternate, oblong, obscurely acute or rounded glabrous leaves 3½–7″ long and 1¼–2½″ wide, all parts of the plant with abundant milky latex. Mature specimens usually have a great number of banyan-like interlocking trunks and reddish-brown bundles of fibrous adventitious roots which are pendent from the branches. The fruits are globose, and nearly 1″ in diameter, usually in pairs in the leaf axils. One of the most common and characteristic local trees, frequently used for living fence posts and for shade near Palmar and Palmar Norte.—Road to the Golfito Dairy, sea level, *Allen 6628.*

FICUS LAPATHIFOLIA (Liebm.) Miq.—Moraceae—*Higueron* (Costa Rica) —*Amate* or *Amate cusho* (Guatemala). (Pl. 5)

A medium-sized or large tree, to about 75 ft. in height, probably always starting as a strangler epiphyte. The leaves are alternate and broadly oblong, usually rounded at the apex and from 4 to 10″ in length, the blades pilose or puberulent beneath, sometimes becoming glabrate with age. Old specimens often develop interlocking trunks or conspicuous flying buttresses, particularly on hillsides.—Puerto Jiménez, *Brenes 12224.*

FICUS OERSTEDIANA Miq.—Moraceae—*Higuillo* (Honduras)—*Matapalo* (Guatemala).

Small or medium-sized trees, to about 45 ft. in height, with alternate, obovate or oblanceolate, acute or obtuse leaves 1¼–4½″ long and ⅝–1¾″ wide. The fruits are among the smallest known for any Central American species, averaging about ¼″ in diameter. The species is often, probably usually, epiphytic.—Puerto Jiménez, *Brenes 12141.*

FICUS TONDUZII Standl.—Moraceae—*Higueron* (Local)—*Chilamate* (Costa Rica)—*Wild fig, Higo,* or *Higuero* (Panama).

Medium-sized to large trees, 50–80 ft. in height, usually with a rounded crown of broad, leathery elliptic-oblong leaves, the blades with 8–10 prominent secondary nerves on the lower surface. The green, globose fruits are 1″ in diameter and mature in late April. The soft white wood is not used.—Frequent in lowland forests near Golfito and Palmar Norte, *Allen 5250.*

FICUS WERCKLEANA Rossberg—Moraceae.

Large, handsome, white-trunked trees, to about 75 ft. in height and 30″ in diameter above the rather prominent basal buttresses. The large leathery leaves are rounded at the base, and have 20–25 pairs of secondary nerves. Frequent in pastures and lowland forests throughout the area.—Santo Domingo de Osa, *Tonduz 9887*—Pastures near Palmar Sur, sea level, *Allen 6708.*

FLACOURTIACEAE—KEY

1. Flowers and fruits produced directly from the bark of the trunk and larger branches as in Cacao. Fruits with conspicuous longitudinal wings .... *Carpotroche*
1. Flowers and fruits not produced as above.
    2. Flowers red, in dense pendulous panicles. Leaves more than 1 ft. in length .... *Tetrathylacium*
    2. Flowers white or green. Leaves much less than 1 ft. long.
        3. Fruits spiny .... *Mayna*
        3. Fruits not spiny.
            4. Leaves with 3–5 longitudinal nerves .... *Hasseltia*
            4. Leaves with a single longitudinal nerve.
                5. Inflorescences elongate racemes, equaling or exceeding the leaves in length.
                    6. Individual flowers about ½″ in diameter. Tall trees, to about 100 ft. .... *Homalium*
                    6. Individual flowers minute, less than ⅛″ in diameter. Small trees, to about 20 ft. .... *Lozania*
                5. Inflorescences sessile fascicles or short racemes or panicles, always much shorter than the leaves.
                    6. Trunks armed with dense fascicles of long, branching, needle-like spines. Floral disk without staminodia-like appendages .... *Xylosma*

6. Trunks never armed with spines. Floral disk with staminodia-like appendages .... *Casearia*

Flor amarilla—See *Tecoma stans* (Nicaragua). Leaves pinnate.
Flor amarilla—See *Zexmenia frutescens* (Guatemala). Leaves simple.
Flor blanca or blanco—See *Plumeria rubra* forma *acutifolia* (Costa Rica and Salvador).
Flor de arco—See *Triplaris melaenodendron* (Nicaragua).
Flor de arito—See *Malvaviscus arboreus* var. *penduliflorus* (Salvador).
Flor de cangrejo—See *Hamelia patens* (Guatemala).
Flor de cruz—See *Randia armata* (Guatemala).
Flor de ensarta—See *Plumeria rubra* forma *acutifolia* (Salvador).
Flor de garrobo—See *Triplaris melaenodendron* (Nicaragua).
Flor de la cruz—See *Plumeria rubra* forma *acutifolia* (Guatemala).
Flor de Mayo—See *Plumeria rubra* forma *acutifolia* (Guatemala). Trees with milky juice. Flowers white.
Flor de Mayo—See *Vochysia ferruginea* (Panama). No milky juice. Flowers orange.
Flor de mico—See *Phyllocarpus septentrionalis* (Honduras and Guatemala).
Flor lila—See *Cornutia grandifolia* (Guatemala).
Fosforito—See *Protium copal* var. *glabrum* (Nicaragua). Fruits not tuberculate.
Fosforito—See *Trichilia tuberculata* (Chiriquí and Bocas del Toro). Fruits tuberculate.
Frangipani—See *Plumeria rubra* forma *acutifolia* (Local and general).
Fria—See *Dialium guianense* (Panama).
Friega plato—See *Solanum verbascifolium* var. *adulterinum* (Salvador and Honduras). Leaves with a single longitudinal nerve.
Friega platos—See *Miconia argentea* (Panama). Leaves with several longitudinal nerves.
Frijolillo—See *Albizzia longepedata* (Honduras). Flowers mimosa-like. Leaf margins entire.
Frijolillo—See *Astronium graveolens* (Honduras). Flowers not mimosa-like. Leaf margins serrate.
Frijolillo—See *Cassia spectabilis* (Honduras). Flowers not mimosa-like. Leaf margins entire.
Fruta de chancha—See *Croton glabellus* (Guatemala).

Fruta de mono—See *Posoqueria latifolia* (Costa Rica). Plant without yellow latex.

Fruta de mono—See *Rheedia madruno* (Panama). Plant with yellow latex.

Fruta de murcielago—See *Posoqueria latifolia* (Panama).

Fruta de pava—See *Ardisia revoluta* (Panama). Leaves alternate.

Fruta de pava—See *Chione costaricense* (Local). Leaves opposite.

Fruta dorada—See *Virola,* all species. (Local and Panama).

Fustan de vieja—See *Aspidosperma megalocarpon* (Guatemala).

Fustic—See *Chlorophora tinctoria* (Bocas del Toro).

Gallina—See *Phyllanthus acuminatus* (Costa Rica).

Gallinazo—See *Jacaranda copaia* (Costa Rica). Flowers blue.

Gallinazo—See *Schizolobium parahybum* (Local). Flowers yellow.

Gallito—See *Triplaris melaenodendron* (Salvador).

Gatillo—See *Ochroma lagopus* (Nicaragua).

Gallote—See *Trophis racemosa* (Panama).

Gargoran—See *Didymopanax Morototoni* (Panama).

Garroche—See *Quararibea guatemalteca* (Local).

Gavilan—See *Schizolobium parahybum* (Local and Nicaragua).

GENIPA CARUTO HBK—Rubiaceae—*Guaitil* (Costa Rica)—*Guaitil blanco, Guaytil blanco, Jagua, Jagua amarilla, Jagua blanca, Jagua colorado, Jagua de montaña,* or *Jagua negra* (Panama)—*Gigualti, Yigualti,* or *Tapaculo* (Nicaragua)—*Irayol, Tambor,* or *Tiñadientes* (Salvador)—*Genipap* (Honduras)—*Irayol* or *Irayol de loma* (Guatemala).

Small or medium-sized trees, to about 40 ft. in height, with opposite, rather leathery oblong or obovate leaves and terminal cymes of yellowish-white flowers, followed by nearly globose, fleshy fruits, which are frequently used by the Indians of Darien Province in Panama to provide a very dark-blue dye commonly seen as body paint during and after ceremonial dances. The wood is hard and strong, and is suitable for tool handles, interior construction, and door frames. A rather rare species of the delta of the Río Terraba, but to be expected as a common plant near Boruca and Buenos Aires.

Genipap—See *Genipa Caruto* (Honduras).

Genisero—See *Enterolobium cyclocarpum* (Costa Rica).

GEONOMA—KEY

1. Trunks solitary.......... *Geonoma binervia*
1. Trunks multiple.

2. Fruits about ½″ in diameter. Canes more than 1″ thick............... *Geonoma congesta*
2. Fruits about ⅛″ in diameter. Canes less than ½″ thick...*Geonoma* sp.—*Allen 6750*

GEONOMA BINERVIA Oerst.—Palmaceae—*Surtuba* (Costa Rica).

Slender, single-trunked palms, to about 15 ft. in height, with relatively large, irregularly pinnatisect fronds and twice-branched, pendulous, reddish or reddish-brown inflorescences.—Occasional in the forested hills above Palmar Norte, mostly at 1,500-2,500 ft., *Moore 6523, 6542, & 6546* & *Allen 6758.*

GEONOMA CONGESTA Wendl. ex Spruce—Palmaceae—*Caña de danta* (Local). (Pl. 4)

Handsome, erect, unarmed palms, with up to about 12–15 canes which average 2½″ in diameter, each bearing at its apex 10–12 pinnate, bifid fronds, the lateral pinnae usually fused into a few broad, confluent blocks, but sometimes much narrower. There are commonly 3–5 branching scapes borne just below the leafy crown, the small black fruits maturing in late March. The canes are sometimes used for house walls, and the fronds for thatch, which may be expected to last for approximately two years, according to report. A very common species of the clay hillsides in the Esquinas Forest at low elevations.—*Allen 6039, 6748, & 6753.*

GEONOMA sp.—Palmaceae.

Slender, stoloniferous palms with 3–5 canes about 6–10 ft. in height and ⅜″ in diameter, each with 6–7 bifid, pinnate fronds, usually composed of 3 pairs of very broad pinnae. Inflorescences 1 or 2 branching scapes about 7″ long, borne below the crown of leaves. The black, nearly globose fruits are about ⅛″ in diameter.—Esquinas Forest, 200 ft., *Moore 6538* & *Allen 6750.*

Gigualti—See *Genipa Caruto* (Nicaragua).

CLIRICIDIA SEPIUM (Jacq.) Steud.—Leguminosae; Papilionatae—*Madera negra* or *Madero negro* (Local)—*Madre de cacao* or *Sangre de drago* (Costa Rica)—*Bala, Balo, Madera negra,* or *Mata raton* (Panama)—*Madre de cacao, Cacagua, Madrial,* or *Madriado* (Honduras)—*Cacahuanance* (Salvador). (Pl. 19)

Small trees, 15 to about 40 ft. in height with alternate odd-pinnate leaves which are usually deciduous during the dry season. The attractive pink flowers are produced in dense axillary racemes in January and February, and are followed by linear, flat, dehiscent pods. The species is often planted in Central America as shade for cacao and the flowers are sometimes fried in egg batter and eaten. The leaves are relished

by cattle and their juice is a home remedy for baldness, but they are reputed to be poisonous to rodents and dogs. The ground bark and/or leaves are mixed with cooked grain and used to poison rats and mice. The heartwood is olive brown, or occasionally lighter in color, or with a purplish cast, individual pieces tending to darken on exposure to the air. The wood has a cucumber-like odor when freshly cut and is hard, heavy, tough, strong, rather coarse-textured, and with an irregular grain. It is sometimes difficult to work, but finishes smoothly and takes a high polish. It has a reputation for being extremely durable in contact with the ground, and is suitable for railroad ties, house posts, culverts, fence posts, and heavy durable construction. The trees are very commonly planted in our area for living fence posts and have the aspect of an introduced plant.—Buenos Aires, *Tonduz 4993*—Santo Domingo de Osa, *Tonduz 9890*.

GLOEOSPERMUM DIVERSIPETALUM Standl. & L. Wms.—Violaceae.

Small forest trees, to about 30 ft. in height, with alternate, elliptic-lanceolate, acuminate leaves which have obscurely serrate margins, particularly on the apical half of the blade. The small white flowers are produced in short axillary fascicles in late April and early May and are followed by 3-valvate, capsular fruits.—Lowland forests near Jalaca, 20 ft., *Allen 5278*.

Goatwood—See *Cassipourea podantha* (Panama).

GODMANIA AESCULIFOLIA (HBK) Standl.—Bignoniaceae—*Corteza de chivo* (Costa Rica)—*Cacaloguiste de flor quemada* (Nicaragua)—*Palo blanco, Palo de agua, Chorequillo,* or *Cortes* (Guatemala)—*Cortes blanco* (Salvador).

Small trees, 15–40 ft. in height, with opposite, long-stalked, digitately compound leaves having 5–9 obovate or oblong-lanceolate, acuminate leaflets and dense terminal panicles of greenish-yellow flowers which appear during the dry season, from January to about May, and are followed by long, slender, cylindrical capsules filled with papery, winged seeds. Known in our area only from the region of Buenos Aires.—*Tonduz 6714*.

GOETHALSIA MEIANTHA (Donn. Sm.) Burret—Tiliaceae—*Guacimo blanco* (Local and Chiriquí)—*Chancho blanco* (Costa Rica). (Pl. 24)

Tall, usually slender trees, 60–90 ft. in height and 16–24″ in trunk diameter above the prominently developed buttresses. The leaves are alternate, with elliptic-oblong, acuminate blades. The small, pale-yellow

flowers are produced in August and September in open terminal or axillary cymes, rendering the trees fairly conspicuous from a distance. The oblong fruits mature from November to January, and consist of 3 winged seeds which are fused together along the midrib. They are much relished by macaws, which congregate in gorgeous, noisy bands in the treetops during the fruiting season. The species often forms small, nearly pure stands on old clearings and on the margins of forest, but individual specimens are occasionally left standing in pastures, where they bear a considerable superficial resemblance to *Luehea Seemannii* ("Guacimo colorado") when not in flower or fruit. The white or grayish wood is light, soft, straight-grained, and easy to work, but tends to saw woolly, sapstains badly, and is not durable. Since it is abundant, it might be used for cheap boxes or paper pulp. One of our most common and characteristic local trees.—Pastures near Palmar Norte, 100 ft., *Allen 5604, 5665, & 5715.*

Gonzalo alves—See *Astronium graveolens* (U. S. trade name).
Gorgojillo—See *Miconia argentea* (Panama).
Gorgojo—See *Miconia argentea* (Panama).
Granadillo—See *Crataeva tapia* (Guatemala). Leaves digitately compound.
Granadillo—See *Platymiscium pinnatum* (Honduras). Leaves pinnate.
Granadillo macho—See *Crataeva tapia* (Salvador).

GRIAS FENDLERI Seem.—Lecythidaceae—*Tabaco* or *Tabacon* (Local)—*Haguey, Jaguey, Membrillo,* or *Sapo* (Chiriquí)— *Membrillo macho* (Panama)—*Irayol* or *Jaguillo* (Honduras).

Medium-sized forest trees, 40–60 ft. in height, with very large, drooping, oblanceolate or spatulate leaves 3–4 ft. in length, which are clustered at the ends of the stout branches. The small white flowers are produced in late March and early April in short, dense fascicles directly from the bark of the trunk and larger branches, and are followed in September by the 1-seeded indehiscent fruits. The trees are very conspicuous in late December, because of the flaccid terminal flushes of pale yellowish-green new leaves. The light, rather hard, yellowish wood is coarse-textured and is not used. A very common and characteristic species of the wet, lowland forests throughout our area.—Jalaca, 50 ft., *Allen 5216*—Lowland forests near Palmar Norte, 100 ft., *Allen 6547.*

Guabillo—See *Alibertia edulis* (Guatemala).
Guacal—See *Crescentia cujete* (Local, Nicaragua, and Salvador).

Guacamayo—See *Andira inermis* (Guatemala). Flowers lavender or purple, in terminal panicles.

Guacamayo—See *Phyllocarpus septentrionalis* (Honduras and Guatemala). Flowers red, in short, axillary clusters.

Guachipelin—See *Diphysa robinioides* (Local and Panama).

Guachipili—See *Albizzia longepedata* (Honduras).

Guachipilin—See *Diphysa robinioides* (Honduras and Guatemala).

Guacimo—See *Guazuma ulmifolia* (Local and Salvador). Fruits round in cross section.

Guacimo—See *Luehea Seemannii* (Panama, Costa Rica, and Honduras). Fruits star-shaped in cross section.

Guacimo blanco—See *Goethalsia meiantha* (Local and Chiriquí). Fruits winged.

Guacimo blanco—See *Guazuma ulmifolia* (Local). Fruits not winged.

Guacimo colorado—See *Luehea Seemannii* (Local, Nicaragua, and Honduras).

Guacimo de ternero—See *Guazuma ulmifolia* (Chiriquí).

Guacimo macho—See *Luehea Seemannii* (Local and Nicaragua).

Guacimo molenillo—See *Luehea Seemannii* (Panama).

Guacimo molinero—See *Luehea Seemannii* (Nicaragua).

Guacuco—See *Casearia aculeata* (Honduras).

Guagara—See *Cryosophila guagara* (Local).

Guaitil—See *Genipa Caruto* (Costa Rica). Leaf petioles about 1/4″ long.

Guaitil—See *Sickingia Maxonii* (Local). Leaf petioles more than 1/2″ long.

Guaitil blanco—See *Genipa Caruto* (Panama).

Guaitil colorado—See *Sickingia Maxonii* (Panama).

Guama—See *Inga punctata* (Honduras).

Guamo—See *Inga multijuga* (Honduras).

Guanacaste—See *Enterolobium cyclocarpum* (Local and Guatemala). Flowers white. Pods ear-shaped.

Guanacaste—See *Schizolobium parahybum* (Guatemala). Flowers yellow. Pods not ear-shaped.

Guano—See *Ochroma lagopus* (Honduras).

Guapinol negro—See *Cynometra hemitomophylla* (Local).

GUAREA—KEY

1. Leaf rachis winged.......... *Guarea aligera*
1. Leaf rachis not winged.
   2. Largest leaflets more than 10″ long. Fruits pear-shaped, usually more than 2″ long.......... *Guarea longipetiola*

2. Largest leaflets usually less than 8″ long. Fruits much less than 2″ long.
   3. Leaflets about three times as long as broad, essentially glabrous, the veins of the lower surface sometimes minutely puberulent........*Guarea Guara*
   3. Leaflets about twice as long as broad, the veins of the lower surface densely pilose........*Guarea Hoffmanniana*

GUAREA ALIGERA Harms—Meliaceae.

Slender trees of varying size, sometimes reaching as much as 75 ft., but often flowering and fruiting when little more than shrubs. The large, pinnate leaves are alternate on the branches, and have a conspicuously winged rachis, with 3–6 pairs of rather leathery, oblong-lanceolate, acute leaflets which are dark glossy green above and paler in color on the lower surface. The individual blades average about 6–9″ long and 2½–3½″ wide, but may sometimes exceed 1 ft. in length. The small flowers are produced in axillary or subterminal panicles, and are followed by dull-red, 1- or 2-celled, obcordate capsular fruits which enclose the scarlet seeds.—Occasional in the forested hills above Palmar Norte at about 1,000–1,800 ft. in elevation, *Allen 5951.*

GUAREA GUARA (Jacq.) P. Wilson—Meliaceae—*Carbonero* (Boruca) —*Manu* (Local)—*Cedro macho* (Panama).

Forest trees of varying size, to about 90 ft. in height, with alternate, pinnate leaves, the 8–20 elliptic to oblong leaflets about 3 times as long as broad and essentially glabrous, the veins of the lower surface sometimes minutely puberulent. The small white flowers are borne in axillary racemes and are followed in November and December by reddish-brown obovoid capsular fruits about ¾″ in diameter.—Esquinas Forest, 100 ft., *Allen 5377.*

GUAREA HOFFMANNIANA C. DC.—Meliaceae—*Manu* (Local).

Medium-sized forest trees 30–50 ft. in height and up to about 24″ in trunk diameter, with alternate, pinnate leaves, the 6–12 elliptic-oblong leaflets about twice as long as broad, with abruptly acuminate apices, the veins of the lower surface densely pilose. The pinkish-tan flowers are produced in short axillary racemes in late March and are followed by reddish-brown capsular fruits. The wood is suitable for furniture and interior cabinet work, but it is unfortunate that it should be known as "Manu," a name usually applied to *Minquartia guianensis.* It has been reported that railroad ties, either of this or some closely related species of *Guarea,* were sold to the United Fruit Company for use in the Coto District, where they lasted less than a year, giving the excellent and highly durable wood of *Minquartia* a bad name in the area.—Forests near Palmar Norte, 100 ft., *Allen 5501.*

GUAREA LONGIPETIOLA C. DC.—Meliaceae—*Cola de pava* or *Cortes negro* (Local)—*Chuchupate* (Chiriquí)—*Carbon* (Black River, Honduras).

Forest trees, 65–100 ft. in height and up to 24–30″ in trunk diameter, with large, alternate, even-pinnate leaves, the individual pinnae broadly oblong in outline with obtuse or shortly acute apices. The small, pale-yellow flowers are produced in September in slender axillary racemes, and are followed in February and March by the woody, reddish-brown obovoid, capsular fruits which average about 1½″ in diameter. The wood is reported to be used for the same purposes as mahogany, to which it is related, and has been considered suitable in Chiriquí Province for flooring, furniture, and general construction.—Esquinas Forest, 250 ft., *Allen 5804 & 6015*—Forested hills near Golfito, 100 ft., *Allen 6562.*

Guarumo—See *Cecropia* sp. (Local and general). Leaves peltate.
Guarumo—See *Pourouma aspera* (Costa Rica). Leaves not peltate.
Guarumo de montaña—See *Pourouma aspera* (Costa Rica, Honduras, and Guatemala).
Guarumo macho—See *Pourouma aspera* (Nicaragua).
Guastomate—See *Ardisia Dodgei* (Local).
Guatemalan flame tree—See *Phyllocarpus septentrionalis* (Honduras, Guatemala, and general).

GUATTERIA—KEY

1. Branches and lower leaf surfaces reddish brown, velvety .... *Guatteria aeruginosa*
1. Branches and lower leaf surfaces glabrous.
   2. Leaves usually less than 5″ long, gradually tapering to an acute apex. Individual carpels of the fruit produced on elongate stipes which average 1½″ in length .... *Guatteria lucens*
   2. Leaves more than 5″ long (usually much more). Stipes of the individual carpels of the fruits much less than 1½″ long.
      3. Leaves linear-lanceolate, with parallel margins .... *Guatteria chiriquiensis*
      3. Leaves elliptic-lanceolate, with strongly curving margins ...... *Guatteria amplifolia*

GUATTERIA AERUGINOSA Standl.—Annonaceae—*Malagueto* (Panama). (Pl. 25)

Trees 15–45 ft. in height and up to 18″ in trunk diameter, with alternate, narrowly oblong leaves 5–7″ long and about 2″ wide, with abruptly and shortly acuminate apices and very short petioles. The veins of the lower surface, and the branchlets are densely covered with a conspicuous reddish-brown tomentum. The small, fleshy flowers are solitary in the leaf axils, and are followed in February by globose heads of small

ellipsoid fruits.—Forested hills above Palmar Norte, 2,500 ft., *Allen 6734.*

GUATTERIA AMPLIFOLIA Tr. & Pl.—Annonaceae—*Anona* (Guatemala).

Shrubs or small trees, to about 20 ft. in height, with large, alternate, elliptic-oblong, shortly acute or acuminate leaves. The flowers are green, with a yellow center, and are produced from the trunk or larger branches. They are followed in June and July by globose heads of small, long-stipitate fruits. Locally common in hillside forests.—Esquinas Forest, 200 ft., *Skutch 5323*—Hills near Palmar Sur, 100 ft., *Allen 5570 & 6679.*

GUATTERIA CHIRIQUIENSIS R. E. Fries—Annonaceae—*Burio* or *Burillo* (Local)—*Malagueto* (Chiriquí). (Pl. 25)

Narrow-crowned forest trees 60–80 ft. in height and 18–24″ in trunk diameter, with alternate, linear, acuminate leaves which average about 9″ in length, and hang in a very distinctive manner, rather reminiscent of the cultivated Ylang-ylang *(Canangium)*, so that the trees can be recognized from a considerable distance. The fleshy green, fragrant flowers are produced from the leaf axils more or less throughout the year and are followed by small globose heads of green or black, stipitate fruits. The wood is apparently not used. One of the most common and characteristic local species.—Jalaca, sea level, *Allen 5217*—Margins of forest near Golfito, 100 ft., *Allen 5233 & 6704.*

GUATTERIA LUCENS Standl.—Annonaceae. (Pl. 25)

Forest trees, 30–65 ft. in height with alternate, elliptic-oblong or lanceolate, acute or acuminate leaves, which are dark glossy green on the upper surface, the solitary greenish flowers being produced in the axils, and followed in February in our area by open, long-stalked heads of yellow fruits. The hard white wood appears to be of good quality, but is not used.—Fairly frequent in the forested hills above Palmar Norte at 1,500–2,500 ft. elevation, *Allen 5906.*

Guava—See *Inga spectabilis* (Costa Rica).

Guava de Castilla—See *Inga spectabilis* (Costa Rica).

Guava de mono—See *Inga marginata* (Panama). Leaf rachis winged.

Guava de mono—See *Inga punctata* (Panama). Leaf rachis not winged.

Guava real—See *Inga spectabilis* (Costa Rica).

Guavo—See *Inga,* various species (Central America). Lower leaf petiole without a very large cup-shaped gland.

Guavo—See *Pithecolobium macradenium* (Local). Lower leaf petiole with a very large cup-shaped gland.

Guavo—See *Quassia amara* (Costa Rica). Small tree with pink flowers. Bark very bitter.

Guavo de montaña—See *Pithecolobium macradenium* (Local).

Guavo machete—See *Inga spectabilis* (Costa Rica).

Guayaba—See *Psidium Guajava* (Local and general).

Guayaba de gusano—See *Psidium Guajava* (Nicaragua).

Guayaba de monte—See *Alibertia edulis* (Guatemala).

Guayaba perulera—See *Psidium Guajava* (Nicaragua).

Guayabillo—See *Alibertia edulis* (Guatemala). Leaves opposite. Flowers not subtended by showy bracts.

Guayabillo—See *Calycophyllum candidissimum* (Guatemala). Leaves opposite. Flowers subtended by showy white bracts.

Guayabillo—See *Casearia* sp. (Guatemala). Leaves alternate. Fruits not winged.

Guayabillo—See *Terminalia amazonia* (Honduras and Guatemala). Leaves clustered at the ends of the branches. Fruits winged.

Guayabito—See *Triplaris melaenodendron* (Nicaragua).

Guayabo—See *Psidium Guajava* (Local and general). Fruits fleshy.

Guayabo—See *Terminalia* sp. (Honduras and Guatemala). Fruits winged, papery.

Guayabo del monte—See *Terminalia amazonia* (Local). Bark not guava-like.

Guayabo del monte—See *Terminalia lucida* (Local). Bark guava-like.

Guayabo de monte—See *Terminalia lucida* (U. S. trade name).

Guayabo de mico—See *Posoqueria latifolia* (Costa Rica).

Guayabo hormiguero—See *Triplaris melaenodendron* (Panama).

Guayabo volador—See *Terminalia amazonia* (Guatemala).

Guayabon—See *Terminalia lucida* (Local and Nicaragua).

Guayacan—See *Swartzia panamensis* (Costa Rica). Leaves pinnate. Flowers yellow.

Guayacan—See *Sweetia panamensis* (Local, Boruca). Leaves pinnate. Flowers white.

Guayacan—See *Tabebuia chrysantha* (Costa Rica). Leaves digitately compound. Flowers yellow.

Guaytil—See *Sickingia Maxonii* (Local).

Guaytil blanco—See *Genipa Caruto* (Panama).

GUAZUMA ULMIFOLIA Lam.—Sterculiaceae—*Guacimo* or *Guacimo blanco* (Local)—*Guacimo de ternero* (Chiriquí)—*Bastard cedar* (Bocas del Toro)—*Cabeza de negrito* (Fruits; Panama)—*Cablote, Caulote, Guacimo,* or *Tapaculo* (Honduras)—*Caca de mico, Chicharron,* or *Guacimo* (Salvador)—*Caulote, Contamal, Pixoy, Tapaculo,* or *Xuyuy* (Guatemala). (Pl. 23)

Common, medium-sized, round-topped trees 40–65 ft. in height and 12–24″ in trunk diameter, with alternate, oblong or ovate, acute or acuminate leaves which have serrate margins. The small, pale-yellow, very fragrant flowers are produced in April and May in small axillary cymes, and are followed in February and March of the next year by oval or nearly globose, black, woody, tuberculate capsules which are round in cross section. The pinkish wood is light, tough and strong, but coarse-textured and not durable. It is little used locally excepting for fuel, but is considered suitable for slack cask staves, tool handles, and gunstocks. It is reported to make an excellent charcoal, which was used in colonial times in Central America for the manufacture of gunpowder. The trees are often left standing in pastures since the leaves and fruits are relished by cattle. The mucilaginous sap is used to clarify syrup in making sugar, and the bark contains a tough fiber sometimes employed for rough cordage. A frequent species of pastures, fence rows, and cutover land throughout the area.—Palmar, *Tonduz 6725* & *Allen 6702*—Terraba Valley, 100 ft., *Pittier 11976* & *Allen 5282.*

GUETTARDA MACROSPERMA Donn. Sm.—Rubiaceae.

Shrubs or small to medium-sized trees, to about 40 ft. in height, with opposite, oblong or elliptic-lanceolate, acute or acuminate leaves which have slender petioles about half the leaf blade in length. The small white tubular flowers are produced in late August and early September in few-flowered axillary cymes, and are followed by subglobose fruits which are reported to be dark red at maturity.—In forest above rocky beaches near the delta of the Río Esquinas, Golfo Dulce, 50 ft., *Allen 5642.*

Guijarro—See *Stemmadenia Alfari* (Costa Rica).

GUILIELMA UTILIS Oerst.—Palmaceae—*Pejibaye* (Costa Rica)—*Pijibay* (Nicaragua)—*Supa* (Sumo, Nicaragua).

Handsome cultivated palms with pinnate fronds and 4–6 or more trunks 20–40 ft. in height, which are heavily armed with broad bands of long, dark-brown or black spines. The small yellow flowers are produced in dense racemes, and are followed by top-shaped red or yellow fruits. These are of a mealy texture and are sweet and edible when cooked, being often sold on the streets in San José and Cartago. The species is not definitely known in the wild, but is probably of South American origin, and has been grown in Costa Rica since pre-Columbian times. The hard black wood is suitable for fishing rods, canes, and archery bows, and the tender terminal "cabbage" of the crown is sweet and

edible. The species is planted throughout our area and is occasionally found along trails near old house sites.

Guiril—See *Sapindus saponaria* (Guatemala).
Guiro—See *Crescentia cujete* (British Honduras).
Gumbolimbo—See *Bursera simaruba* (Honduras).

GUSTAVIA ANGUSTIFOLIA Benth.—Lecythidaceae.

Small, usually unbranched trees, 9 to about 12 ft. in height, with very large, conspicuous, spatulate leaves which are clustered at the apex of the trunk or branches. The inflorescence is apparently terminal, and probably nearly sessile, but has been seen but once, as young fruiting material in the Esquinas Forest. A rather infrequent species, sometimes forming small local colonies of a few specimens in climax forest at low elevations.—Río Piedras Blancas, 100 ft., *Allen 6724*.

Gutigamba—See *Stemmadenia Donnell-Smithii* (Guatemala).

GUTTIFERAE—KEY

1. Flowers blood red. Lower trunk often with stilt roots.......*Symphonia*
1. Flowers white, yellow, or pale orange. Lower trunk never with stilt roots.
  2. Inflorescences terminal.
    3. Leaves brown or pale greenish tan on the lower surface. Sap turns red on exposure to the air.......*Vismia*
    3. Leaves green on the lower surface. Sap not red.......*Tovomitopsis*
  2. Inflorescences axillary.
    3. Leaves more than 1 ft. long, with many very prominent veins on the lower surface.......*Marila*
    3. Leaves much less than 1 ft. in length, the veins inconspicuous.
      4. Lateral veins of the leaves very fine, and so closely spaced that there is no interval between them.......*Calophyllum*
      4. Lateral veins of the leaves much more widely spaced, with a fine but definite cross reticulation in the interval between them.......*Rheedia*

Habillo—See *Hura crepitans* (Local and general).
Habin—See *Lonchocarpus guatemalensis* (Guatemala).
Haguey—See *Grias Fendleri* (Chiriquí).
Halal—See *Dracaena americana* (Guatemala).
Half crown—See *Mouriria parvifolia* (British Honduras).

HAMELIA—KEY

1. Leaf blades usually less than 4″ long, densely puberulent on the petioles and lower surface (lens!).......*Hamelia patens*
1. Leaf blades usually more than 6″ long, completely glabrous on the petioles and lower surface.......*Hamelia magnifolia*

HAMELIA MAGNIFOLIA Wernham—Rubiaceae—*Zorillo colorado* (Costa Rica).

Attractive shrubs or small trees, to about 18 ft. in height, with opposite, long-petiolate, broadly elliptic-oblong, shortly and abruptly acute or acuminate leaves, which have red veins and petioles. The slender yellow tubular flowers are produced in broad terminal cymes, and are followed by small red fruits. Occasional in forests and in old clearings.—Santo Domingo de Osa, *Tonduz 10091*—Margins of forest near Palmar Norte, 50 ft., *Allen 5259*—Forested hills above Palmar Norte, 2,500 ft., *Allen 5901*—Esquinas Forest, 50 ft., *Allen 6243.*

HAMELIA PATENS Jacq.—Rubiaceae—*Añileto, Azulillo, Coralillo, Palo camaron, Pissi, Zorillo,* or *Zorillo real* (Costa Rica)—*Red berry* or *Uvero* (Panama)—*Coralillo* or *Canilla de venado* (Nicaragua)—*Achiotillo colorado, Coral, Coralillo,* or *Coloradillo* (Honduras)—*Chichipince, Chichipinte, Coralillo,* or *Sisipince* (Salvador)—*Coralillo* (British Honduras)—*Canudo, Chamah, Chac-ixcanan, Chichipin, Clavito, Cuetillo, Flor de cangrejo, Hierba de erisipela, Hierba del cancer, Ixcanan, Ixcanan amarillo, Sicunken, Sisipince,* or *Xcanan* (Guatemala).

Shrubs or small trees, 6 to about 20 ft. in height, with verticillate clusters of 4 elliptic-lanceolate, short-acuminate leaves at each node. The small, orange-red, tubular flowers are produced in terminal cymes from about March until May, and are followed by small, dark-red, juicy edible fruits. Common along trails, sometimes forming small thickets.—Santo Domingo de Osa, *Tonduz 10052*—Valley of the Río Terraba, near La Presa, 50 ft., *Allen 5283.*

HAMPEA—KEY

1. Leaves with 5 distinct longitudinal nerves.................................*Hampea platanifolia*
1. Leaves with 3 longitudinal nerves.................................*Hampea Allenii*

HAMPEA ALLENII Standl. & L. Wms.—Bombacaceae—*Burio* (Local). (Pl. 18)

Small or medium-sized trees, 15–65 ft. in height, with alternate, long-petiolate leaves, the broadly ovate, acute blades paler in color on the lower surface and averaging about 7″ in length. The small but attractive white flowers have masses of yellow stamens, and are produced in many, short, few-flowered fascicles from the leaf axils in late August and early September, and are followed in November by the 3-celled, brown-puberulent capsular fruits which contain jet black seeds attached to a fleshy white aril. Common on the margins of forest, particularly along the road leading to the inner dairy pastures near Golfito.—*Allen 5618 & 6610.*

HAMPEA PLATANIFOLIA Standl.—Bombacaceae.

Trees 12–20 ft. in height, usually with multiple trunks. Leaves alternate, with long slender petioles, the very broad, palmately lobed blades about 9–14″ long and nearly as wide, with 5 radiating longitudinal nerves. Flowers white, in short axillary clusters, the individuals about 1¼–1½″ long, with strongly reflexed petals and attractive golden-yellow anthers. They are followed in late October and early November by ellipsoid, 3-celled capsular fruits which are densely covered with a whitish stellate pubescence. Common on cutover land, particularly in the Coto Valley.—Area between the Estero Azul and the Río Sierpe, sea level, *Allen & Chittenden 6599.*

Harino—See *Calycophyllum candidissimum* (Panama). Leaves simple.
Harino—See *Dipterodendron costaricense* (Panama). Leaves pinnate.

HASSELTIA—KEY

1. Leaves with 3 longitudinal veins .... *Hasseltia floribunda*
1. Leaves with 5 longitudinal veins .... *Hasseltia quinquenervia*

HASSELTIA FLORIBUNDA HBK—Flacourtiaceae—*Raspa lengua* (Panama)—*Muñeca* (Nicaragua).

Shrubs or small trees, to about 25 ft. in height, with alternate, short-petiolate leaves, the elliptic oblong blades with 3 longitudinal nerves and coarsely serrate margins. The small white flowers are produced in February and March in large terminal panicles, followed by small globose fruits. The creamy white wood is described as being fairly light, rather hard, with a fine, straight grain, and is easy to work, but has no local uses.—Boruca, 1,500 ft., *Tonduz 6864*—Diquis, 1,800 ft., *Pittier 12148*—Forested hills above Palmar Norte, 1,800 ft., *Allen 5956 & 6743.*

HASSELTIA QUINQUENERVIA Standl. & L. Wms.—Flacourtiaceae—*Quebracho blanco* (Local).

Medium-sized forest trees, 40–75 ft. in height, with alternate, long-petiolate leaves, the blades elliptic-oblong, with 5 longitudinal nerves and entire margins. The small, fragrant white flowers are somewhat variable in size, and are produced in large, often rather showy terminal panicles, rendering the trees conspicuous from a distance. The flowering season is from February to May, and the clusters of small red, attractive fruits mature in August and September. The species is occasional on the margins of forest, particularly along the Golfito airstrip and along the railroad line between Kilometer 5 and Kilometer 15. Quite a few young plants were being grown in the Golfito nursery in 1952 for decorative

use in the area.—Esquinas Forest, *Allen 5240*—Forests near Palmar Norte, *Allen 5509*—Forests near Golfito, 100 ft., *Allen 6247.*

Hediondilla—See *Solanum verbascifolium* (Guatemala).

HEDYOSMUM MEXICANUM Cordemoy—Chloranthaceae—*Vara blanca* (Costa Rica)—*Macetero, Ocze, Mazorco, Onc, Onj, Palo de agua, Sandio, Te azteco, Te de monte,* or *Te maya* (Guatemala).

Shrubs or small trees, 10 to about 30 ft. in height, with rather thick, very brittle branches and opposite, ovate to oblong-ovate, serrate leaves, which have a strong aromatic odor when crushed. The bases of the petioles are joined, and together form a tubular sheath enveloping the stem. The flowers are dioecious, the staminate form in catkin-like spikes and the pistillate in relatively short heads which become very fleshy at maturity, with many whitish, 3-angled drupaceous fruits which are edible. The leaves are reported by Standley to be used in Guatemala for the preparation of a tealike beverage. A very common species of wet highland forests, old specimens often having conspicuous stilt roots.—Cañas Gordas, 3,500 ft., *Pittier 11114 & 11113.*

HEISTERIA—KEY

1. Calyx circular, undivided. Fruits black at maturity........................*Heisteria longipes*
1. Calyx deeply lobed. Fruits white at maturity..................................*Heisteria concinna*

HEISTERIA CONCINNA Standl.—Olacaceae—*Naranjillo* (Costa Rica).

Small, bushy trees, 20–35 ft. in height, with alternate, elliptic-lanceolate, acute or acuminate, rather leathery leaves 2½–5″ long and 1–2½″ wide. The minute flowers are produced in dense sessile, axillary fascicles and are followed by white ellipsoidal drupes, which are subtended by a rather showy, deeply lobed red calyx.—Area between the Estero Azul and the Río Sierpe, sea level, *Allen & Chittenden 6598.*

HEISTERIA LONGIPES Standl.—Olacaceae—*Coloradito* (Local).

Small forest trees, 20–30 ft. in height, with alternate, elliptic-oblong, acute leaves which average 3 or 4″ in length. The round, thin calyx of the inconspicuous axillary flowers becomes much enlarged and red, and is persistent, subtending the solitary green or black drupaceous fruits. A common species, found throughout the area.—Forests near Palmar Norte, 100 ft., *Allen 5574.*

HELIOCARPUS APPENDICULATUS Turcz.—Tiliaceae—*Burio* (Costa Rica)—*Pestaño de mula* or *Balsa* (Nicaragua)—*Majao, Mecate de agua,* or *Damajao colorado* (Honduras)—*Cajeton* or *Majauha* (Guatemala). (Pl. 29)

Small or medium-sized trees 20–75 ft. in height, with alternate, long-petiolate leaves which are usually broadly ovate, or occasionally obscurely 3-lobed in outline. The small greenish-yellow flowers are produced in terminal cymes in December and January and are followed by conspicuous pink or red bristly fruits. The species is very common in the forested hills along the Golfito air field, and is quite showy in January during the fruiting season. The bark contains a tough fiber which is sometimes used for the manufacture of coarse twine or rope in northern Central America. The wood is very light, and the logs are sometimes used for rafts as a substitute for balsa.—Forested hills near Golfito, 50 ft., *Allen 6343 & 6691.*

HEMITELIA CHORICARPA Maxon—*Tree fern.*

One of the common species of tree ferns, averaging 8–12 ft. in height, found in the Esquinas Forest and elsewhere throughout the area.—*Skutch 5307.*

HENRIETTELLA TUBERCULOSA Donn. Sm.—Melastomaceae.

Shrubs or small trees, to about 18 ft. in height, with opposite, lanceolate, acuminate leaves and sessile fascicles of small white flowers, which are produced from the nodes of the leafless parts of the branches in late January and early February.—Esquinas Forest, 250 ft., *Allen 5821* & *Skutch 5400.*

HERNANDIACEAE—One genus, *Hernandia.*

HERNANDIA DIDYMANTHA Donn. Sm.— Hernandiaceae.

Tall forest trees, to about 100 ft. in height with alternate, long-petiolate leaves, the elliptic-oblong to ovate-oblong blades with acuminate apices and rounded bases. The small fragrant white flowers are subtended by many greenish-white involucral bracts, and are produced in late February in slender axillary cymes which equal or exceed the leaves in length. The fruits are enclosed in an inflated, fleshy saclike involucel. The soft white wood is very light, somewhat resembling balsa in general appearance and texture.—Occasional in lowland forests near the Tinoco Station, *Allen 5981.*

HERRANIA PURPUREA (Pittier) R. E. Schultes—Sterculiaceae—*Cacao del monte* (Local)—*Cacao de mico* or *Cacao de ardilla* (Costa Rica)—*Cacao mani* (Chiriquí)—*Cacao cimarron* or *Chocolatillo* (Panama)—*Wild cacao* (Bocas del Toro and Limón).

Slender shrubs or small trees, to about 12 ft. in height, with alternate,

long-petiolate, digitately 5-parted leaves, the very large obovate-lanceolate leaflets with acuminate apices. The dark-red, rather fleshy flowers are about $\frac{3}{4}''$ in diameter and are produced in small, sessile clusters directly from the bark of the lower trunk, the flowering season being in late February and early March. The yellow, longitudinally ridged pods average about 3″ in length, and are covered with minute translucent, mildly irritant hairs. The fruits mature from March until about May and are filled with a sweet, edible pulp. The seeds are reported by Pittier as having been used by the Bribri Indians for the preparation of a rather bitter drink. The species is often listed as *Theobroma purpureum* Pittier, but would seem to have ample characters to warrant segregation in a distinct genus. Occasional in wet woodlands throughout the area.—Palmar Norte, 100 ft., *Allen 5516.*

HIBISCUS TILIACEUS L.—Malvaceae—*Majagua* (Local)—*Algodoncillo* or *Majagua* (Panama)—*Mahoe* (Bocas del Toro)—*Majao* or *Majagua* (Honduras). (Pl. 16)

Shrubs or bushy trees, 12–30 ft. in height, common to sea beaches and the margins of mangrove swamps where they often form small, nearly pure stands. The alternate, long-petiolate leaves are broadly heart-shaped in outline, and are usually more or less whitish on the lower surface. The large, attractive flowers are produced during most of the year, and are at first rich yellow, turning reddish brown later in the day. The soft, white or sometimes purplish wood is light, firm, straight-grained, easy to work, and fairly durable, but is seldom used excepting as floats or as a substitute for cork. The bark contains a tough fiber which has been used for making rope, mats, hammocks, and coarse cloth, and has been found to compare favorably with jute for general utility. It has in addition the unusual quality of becoming stronger when wet. The species is common along sea beaches, particularly in the deltas of the Río Coto and Río Esquinas, and along the railroad line near Golfito.—*Allen 5640.*

Hierba del cancer—See *Hamelia patens* (Guatemala).
Hierba de erisipela—See *Hamelia patens* (Guatemala).
Hierba mala—See *Euphorbia continifolia* (Guatemala).

HIERONYMA TECTISSIMA Standl. & L. Wms.—Euphorbiaceae—*Platano, Pilon,* or *Zapatero* (Local). (Pl. 29)

Medium-sized or large forest trees 65–120 ft. in height and from 1–4 ft. in diameter above the moderately developed buttresses. The alternate,

elliptic-oblong, acute leaves have slender petioles, the blades averaging about 5″ in length in mature specimens, but considerably larger when the trees are young. The small, greenish-yellow, very fragrant, dioecious flowers are produced in axillary panicles, apparently at intervals throughout the year, and are followed by the small, black, juicy fruits, many of which seem, in spite of the erratic flowering season, to mature in February and March. The trunks have a characteristic black sap when cut, and the trees are often easy to spot from a distance since they nearly always have a few bright-red, senescent leaves scattered through the crown, a peculiarity also characteristic of the closely allied *Hieronyma alchorneoides,* which is known as "Curtidor" in Honduras. The hard, heavy, coarse-textured wood is pink or reddish brown in color and is used locally as a secondary timber for piling, fence posts, railroad ties, bridge boards, and miscellaneous heavy construction, but is considered inferior to "Nispero colorado" *(Pouteria heterodoxa)* or "Mora" *(Chlorophora),* since in an untreated state it is not very durable in contact with the ground and sometimes tends to split or warp badly in the process of seasoning. It has been used in Bocas del Toro for heavy beams and engine beds where protected from the weather. The Engineering department of the United Fruit Company in the Armuelles Division have found that railroad ties of this species can be heated and then soaked in a standard penta-diesel solution, and that individual ties will absorb 5 or 6 pounds of the preservative, with a life expectancy of 10–15 years as compared with 3–5 years for untreated material. The trees are very common in the hillside forests throughout the area, up to about 2,000 ft. in elevation.—Hills above Palmar Norte, 1,000 ft., *Allen 5865 & 6000* —Esquinas Forest, 100 ft., *Allen 6003.*

Higo—See *Ficus costaricana* (Local and general). Leaves less than 6″ long.

Higo—See *Ficus Tonduzii* (Panama). Leaves more than 6″ long.

Higuera—See *Oreopanax xalapense* (Costa Rica).

Higuerilla—See *Ricinus communis* (Costa Rica).

Higuerillo blanco—See *Ricinus communis* (Local). Green leaved form.

Higuerillo rojo—See *Ricinus communis* (Local). Red leaved form.

Higuero—See *Ficus costaricana* (Honduras). Leaves less than 6″ long.

Higuero—See *Ficus Tonduzii* (Panama). Leaves more than 6″ long.

Higuero—See *Ricinus communis* (Central America generally). Leaves peltate.

Higueron—See *Ficus costaricana* (Costa Rica). Receptacles sessile. Plants often epiphytic.

Higueron—See *Ficus lapathifolia* (Costa Rica). Receptacles pedunculate. Plants often epiphytic.

Higueron—See *Ficus Tonduzii* (Local). Large tree, never epiphytic.

Higueron colorado—See *Ficus costaricana* (Costa Rica).

Higuillo—See *Ficus costaricana* (Honduras). Fruits sessile.

Higuillo—See *Ficus Oerstediana* (Honduras). Fruits pedunculate.

HIPPOCASTANACEAE—One genus, *Billia.*

HIPPOMANE MANCINELLA L.—Euphorbiaceae—*Manzanillo* or *Manzanillo de la playa* (Local and general). (Pls. 7 & 16)

Attractive, gray-barked, round-topped trees, 40–60 ft. in height, common to sea beaches in Central America and the West Indies, often forming nearly pure stands. The glossy-green leaves are alternate, with slender petioles, the blades ovate or elliptic-ovate in outline, with acute or shortly acuminate apices and rounded bases. The small greenish flowers are produced in terminal spikes and are followed by small green or yellowish fruits about 1¼″ in diameter, which superficially resemble crab apples, but are poisonous. All parts of the plant contain a milky juice which is caustic and highly poisonous, causing severe swelling, burning, blistering, and inflammation on contact with the skin and temporary blindness if rubbed into the eyes. The yellowish-brown wood resembles walnut, and has been used since colonial times for the manufacture of fine furniture. The cutting of the trees and the working of the wood must however be undertaken with some caution, the common practice being to char the trunks before felling. The wood cannot be safely used for fuel, since the smoke is intensely irritant to the eyes. Occasional along sea beaches in the Golfo Dulce area.—Puerto Jiménez, *Tonduz 10049.*

HIRTELLA—KEY

1. Flowers purple. Branches and rachis of the inflorescence often with a conspicuous reddish tomentum. Leaves rounded or obtuse at the base................*Hirtella americana*
1. Flowers white, with purple stamens. Branches and rachis of the inflorescence usually not as above. Leaves acute at the base..*Hirtella triandra*

HIRTELLA AMERICANA L.—Rosaceae—*Pelo de indio* (Nicaragua)—*Pasta* (Honduras)—*Aceituno* or *Aceituno peludo* (Guatemala)—*Pigeon plum* or *Wild cocoplum* (British Honduras).

Shrubs or small trees to about 12 ft. in height, with alternate, elliptic-oblong, acute leaves. The small, rose-purple flowers are produced in simple racemes or narrow panicles and are followed by 1-seeded, obovoid,

drupaceous fruits.—The species is known in our area only from the vicinity of Buenos Aires, *Tonduz 3982, 3824, 6510, & 6708.*

HIRTELLA TRIANDRA Swartz—Rosaceae—*Camaroncillo, Carapato,* or *Chicharron* (Panama)—*Wild cocoplum* or *Wild pigeon plum* (British Honduras).

Shrubs or small trees, sometimes as much as 45 ft. in height, with alternate, lanceolate or elliptic-lanceolate leaves with acute or acuminate apices and wedge-shaped, very shortly petiolate bases. The small white flowers have prominent purple stamens and are produced in slender terminal racemes.—Esquinas Forest, *Skutch 5388.*

Hog plum—See *Spondias mombin* (Bocas del Toro and Limón).
Hoja blanca—See *Solanum verbascifolium* var. *adulterinum* (Honduras).
Hoja chigue—See *Curatella americana* (Costa Rica).
Hoja de duende—See *Bactris Baileyana* (Local).
Hoja de pasmo—See *Miconia impetiolaris* (Costa Rica).
Hoja de zope—See *Cornutia grandifolia* (Guatemala).
Hoja tinta—See *Trophis racemosa* (Honduras).
Holillo—See *Raphia taedigera* (Bluefields, Nicaragua).
Hom—See *Crescentia cujete* (Guatemala).

HOMALIUM EURYPETALUM Blake—Flacourtiaceae. (Pl. 29)

Forest trees, 30 to about 100 ft. in height and 1–2 ft. in trunk diameter above the moderately developed buttresses. The alternate, elliptic-lanceolate leaves are shortly acute at the apex and have obscurely crenate margins. The greenish-white flowers are produced in great profusion in slender axillary racemes or panicles, lasting from early March through most of April, the trees being fairly conspicuous when seen from a distance. The pale-brown wood is moderately hard, heavy, strong, and probably fairly durable, and would appear to be suitable for interior construction and cabinet work, but is not used locally. The species is frequent in swampy forests near Tinoco Station and the Laguna de Sierpe.—Coto, *Manuel Valerio 387*—Laguna de Sierpe, *Pittier 6817*—Tinoco Station, *Allen 6041.*

Hombre grande—See *Quassia amara* (Costa Rica).
Hombron—See *Quassia amara* (Costa Rica).
Hormigo—See *Platymiscium pinnatum* (Honduras). Leaves pinnate.
Hormigo—See *Triplaris melaenodendron* (Costa Rica and Guatemala). Leaves simple.

Horse cassia—See *Cassia grandis* (Honduras).

HUBERODENDRON ALLENII Standl. & L. Wms.—Bombacaceae—*Popon-joche* (?) (Local). (Fig. 13 and Pl. 5)

Giant forest trees, to about 160 ft. in height and 3–4 ft. in diameter above the tremendous plank buttresses, which often ascend the trunk for 40 or 50 ft. The alternate, long-petiolate leaves have ovate or oblong-elliptic, acute or shortly acuminate blades 3–5″ in length, and tend to be clustered at the ends of the branchlets. The brown, woody, 5-valvate capsules are about 9″ long, and are pendulous, resembling slender calabashes or sapotes when seen from a distance. These fruits mature in March and April and are filled with relatively large, brown, winged seeds which superficially resemble those of the Central American Mahogany *(Swietenia)*. The light, firm, pale-brown wood has no local uses. It is suspected that the local vernacular name may be a corruption of "Pumpunjuche" applied to the large-fruited *Pachira aquatica* in Nicaragua.—Occasional in the hillside forests, Esquinas Forest, 250 ft., *Allen 6014*.

Huesillo—See *Casearia arguta* (Costa Rica). Leaves simple.

Huesillo—See *Cupania guatemalensis* (Costa Rica). Leaves pinnate, with 6–10 leaflets.

Huesillo—See *Talisia nervosa* (Local). Leaves pinnate, with 10–14 leaflets.

Huesito—See *Cassipourea podantha* (Panama). Leaves simple.

Huesito—See *Cupania guatemalensis* (Salvador). Leaves pinnate. Inflorescences terminal.

Huesito—See *Trichilia acutanthera* (Chiriquí). Leaves pinnate. Inflorescences axillary.

Huevo de mono—See *Posoqueria latifolia* (Panama).

Huevos de caballo—See *Stemmadenia Donnell-Smithii* (Local).

Huevos de gato—See *Pterocarpus officinalis* (Panama).

Huevo de toro—See *Cymbopetalum costaricense* (Honduras).

Huiril—See *Sapindus saponaria* (Guatemala).

Huiscoyol—See *Bactris* sp. var.

Huitite—See *Xylosma excelsum* (Local).

Hule—Reported from the area, but not seen. Possibly *Castilla nicoyensis* O. F. Cook.

Hule blanco—See *Castilla fallax* (Costa Rica).

Hule macho—See *Castilla fallax* (Costa Rica).

Fig. 13. HUBERODENDRON ALLENII. 1 — Branch with leaves. 2 — Cross section of capsule, showing arrangement of seeds. 3 — Individual seed.

HUMIRIACEAE—Key

1. Leaves abruptly acute at the apex, the margins of the blades minutely serrate.......... ....*Saccoglottis*
1. Leaves rounded at the apex, the margins of the blades entire........................*Vantanea*

Hura crepitans L.—Euphorbiaceae—*Habillo* or *Jabillo* (Local and general)—*Javillo* (Costa Rica)—*Nune* (Chiriquí)—*Sandbox* or *White cedar* (Bocas del Toro and Limón)—*Jabillo, Javillo,* or *Tronador* (Panama)—*Possum-wood* (U. S. trade name). (Fig. 14)

Trees of varying size, often reaching 100 ft. or more in height and up to 6 ft. in trunk diameter, easily recognized by the pale-gray bark which is heavily armed with hard, sharp, conic spines. The leaves are alternate, with elongate, slender petioles, the broadly ovate blades with abruptly acuminate apices and rounded, subcordate bases which are characterized by 2 prominent glands. The male and female flowers are borne on separate inflorescences, the relatively small and individually inconspicuous staminate form on a long-stalked conical spike, while the dark-red, rather showy, star-shaped pistillate form is solitary in the leaf axils or at the base of the staminate inflorescence. There is some indication that the trees may flower and fruit at irregular times, but most of the flowers seem to be produced in October and November and are followed by the depressed-globose, many-celled, woody, pumpkin-like fruits, which average about 3″ in diameter and mature from November until about February. When fully ripe these capsules explode with a loud report, scattering the relatively large, oily seeds, which are much relished by monkeys and macaws. The watery, whitish sap of the tree is irritant and caustic, and will produce inflammation and blistering of the skin in susceptible individuals. It is often mixed with cornmeal or sand and used as a barbasco for catching fish. The seeds have an agreeable taste, but are violently purgative and may be poisonous if eaten in quantity, being used in Mexico to kill coyotes and other noxious animals. The light, rather soft but firm wood is variable in color from white to yellow, pale olive gray or brown. It is suitable for boxes, crates, concrete forms, rough carpentry, cheap furniture, plywood corestock, and probably paper pulp. Men engaged in felling the trees are liable to swelling of the face and hands because of the action of the acrid sap, and commonly remove a broad band of bark to reduce spattering. One of the most common and characteristic trees of the area, being found from sea level to about 2,000 ft. in elevation.—Boruca, *Tonduz 4802*—Esquinas Forest, 100 ft., *Allen 6609*—Pastures near Golfito, sea level, *Allen 6621.*

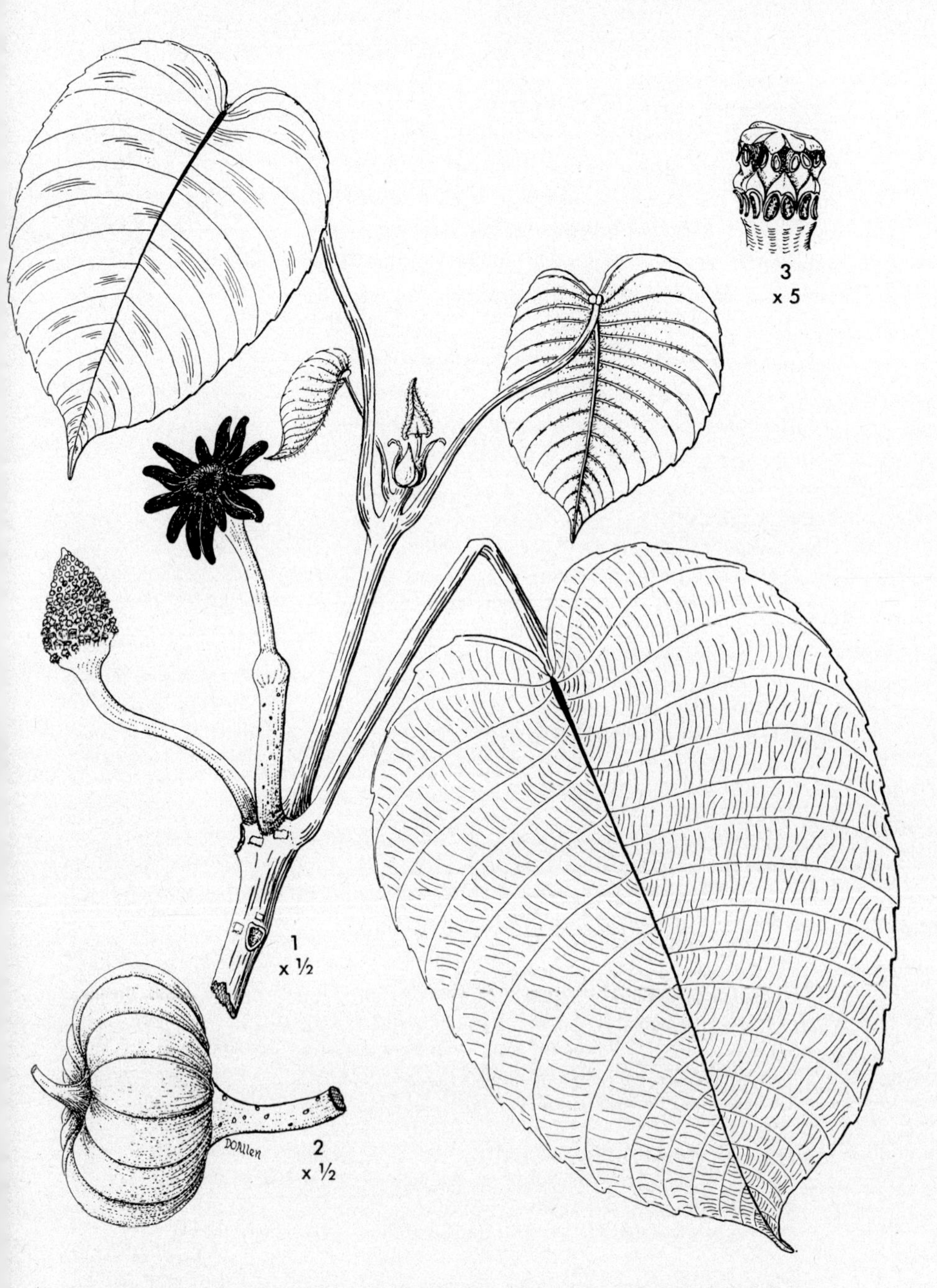

Fig. 14. Hura crepitans. 1 — Branch with staminate inflorescence (left) and pistillate flower. 2 — Mature fruit. 3 — Enlargement of stamen column, showing anthers.

HYOSPATHE LEHMANNII Burret—Palmaceae.

Unarmed, strongly stoloniferous, pinnate-leaved palms, with up to 20–25 slender green canes 1–2″ in diameter and 9–15 ft. in height. Spathes 2, the flowers pinkish in bud. Fruiting spadices borne below the crown of leaves, with seeds about ⅜″ long and 3/16″ wide.—Locally common in the forested hills above Palmar Norte at about 2,500 ft. in elevation, *Moore 6544* & *Allen 6764.*

Iguano—See *Dipterodendron costaricense* (Costa Rica).
Ilcaax—See *Dracaena americana* (Guatemala).
Indio—See *Schizolobium parahybum* (Panama).
Indio desnudo—See *Bursera simaruba* (Local and general).

INGA—KEY

1. Leaf rachis not winged.
   2. Leaflets 2–4 pairs.
      3. Flowers large, the tubular corolla (not including exserted stamens) more than ½″ in length.
         4. Leaflets essentially glabrous. Pods flat........ *Inga spectabilis*
         4. Leaflets pubescent, at least on the petioles and veins of the lower surface. Pods not flat........ *Inga sapindoides*
      3. Flowers small, the tubular corolla less than ½″ in length.
         4. Flowers in short, compact heads 1″ or less in length at the ends of the spikes.
            5. Leaflets usually 2 pairs........ *Inga punctata*
            5. Leaflets 3–4 pairs.
               6. Flowering heads less than ½″ in length. Leaflets usually 3 pairs........ *Inga quaternata*
               6. Flowering heads about 1″ long. Leaflets 4 pairs........ *Inga Ruiziana*
         4. Flowers distributed uniformly along the spike for 2 or 3″, never in short, compact heads.
            5. Leaflets usually less than 3″ long........ *Inga laurina*
            5. Leaflets about 6″ long........ *Inga coruscans*
   2. Leaflets 5–10 pairs.
      3. Leaflets and rachis usually with a minute but conspicuous rufous puberulence. Calyx usually more than ¼″ long........ *Inga multijuga*
      3. Leaflets and rachis not rufous (nearly glabrous). Calyx about ⅛″ long........ *Inga Ruiziana*
1. Leaf rachis winged, at least in part.
   2. Flowers in short, compact heads at the ends of the spikes.
      3. Calyx pubescent, less than ½″ long.
         4. Leaflets usually 2 pairs, the terminal leaflets usually more than 8″ long........ *Inga spectabilis*
         4. Leaflets usually 3 pairs, the terminal leaflets usually less than 6″ long........ *Inga sapindoides*
      3. Calyx glabrous, usually more than ½″ long........ *Inga portobellensis*

2. Flowers distributed uniformly along the spike for 3 or 4″, never in compact heads... *Inga marginata*

INGA CORUSCANS HBK—Leguminosae; Mimosoideae.

Trees about 60 ft. in height, with alternate, pinnate leaves which have a slender, terete rachis and 2 or 3 pairs of lanceolate or elliptic-lanceolate, acuminate, glabrous leaflets and flat, linear or slightly undulate pods 5-6″ in length, which are produced in December.—Frequent along the Río Esquinas near Kilometer 42, sea level, *Allen 5433.*

INGA LAURINA (Sw.) Willd.—Leguminosae; Mimosoideae—*Guavo* (Panama)—*Cujincuil* or *Paternillo* (Salvador)—*Palal* (Guatemala).

Trees 50–75 ft. in height with alternate, pinnate leaves. Leaflets usually 4 on a slender terete rachis. The small white flowers are produced on slender axillary spikes and are followed by slender flattened pods about 4–5″ long.—Esquinas Forest, 250 ft., *Allen 6345.*

INGA MARGINATA Willd.—Leguminosae; Mimosoideae—*Cuajiniquil* (Costa Rica)—*Guava de mono* (Panama)—*Sweetwood* (Bocas del Toro). (Fig. 15)

Medium-sized, often round-topped, spreading trees 18–45 ft. in height, with alternate, pinnate leaves, the rachis alate, the wings broader toward the apex and bearing a small, cuplike gland at the base of each of the 2 pairs of lanceolate, acuminate, glabrous leaflets. The small white flowers are produced twice a year in great profusion, each blooming season of several closely successive flowerings which coincide with the conspicuous brownish flushes of new leaves. These appear during heavy rains in May, and again in November and December, the flowers having a very pleasant, characteristic fragrance, rather reminiscent of that of sweet clover. The pods are linear, about 5″ in length, becoming thickened about the seeds at maturity. Fairly frequent in lowland forests throughout the area.—Pastures near Golfito, sea level, *Allen 6242*—Forest near Palmar Norte, 100 ft., *Allen 6338*—Tinoco Station, sea level, *Allen 6615.*

INGA MULTIJUGA Benth.—Leguminosae; Mimosoideae—*Guamo* (Honduras)—*Cuje* or *Nacaspiro* (Guatemala).

Trees 12–35 ft. in height, with alternate, pinnate leaves, the densely ferruginous-tomentulose rachis not winged, the 5–10 pairs of ovate or oblong-elliptic, acute or acuminate leaflets densely rough-pubescent on the upper surface and densely velutinous-pubescent below, each pair of leaflets having a small, sessile cupular gland at the base. The small whitish flowers are produced in short, pedunculate spikes and are followed by slender flat or subterete, densely brown-tomentulose pods, in which the edges are conspicuously thickened at maturity.—Puerto Jiménez,

Fig. 15. Inga marginata. 1 — Flowering branch. 2 — Seed pod.

Brenes 12164 A—Playa Blanca, *Manuel Valerio 469*—Boca Zacate, Delta del Diquis, *Tonduz 6793.*

INGA PORTOBELLENSIS Beurling—Leguminosae; Mimosoideae.

Medium-sized trees, with alternate, pinnate leaves, the winged rachis with an elevated cupular gland at the base of each of the two pairs of obovate-oblong, abruptly acuminate, glabrous leaflets. The showy flowers have very large calyxes and are produced in solitary pedunculate heads from the leaf axils, and are followed by the flat woody pods, which are thickened on the margins and average about 6″ in length.—Santo Domingo de Osa, *Tonduz 9879.*

INGA PUNCTATA Willd.—Leguminosae; Mimosoideae—*Cuajiniquil* or *Guavo* (Costa Rica)—*Bribri* or *Guavo* (Chiriquí and Bocas del Toro)—*Guava de mono* (Panama)—*Guavo* (Nicaragua)—*Cuajiniquil* or *Guama* (Honduras)—*Pepeto* (Salvador)—*Bitze, Caspiro, Cuajiniquil, Cuajiniquil blanco, Nacaspiro,* or *Ixcapirol* (Guatemala).

Medium-sized, round-topped trees 24–36 ft. in height, with alternate, pinnate leaves, the terete rachis with 2 unequal pairs of lanceolate, acuminate, nearly glabrous leaflets, the terminal pair 4 to about 9″ in length. The small fragrant white flowers are produced in open axillary panicles in late August, and are soon followed by the thin, flat, linear pods, which are about 6–7″ long, with somewhat thickened margins.—Frequent in swampy pastures throughout the Golfito Valley, *Allen 5620 & 6629*—Palmar, *Tonduz 6756.*

INGA QUATERNATA Poeppig—Leguminosae; Mimosoideae—*Bribri* (Bocas del Toro).

Small trees, 10–25 ft. in height, with rather large, alternate, pinnate leaves, the terete rachis with 3 pairs of oblong-elliptic, acute, nearly glabrous leaflets, each pair having a small gland at the base. The small whitish flowers are produced in terminal or subterminal paniculate umbels, followed by linear, oblong, flat, prominently margined pods which average about 6″ in length.—Santo Domingo de Golfo Dulce, *Tonduz 10032*—Tinoco Station, sea level, *Allen 6618.*

INGA RUIZIANA G. Don—Leguminosae; Mimosoideae—*Bribri* (Chiriqui and Bocas del Toro).

Rather slender, round-topped trees, 25–60 ft. in height, with large alternate, pinnate leaves, the terete rachis densely puberulent, with small shallow sessile cupular glands at the base of each of the 6 pairs of elliptic-lanceolate, acute or shortly acuminate leaflets which are nearly glabrous on the upper surface, the lower side having prominent reticulate veins. The rather large, attractive, white, fragrant flowers are pro-

duced in early February in short terminal or subterminal, congested, pedunculate heads, which are soon followed by the flat, more or less linear pods, which have thickened margins and average about 5–6″ in length.—Esquinas Forest, in the low wet area below the residence, *Allen 5842*—Pastures near Golfito, sea level, *Allen 6622 & 6631.*

INGA SAPINDOIDES Willd.—Leguminosae; Mimosoideae—*Bribri* or *Guavo* (Panama).

Slender trees, 25–50 ft. in height, with alternate, pinnate leaves, the rachis narrowly winged (sometimes only near the apex) and with 3 pairs of elliptic-lanceolate or oblong-lanceolate, acute or acuminate leaflets which are rufous-tomentulose on both surfaces. The rather large, attractive flowers are produced in axillary, pedunculate spikes and are at first white, becoming pale tan with age. They appear in late December and early January and are followed by thick, linear-oblong, 4-angled, glabrous pods which average 6–9″ in length. Occasional in pastures and on the margins of forest throughout the area.—Río Ceibo, near Buenos Aires, *Tonduz 4977*—Santo Domingo de Osa, *Tonduz 10030*—Palmar Norte, 100 ft., *Allen 5735.*

INGA SPECTABILIS (Vahl) Willd.—Leguminosae; Mimosoideae—*Guava, Guava de Castilla, Guava real,* or *Guavo machete* (Costa Rica)—*Bribri guavo* or *Monkey tambrin* (Panama).

Spreading trees with rounded crowns, usually 20–45 ft. in height, with large, alternate, pinnate leaves, the terete or narrowly winged rachis with 2 pairs of broadly elliptic leaflets. The flowers are relatively large, in dense heads at the ends of the terminal, spicate inflorescences, and are followed by very large, flat pods which are about 1 ft. in length. Planted in townsites and occasional in pastures.—Palmar Norte, 50 ft., *Allen 6710*—Buenos Aires, *Tonduz 3826*—Boruca, *Tonduz 4765.*

Inkwood—See *Sickingia Maxonii* (Limón).

Inup—See *Ceiba pentandra* (Guatemala).

Ira—See *Ocotea Ira* (Costa Rica).

Ira chiricana—See *Vantanea Barbourii* (Costa Rica).

Ira rosa—Probably *Virola* sp. (Local).

Irayol—See *Genipa Caruto* (Guatemala). Flowers and fruits not produced from the trunk.

Irayol—See *Grias Fendleri* (Honduras). Flowers and fruits produced from the trunk.

Irayol de loma—See *Genipa Caruto* (Guatemala).

IRIARTEA GIGANTEA Wendl. ex Burret—Palmaceae—*Stilt palm* (Local and general)—*Chonta negra* (Boruca). (Pl. 9)

Strikingly handsome, single-trunked, unarmed palms, 30–90 ft. in height, with low, compact clusters of stilt roots which seldom exceed 3–4 ft. in height and which may be completely lacking in juvenile specimens. Individual pinnae strongly wedge-shaped and broadest near the apex, those of juvenile fronds being all on one plane, while in mature plants they are produced in 2 ranks which are at about a 30° angle with one another on the rachis, giving the fronds a tousled appearance. The slender, pendulous, unopened spathes are hornlike in appearance and are borne on the trunk just below the crownshaft. The pale-yellow flowers are produced on elongate strands, those of the staminate form with about 14 stamens. The fruits are globose, about 1″ in diameter, with a lateral embryo. The wood in old specimens is black and exceedingly hard, like that of *Astrocaryum Standleyanum.* The trees occur as isolated specimens or in magnificent groves along the forest trail from Palmar Norte to Buenos Aires at about 2,500 ft. elevation.—*Moore 6524 & 6555* & *Allen 6745 & 6763.*

Ishtaten—See *Avicennia nitida* (Salvador).
Isna—See *Mouriria parvifolia* (Honduras).
Isote—See *Dracaena americana* (Honduras).
Istaten—See *Avicennia nitida* (Salvador).
Ixcanan—See *Hamelia patens* (Guatemala).
Ixcanan amarillo—See *Hamelia patens* (Guatemala).
Ixcapirol—See *Inga punctata* (Guatemala).
Ixec-subin—See *Lonchocarpus guatemalensis* (Guatemala).
Ixim-che—See *Casearia arguta* (Guatemala).

IXORA NICARAGUENSIS Standl.—Rubiaceae.

Shrubs or bushy trees to 20–30 ft. in height, with opposite, lanceolate or elliptic-lanceolate, acute or acuminate leaves 2–6″ long and 1–2″ wide. The small white flowers are produced in October in short terminal cymes or panicles and are followed by small red fruits. Locally very common, often one of the most characteristic understory species in wet lowland forest.—Area between the Estero Azul and the Río Sierpe, sea level, *Allen & Chittenden 6595.*

Izote—See *Dracaena americana* (Honduras).
Izote de montaña—See *Dracaena americana* (Guatemala).
Jabillo—See *Hura crepitans* (Local and general).

Jaboncillal—See *Sapindus saponaria* (Guatemala).

Jaboncillo—See *Sapindus saponaria* (Local and general).

JACARANDA—KEY

1. Inflorescences terminal. Secondary leaf petioles not winged............*Jacaranda copaia*
1. Inflorescences produced from the upper axils of the current new growth. Secondary leaf petioles narrowly winged............*Jacaranda lasiogyne*

JACARANDA COPAIA (Aubl.) D. Don—Bignoniaceae—*Gallinazo* (Costa Rica)—*Palo de buba* (Panama).

Tall, slender forest trees, with narrow, rounded crowns of large, opposite bipinnate leaves. The attractive dark-blue flowers are produced in large terminal panicles in March and April, rendering the trees very conspicuous from a distance. The species is apparently very rare in our area, two or three specimens having been seen in the hills near Golfito, one along a small stream near Kilometer 35 and one near the Esquinas rest pasture. An infusion of the bark is used for the treatment of skin diseases. The firm white wood is straight-grained, easy to work, and is suitable for boxes and interior carpentry.

JACARANDA LASIOGYNE Bur. & K. Schum.—Bignoniaceae.

Forest trees 65–90 ft. in height and 18-36″ in trunk diameter, with opposite, bipinnate leaves, the small, obliquely oblong, apiculate leaflets paler in color on the lower surface. The attractive dark-blue flowers are produced in early April in short subterminal panicles and are followed in January and February by oblong, flattened woody pods which have very strongly undulate margins. The nearly white, moderately hard, heavy wood is fairly straight-grained, strong, and easy to work, but is not durable in contact with the ground. It is not used in our area, probably due to the low-branching habit and often poor form of the trunks. Fairly frequent in hillside forests above Palmar Norte up to about 1,500 ft. in elevation.—Curres de Diquis, *Pittier 12050*—Santo Domingo de Osa, *Tonduz 7102*—Hills near Palmar Norte, *Allen 5793 & 6740*.

JACARATIA DOLICHAULA (D. Sm.) I. M. Johnston—Caricaceae—*Papayillo de venado* (Local)—*Papaya de monte, Papayillo,* or *Palo de barril* (Costa Rica)—*Wild cucumber* or *Papaya* (Panama).

Slender or corpulent, unarmed trees, 20–40 ft. in height, with alternate, 3- to 5-parted, digitately compound leaves which are deciduous during the dry season. The slender, tubular, white flowers are produced when the trees are bare of leaves, and are followed by fleshy fruits about the size of a lemon. The soft, spongy wood is not used. Bark cylinders, probably from this species, are reported by Standley to be used as bins for

storing corn and rice in Guanacaste.—Occasional in the hills near Palmar Norte, up to about 1,800 ft., *Allen 5927*.

Jagua—See *Genipa Caruto* (Panama).
Jagua amarilla—See *Chimarrhis latifolia* (Chiriquí). Flowers less than ½″ in diameter.
Jagua amarilla—See *Genipa Caruto* (Panama). Flowers more than ¾″ in diameter.
Jagua blanca—See *Genipa Caruto* (Panama).
Jagua colorado—See *Genipa Caruto* (Panama).
Jagua de montaña—See *Genipa Caruto* (Panama). Flowers more than ¾″ in diameter.
Jagua de montaña—See *Sickingia Maxonii* (Panama). Flowers less than ½″ in diameter.
Jagua negra—See *Genipa Caruto* (Panama).
Jaguey—See *Grias Fendleri* (Chiriquí).
Jaguillo—See *Grias Fendleri* (Honduras).
Jarina—See *Enterolobium cyclocarpum* (Costa Rica). Fruit a coiled, ear-shaped pod.
Jarino—See *Dipterodendron costaricense* (Costa Rica). Fruit a 2- to 3-celled capsule.
Jasmin cimarron—See *Randia armata* (Honduras).
Jasmin de arbol—See *Posoqueria latifolia* (Nicaragua).

JATROPHA CURCAS L.—Euphorbiaceae—*Coquillo, Coquito,* or *Tempate* (Costa Rica)—*Arbol santo* or *Piñon* (Honduras and Panama)—*Yupur* (Guatemala).

Shrubs or small trees, to about 15–18 ft. in height, with alternate, palmately lobed or broadly ovate leaves having 3–5 longitudinal nerves and slender petioles which about equal the blades in length. The greenish-yellow flowers are produced in slender terminal pedunculate cymes and are followed by 3-celled ellipsoid fruits about 1¼″ in length. The large oily seeds are strongly purgative and even poisonous if eaten in quantity. The plants are occasional in hedgerows and townsites throughout the area.

Javillo—See *Hura crepitans* (Costa Rica and Panama).
Jaya—See *Carpotroche platyptera* (Guatemala).
Jelinjoche—See *Pachira aquatica* (Costa Rica).
Jenequite—See *Bursera simaruba* (Honduras).
Jicarillo—See *Cochlospermum vitifolium* (Honduras). Leaves alternate.

Jicarillo—See *Posoqueria latifolia* (Nicaragua). Leaves opposite. Lateral branchlets not armed with spines.
Jicarillo—See *Randia armata* (Salvador). Leaves opposite. Lateral branchlets armed with short spines.
Jicara—See *Crescentia cujete* (British Honduras).
Jicaro—See *Crescentia cujete* (Local and general).
Jicaro de cuchara—See *Crescentia cujete* (Salvador).
Jicote—See *Bursera simaruba* (Honduras).
Jinicuite—See *Bursera simaruba* (Honduras).
Jiñocuavo—See *Bursera simaruba* (Local).
Jiñote—See *Bursera simaruba* (Costa Rica).
Jiote—See *Bursera simaruba* (Honduras and Guatemala).
Jira—See *Socratea durissima* (Panama).
Jobillo—See *Astronium graveolens* (Guatemala).
Jobito—See *Spondias purpurea* (Local).
Jobo—See *Licania arborea* (Salvador). Leaves simple.
Jobo—See *Spondias mombin* (Local and general). Large trees, fruits yellow.
Jobo—See *Spondias purpurea* (Panama). Small trees, fruits red.
Jobo jocote—See *Spondias mombin* (Guatemala).
Jobotillo—See *Phyllanthus acuminatus* (Panama).
Jocomico—See *Rheedia edulis* (Nicaragua).
Jocote—See *Spondias mombin* (Nicaragua). Large tree, fruits yellow.
Jocote—See *Spondias purpurea* (Local and general). Small tree, fruits red.
Jocote de jobo—See *Spondias mombin* (Nicaragua).
Jocote de mico—See *Simarouba glauca* (Guatemala).
Jocote jobo—See *Spondias mombin* (Guatemala and Nicaragua).
Jocote marañon—See *Anacardium occidentale* (Honduras and Guatemala).
Jocote montero—See *Spondias mombin* (Nicaragua).
Jocotillo—See *Trichilia acutanthera* (Salvador).
Joint bush—See *Miconia laevigata* (Bocas del Toro).
John crow wood—See *Cespedesia macrophylla* (Bocas del Toro).
Jolillo—See *Raphia taedigera* (Bluefields, Nicaragua).
Jolte—See *Miconia argentea* (Guatemala).
Jorco—See *Rheedia edulis* (Local). Fruits smooth, orange.
Jorco—See *Rheedia madruno* (Local). Fruits rough, bright yellow.
Ju—See *Persea americana* (Guatemala).
Juan de la verdad—See *Aegiphila martinicensis* (Panama).

Juche—See *Plumeria rubra* forma *acutifolia* (Costa Rica).

Juco—See *Trema micrantha* (Costa Rica).

Jucumico—See *Simarouba glauca* (Guatemala).

Jujul—See *Ochroma lagopus* (Guatemala).

Kenna—See *Mouriria parvifolia* (Panama).

Kib—See *Trema micrantha* (Guatemala).

Kicob—See *Theobroma Cacao* (Guatemala).

Kicou—See *Theobroma Cacao* (Guatemala).

Kiixche—See *Zanthoxylum procerum* (Guatemala).

Kinim—See *Spondias mombin* (Guatemala).

La—See *Urera caracasana* (Guatemala).

LACMELLEA PANAMENSIS (Woods.) Monachino—Apocynaceae—*Cerillo, Espinudo,* or *Lagarto negro* (Local).

Common forest trees of moderate size, averaging 40–65 ft. in height and 12–18″ in diameter, the trunk armed with many low, broad-based conical or subpyramidal spines. The somewhat leathery, opposite leaves are elliptic-oblong in outline, with short petioles and acute or acuminate apices, and are usually about 3–4″ in length. The slender, yellow, tubular, fragrant flowers are produced in few-flowered axillary cymes from April until about June and are followed in February of the next year by ovoid, orange, fleshy, 1- or 2-seeded fruits which are slightly over 1″ in diameter. They are reported to be edible. All parts of the tree contain an abundant white sticky latex which tastes like milk or cream flavored with sugar and vanilla extract. The *Apocynaceae* in general are rather notoriously poisonous, but the latex of several allied South American species of *Lacmellea* can be used for a sort of chicle, for adulterating rubber, and as a substitute for milk in tea and coffee. The present species is often listed in floras and handlists as *Zschokkea panamensis* Woodson. One of the most common and characteristic species of our hillside forests.—Hills above Palmar Norte, 250–1,800 ft., *Allen 5254 & 5935.*

LADENBERGIA—KEY

1. Corolla tube stout, about 1½ times as long as the expanded lobes........ *Ladenbergia Brenesii*
1. Corolla tube very long and slender, about 3–4 times as long as the expanded lobes........ *Ladenbergia chariantha*

LADENBERGIA BRENESII Standl.—Rubiaceae—*Quina* or *Agujilla* (Costa Rica). (Pl. 30)

Forest trees of very variable size and aspect, sometimes flowering when only about 15 ft. in height, but usually much larger, with mature speci-

mens about 90 ft., with very broadly elliptic, opposite, leathery leaves 6–10″ long and 3½–5½″ wide, the blades obtuse or subacute at the apex and rounded at the base, the veins and petioles red in color. The white flowers are produced in narrow terminal paniculate cymes. The species is rather frequent in the dense forests near the crests of the Cerros de Retinto, sometimes forming nearly pure stands.—Hills near Palmar Norte, 2,500 ft., *Allen & Chittenden 6735.*

LADENBERGIA CHARIANTHA Standl. & L. Wms.—Rubiaceae.

Slender forest trees, to about 90 ft. in height, with opposite, broadly elliptic or obovate, leathery leaves 8–12″ long and 4–6″ wide. The fragrant white flowers are borne in short terminal paniculate cymes, the very slender corolla tube about 3–4 times as long as the expanded lobes.—Esquinas Forest, 30 ft., *Skutch 5324.*

LAFOENSIA PUNICIFOLIA DC.—Lythraceae—*Palomo* (Local)—*Amarillo fruto* or *Amarillo de fruta* (Panama)—*Cuyapo* or *Trompillo* (Salvador) —*Palo de culebra* (Guatemala).

Medium-sized or large forest trees, up to about 80 ft. in height and 1 4½ ft. in diameter, with dark-brown, scaling bark. The leaves are opposite, with oblong-lanceolate, acuminate blades. The relatively large, campanulate, greenish-yellow flowers darken in color with age, and have a mass of conspicuous, spirally coiled, dark-red stamens and a long, prominently exserted pistil. They are produced in short, terminal racemes during the latter part of the rainy season from September until about November, and are followed by ovoid, capsular fruits about 1¼″ long, which mature their many-winged seeds in January and February. The strong, yellow, fine-grained wood is fairly hard, heavy and tough; works easily, finishes smoothly, takes a good polish, and is reported to be very durable. It is used in considerable quantity in the Canal Zone for cabinet work and interior boat work.—The species is occasional in the hillside forests near Palmar Norte, *Allen 5659 & 6550.*

Lagartillo—See *Alibertia edulis* (Panama). Leaves simple.
Lagartillo—See *Zanthoxylum procerum* (Costa Rica). Leaves pinnate.
Lagarto—See *Zanthoxylum procerum* (Guatemala).
Lagarto negro—See *Lacmellea panamensis* (Local).

LAGUNCULARIA RACEMOSA (L.) Gaertn.—Combretaceae—*Mangle marequita* (Local)—*Palo de sal* (Costa Rica)—*Mangle blanco* (Panama)—*Sincahuite* (Salvador)—*White mangrove* (British Honduras)—*Mangle blanco* (Honduras). (Pl. 20)

Common shrubs or trees of the mangrove swamps, to about 60 ft. in height and 30″ in diameter, the gray, furrowed trunks without stilt roots or the finger-like pneumatophores found in *Rhizophora* and *Avicennia.* The opposite, oblong-elliptic, leathery leaves are obtuse at the apex, the terete petiole with 2 large glands. The small greenish-white flowers are produced in May in clustered or paniculate spikes and are followed from July to about September by the oblong or obovoid, leathery, 10-ribbed fruits. The heartwood is variable in color from reddish to olive brown and is hard, heavy, tough, strong, and highly durable, but is difficult to work and is seldom used excepting for fuel. It is considered suitable for house and fence posts. The bark and dried leaves contain from 10 to 17% tannin, and might be exploited together with the bark of *Rhizophora.* —Delta of the Río Esquinas, on Golfo Dulce, *Allen 5627.*

Lana—See *Ochroma lagopus* (Panama to Guatemala).
Lancewood—See *Calycophyllum candidissimum* (U. S. trade name).
Lanilla—See *Ochroma lagopus* (Guatemala).

LAURACEAE—KEY

1. Leaves very large, usually more than 1 ft. in length.
 2. Veins of the lower leaf surface conspicuously ferrugineous-tomentose .... *Ocotea mollifolia*
 2. Veins of the lower leaf surface glabrous.
  3. Leaf blades broadest above the middle, the base conspicuously decurrent, with recurved margins .... *Ocotea rivularis*
  3. Leaf blades broadest at about the middle, the base rounded, never decurrent or with recurved margins .... *Ocotea nicaraguensis*
1. Leaves less than 1 ft. in length.
 2. Leaf blades with 3 conspicuous longitudinal nerves .... *Phoebe costaricana*
 2. Leaf blades with a single longitudinal nerve.
  3. Leaf blades strongly decurrent at the base.
   4. Leaves oblanceolate, more than 7″ long .... *Ocotea Ira*
   4. Leaves lanceolate, less than 6″ long .... *Ocotea Williamsii*
  3. Leaf blades rounded or cuneate at the base.
   4. Leaf blades narrowly linear-lanceolate, with parallel margins, usually more than 8″ long .... *Ocotea pergamentacea*
   4. Leaf blades never with parallel margins, mostly less than 8″ long.
    5. Floral pedicels with a conspicuous pair of linear bracts .... *Persea pallida*
    5. Floral pedicels without bracts.
     6. Leaf petioles at least ¼ the length of the blades.
      7. Fruits about ¼″ in diameter .... *Persea Skutchii*
      7. Fruits 3″ or more in diameter .... *Persea americana*
     6. Leaf petioles much less than ¼ the length of the blades.
      7. Axils of the veins of the lower leaf surface with conspicuous tufts of hairs .... *Nectandra perdubia*

7. Axils of the veins of the lower leaf surface without tufts of hairs.
  8. Leaf blades mostly obtuse or obtusely acute at the apex..*Ocotea veraguensis*
  8. Leaf blades mostly acute to caudate-acuminate at the apex.
    9. Fruits nearly globose, subtended by a very shallow cupule..*Nectandra latifolia*
    9. Fruits ellipsoidal or conical, subtended by a deep, acorn-like cupule.
      10. Cupules with double margins.
        11. Lower margin of cupule conspicuously flaring and undulate. Apex of fruits flattened. Flowers yellow...*Licaria Cufodontisii*
        11. Lower margin of the cupule not flaring or undulate. Apex of the fruits conical. Flowers white..............*Licaria Cervantesii*
      10. Cupules with simple margins.
        11. Leaf blades mostly narrowly lanceolate, tapering gradually to an acuminate tip. Flower buds glabrescent. Flowering season in December...............................*Nectandra salicifolia*
        11. Leaf blades mostly elliptic-lanceolate, abruptly acuminate. Flower buds densely pilose. Flowering season in April...*Nectandra nervosa*

Laurel—See *Cordia alliodora* (Local and general). Forks of twigs with hollow, gall-like swellings.

Laurel—See *Nectandra salicifolia* (British Honduras). Forks of twigs without gall-like swellings.

Laurel blanco—See *Cordia alliodora* (Panama to Guatemala).

Laurel macho—See *Cordia alliodora* (Nicaragua).

Laurel negro—See *Cordia alliodora* (Costa Rica and Nicaragua). Flowers less than ½″ in diameter.

Laurel negro—See *Cordia gerascanthus* (Honduras). Flowers more than ¾″ in diameter.

Laureño—See *Cassia reticulata* (Panama).

Lech—See *Calophyllum braziliense* var. *Rekoi* (Guatemala).

Leche—See *Calophyllum braziliense* var. *Rekoi* (Guatemala). Leaf petiole without glands.

Leche—See *Sapium jamaicense* (Local). Leaf petiole with 2 small glands.

Leche amarilla—See *Calophyllum braziliense* var. *Rekoi* (Guatemala). Flowers white.

Leche amarilla—See *Symphonia globulifera* (Honduras and Guatemala). Flowers red.

Lechero—See *Brosimum utile* (Costa Rica).

Lechosa—See *Trophis racemosa* (Bocas del Toro).

LECYTHIDACEAE—Key

1. Leaves very large, usually more than 2 ft. in length, conspicuously clustered at the ends of the branches.
   2. Flowers and fruits produced directly from the bark of the trunk and larger branches. Leaves conspicuously drooping .... *Grias*
   2. Flowers and fruits produced from the ends of the branches. Leaves erect or arching, never conspicuously drooping .... *Gustavia*
1. Leaves usually less than 1 ft. in length, never conspicuously clustered at the ends of the branches.
   2. Flowers pale yellow. Leaves abruptly caudate-acuminate at the apex ... *Eschweilera*
   2. Flowers pink. Leaves broadly rounded or abruptly and very shortly acute at the apex .... *Couratari*

LEGUMINOSAE—Key

1. Trunk or branches armed with thorns or spines.
   2. Leaflets 3. Thorns not in pairs. Flowers red .... *Erythrina costaricensis*
   2. Leaflets many. Thorns or spines in pairs. Flowers yellow.
      3. Thorns large and conspicuous, hollow, usually inhabited by stinging ants.
         4. Thorns white or pale yellow .... *Acacia spadicigera*
         4. Thorns dark brown or black.
            5. Flowers in slender spikes .... *Acacia costaricensis*
            5. Flowers in globose heads .... *Acacia melanoceras*
      3. Thorns small and inconspicuous, not hollow or inhabited by ants .... *Acacia farnesiana*
1. Trunk or branches not armed with thorns or spines.
   2. Leaves simple.
      3. Leaves deeply bilobed .... *Bauhinia ungulata*
      3. Leaves not bilobed .... *Swartzia simplex*
   2. Leaves compound, composed of 2 or more distinct leaflets.
      3. Leaves once pinnate (having a central axis and 2 or more leaflets).
         4. Pinnate leaves opposite on the branches .... *Platymiscium pinnatum*
         4. Pinnate leaves alternate on the branches.
            5. Leaflets less than 5 in number.
               6. Leaflets a single pair.
                  7. Leaflets more than 6″ long. Flowers red. Small trees, to about 20 ft. in height .... *Calliandra*
                  7. Leaflets less than 6″ long. Flowers not red. Tall trees, to 120 ft.
                     8. Leaflets less than 3″ long. Wood purple. Seed pods conspicuously flattened .... *Peltogyne purpurea*
                     8. Leaflets more than 3″ long. Wood brown. Pods not flattened .... *Cynometra hemitomophylla*
               6. Leaflets usually 3 or 4 in number.
                  7. Leaflets 3.
                     8. Flowers red, more than 2″ long. Seeds red ..... *Erythrina costaricensis*
                     8. Flowers not red, much less than 2″ long. Seeds not red.
                        9. Flowers whitish, in elongate spikes .... *Inga laurina*

9. Flowers pink, in short, sessile fascicles...*Pithecolobium longifolium*
7. Leaflets 4.
8. Leaf rachis with a small gland at the base of at least the terminal pair of leaflets. Seed pods long and narrow, with nearly parallel margins.
9. Terminal pairs of leaflets more than 8″ long.
10. Leaf rachis conspicuously winged...*Inga portobellensis*
10. Leaf rachis not winged...*Inga spectabilis*
9. Terminal pair of leaflets less than 8″ long.
10. Flowers pink, in short, sessile clusters. Trees confined to river banks...*Pithecolobium longifolium*
10. Flowers white, in elongate spikes.
11. Leaf petiole narrowly winged...*Inga marginata*
11. Leaf petiole not winged.
12. Terminal flowering portion of the spike usually short and compact, less than ¾″ long...*Inga punctata*
12. Flowering portion of spike more than 2″ long.
13. Leaflets usually less than 3″ long...*Inga laurina*
13. Leaflets about 6″ long...*Inga coruscans*
8. Leaf rachis without a gland at the base of the pairs of leaflets. Seed pods not long and narrow.
9. Flowers in terminal or axillary spikes. Leaves mostly acuminate at the apex. Fruits more than 7″ long...*Mora oleifera*
9. Flowers in terminal spreading paniculate racemes. Leaves mostly broadly acute at the apex. Fruits less than 4″ long...*Prioria copaifera*
5. Leaflets 5 or more.
6. Leaf rachis winged, at least in part.
7. Leaflets 6. Flowers white...*Inga sapindoides*
7. Leaflets 5. Flowers yellow...*Swartzia picramnoides*
6. Leaf rachis not winged.
7. Flowers mimosa-like. Leaf rachis with a small gland or pair of glands at the base of at least the terminal pair of leaflets.
8. Flowers white, in conspicuous globose heads at the ends of slender pendulous scapes which equal the leaves in length...*Pithecolobium glanduligerum*
8. Flowers not in globose heads, not pendulous, the scapes not equaling the leaves in length.
9. Leaflets 3 pairs or less.
10. Flowering portion of the spike short and condensed, usually less than 1″ long...*Inga quaternata*
10. Flowering portion of the spike more than 2″ long.
11. Terminal pair of leaflets up to about 4½″ long...*Inga laurina*
11. Terminal pair of leaflets about 6″ long...*Inga coruscans*
9. Leaflets 4 to 10 pairs.
10. Calyx usually more than ¼″ long...*Inga multijuga*
10. Calyx usually about ⅛″ long...*Inga Ruiziana*

7. Flowers not mimosa-like. Leaf rachis without glands at the base of the leaflets.
  8. Inflorescences terminal.
    9. Leaves even-pinnate, with 2 terminal leaflets.
      10. Terminal leaflets more than 5″ long.
        11. Flowers bright yellow. Small bushy trees, to about 18 ft., common to pastures and house sites............*Cassia reticulata*
        11. Flowers pinkish lavender. Tall forest trees, to about 140 ft.... ....*Tachigalia versicolor*
      10. Terminal leaflets less than 3″ long.....................*Cassia spectabilis*
    9. Leaves odd-pinnate, with a single terminal leaflet.
      10. Leaflets subacute and distinctly notched at the apex................ ....*Sweetia panamensis*
      10. Leaflets acute or acuminate at the apex, never notched.
        11. Flowers olive green or white.
          12. Flowers olive green. Fruits not winged............................ ....*Dialium guianense*
          12. Flowers white. Fruits winged............*Myroxylon balsamum*
        11. Flowers yellow, pink, lavender, or purple.
          12. Flowers yellow.
            13. Outer surface of calyx densely pubescent. Fruits very thin, broadly winged........................*Pterocarpus Hayesii*
            13. Outer surface of the calyx nearly glabrous. Fruits thick, narrowly winged...............*Pterocarpus officinalis*
          12. Flowers not yellow.
            13. Flowers pink.
              14. Calyx very deeply lobed. Flowers not showy from a distance ....*Dussia macrophyllata*
              14. Calyx shallowly lobed. Flowers very showy from a distance.......................................*Dussia mexicana*
            13. Flowers lavender or purple.
              14. Calyx margins deeply lobed. Seeds bright red and black....................................*Ormosia panamensis*
              14. Calyx margins shallowly lobed or entire. Seeds not red and black.
                15. Leaflets broadly elliptic-oblong, abruptly and shortly acute at the apex. Sap dries blood red.........*Lonchocarpus sericeus* var. *glabrescens*
                15. Leaflets narrowly linear-lanceolate, abruptly acuminate at the apex. Sap does not dry red ....*Andira inermis*
  8. Inflorescences axillary.
    9. Mature leaflets less than 3″ long.
      10. Mature leaflets broadly rounded at the apex.
        11. Flowers pink. Pods dark brown, woody, up to about 1 ft. in length..*Cassia grandis*
        11. Flowers yellow. Pods pale tan, papery, about 3″ long......... ....*Diphysa robinioides*

10. Mature leaflets acute or acuminate at the apex.
  11. Flowers blood red........................*Phyllocarpus septentrionalis*
  11. Flowers pink or purple.
    12. Flowers pale pink. Trees planted for living fence posts... ....*Gliricidia sepium*
    12. Flowers purple. Trees not planted for fence posts........... ....*Lonchocarpus minimiflorus*

9. Mature leaflets more than 3″ long.
  10. Leaflets broadly rounded at the apex. Terminal leaflets 2 in number..*Cassia reticulata*
  10. Leaflets subacute to acuminate at the apex. Terminal leaflet 1.
    11. Flowers yellow or orange.
      12. Flowers in elongate, pendulous racemes up to 16″ in length. Fruits very large, opening at maturity and containing several seeds...........................*Swartzia panamensis*
      12. Flowers in relatively short, erect racemes. Fruits single-seeded, not opening at maturity.
        13. Outer surface of the calyx densely brown-tomentose. Fruits thin, broadly winged.........*Pterocarpus Hayesii*
        13. Outer surface of the calyx nearly glabrous. Fruits conspicuously thickened, the margins narrowly winged..*Pterocarpus officinalis*
    11. Flowers pink, lavender, or purple.
      12. Floral calyx deeply lobed, or with conspicuous marginal teeth. Leaves pubescent or tomentulose on the lower surface, at least on the veins. Fruits fleshy.
        13. Calyx divided nearly to the base. Flowers pinkish lavender, not showy.....................*Dussia macrophyllata*
        13. Calyx tubular, with short, distinct marginal teeth. Flowers pink, showy...........................*Dussia mexicana*
      12. Floral calyx not deeply lobed (margins entire or very obscurely lobed). Leaves essentially glabrous. Fruits not fleshy.
        13. Leaves minutely puberulent on the lower surface. Sap turns blood red when exposed to the air........... ....*Lonchocarpus sericeus* var. *glabrescens*
        13. Leaves completely glabrous. Sap not red.
          14. Fruits conspicuously thickened along one margin.................................*Lonchocarpus guatemalensis*
          14. Margins of the fruits not thickened...................... ....*Lonchocarpus latifolius*

3. Leaves bipinnate (having a central nonwoody axis with several lateral pinnae).
  4. Leaves with a single pair of often unequal pinnae, each composed of 2–3 leaflets. Common to rocky river banks.......................*Pithecolobium longifolium*
  4. Leaves with several pairs of equal pinnae, each with more than 3 leaflets.
    5. Individual leaflets more than ¾″ long.

6. Lower leaf rachis with a very large, conspicuous, vase-shaped gland...... .... *Pithecolobium macradenium*
6. Leaf rachis without a vase-shaped gland.
   7. Individual leaflets rounded at the apex.
      8. Leaflets narrowly oblong, with parallel margins. Flowers bright yellow........ *Schizolobium parahybum*
      8. Leaflets broadly elliptic, with curving margins. Flowers pale tan........ .... *Albizzia longepedata*
   7. Individual leaflets acute at the apex. Flowers white, in showy globose heads........ *Pithecolobium glanduligerum*

5. Individual leaflets minute, less than ½″ long.
6. Leaflets with a small, pale, conspicuous gland at the base. Mature fruits red, terete........ *Pithecolobium austrinum*
6. Leaflets without a conspicuous gland at the base. Mature fruits not red or terete.
   7. Fruits linear, with parallel margins.
      8. Pods about 1½″ wide. Tall trees of the hillside forests........ .... *Lysiloma guanacastense*
      8. Pods less than 1″ wide. Small or medium-sized trees, usually confined to stream courses........ *Albizzia filicina*
   7. Fruits coiled to form a circular, earlike pod........ .... *Enterolobium cyclocarpum*

Lemonwood—See *Calycophyllum candidissimum* (U. S. trade name).
Lengua de buey—See *Vernonia patens* (Panama).
Lengua de vaca—See *Curatella americana* (Salvador). Leaves oval or elliptic, with undulate margins.
Lengua de vaca—See *Vernonia patens* (Panama). Leaves lanceolate, often rugose.
Lian—See *Croton glabellus* (Honduras).

LICANIA—Key

1. Leaves broadly rounded at the apex........ *Licania arborea*
1. Leaves acute at the apex........ *Licania operculipetala*

Licania arborea Seem.—Rosaceae—*Alcornoque* or *Roble blanco* (Costa Rica)—*Raspa* (Panama)—*Roble* (Nicaragua)—*Canilla de mula* or *Jobo* (Salvador)—*Caca de niño, Encino, Roble,* or *Zuncilla* (Guatemala).

Medium-sized or sometimes very large trees, 40–120 ft. in height and 1–3 ft. in trunk diameter, with pale-tan bark, which scales off in rather large, irregular flakes. The alternate, broadly oval or oblong, mature leaves are thick and leathery in texture and rounded at the base and apex, the lower surface being conspicuously paler in color. Trees on ridges become completely deciduous during the late dry season, but specimens on deeper soil seem to be evergreen. Tender flushes of new foliage are often

a striking, dark metallic blue, with pure white undersurfaces. The small, yellowish flowers are produced in terminal panicles in June and are followed in July by the ovoid, 1-seeded fruits which average about 1″ in length. The brownish wood is hard and heavy and is variously reported to be of good quality (Standley, *Flora of Costa Rica*) or nondurable and little used (Marker, Barbour *et al.*, *Forests of Costa Rica*). The seeds are very rich in oil and are used for making soap and candles and are strung on sticks and burned for illumination in Mexico.—Very large specimens are frequent in the hills near Palmar Norte, up to about 1,800 ft. in elevation, *Allen 6741*.

LICANIA OPERCULIPETALA Standl. & L. Wms.—Rosaceae—*Sapotillo* (Local).

Tall, handsome forest trees with rounded crowns of glossy foliage, mature specimens averaging about 120 ft. in height and 3 ft. in diameter above the moderately developed buttresses. The leathery, linear, short-petiolate leaves are alternate, with acute apices, and average about 6″ in length. The small, greenish-yellow, very fragrant flowers are produced in short terminal panicles in late March. The fruits have not been seen. The hard, moderately heavy, pinkish-tan wood is apparently not used locally. —Fairly frequent in the forested hills near Palmar, 3 fine specimens being conspicuous in the late afternoon light back of the camp cars at the river crossing, *Allen 6032*.

LICARIA—KEY

1. Lower margin of the cupule conspicuously flaring and crisped. Apex of the fruits flattened. Flowers dark yellow .... *Licaria Cufodontisii*
1. Lower margin of cupule not flaring, crisped, or undulate. Apex of fruits conical. Flowers white .... *Licaria Cervantesii*

LICARIA CERVANTESII (HBK) Kostermans—Lauraceae—*Yayo* or *Llayo* (Local).

Forest trees about 65 ft. in height with alternate, short-petiolate leaves, the lanceolate or elliptic-lanceolate blades acute or obtusely acuminate at the apex and 5–6½″ long and 1¼–2¾″ wide. The small white flowers are produced in February in short, slender, terminal racemes and are followed by conspicuous, acorn-like fruits up to 1¼–1½″ long, the broad, woody cupule with a double margin.—Forested hills near Palmar Norte, *Allen 5658 & 5861*.

LICARIA CUFODONTISII Kostermans—Lauraceae. (Pl. 27)

Small, rather bushy trees about 40 ft. in height with alternate, lanceolate or elliptic-lanceolate, often shortly caudate-acuminate leaves and

terminal or axillary, pyramidal panicles of minute dark-yellow flowers. The fleshy cupules of the acorn-like fruits have double or triple margins, the lower margin conspicuously flaring and undulate. A fairly common species of pastures and the margins of woodland.—Puerto Jiménez, *Cufodontis 187* & *Brenes 12262*—Pastures near Palmar Norte, *Allen 5651 & 6654*.

LILIACEAE—One genus, *Dracaena*.

LIMONCILLO—KEY

1. Leaves pinnate.
   2. Inflorescences axillary........ *Zanthoxylum procerum* (Guatemala)
   2. Inflorescences terminal........ *Sapindus saponaria* (Panama)
1. Leaves simple.
   2. Leaves alternate.
      3. Flowers in subterminal cymes........ *Ximenia americana* (Guatemala)
      3. Flowers in axillary fascicles........ *Casearia aculeata* (Guatemala)
   2. Leaves opposite.
      3. Inflorescences terminal........ *Randia armata* (Costa Rica)
      3. Inflorescences axillary.
         4. Flowers yellow. Shrubs or small trees, to about 18 ft........ *Siparuna patelliformis* (Costa Rica)
         4. Flowers white. Medium-sized or tall trees, to about 90 ft........ *Cassipourea podantha* (Panama)

LINOCIERA PANAMENSIS Standl.—Oleaceae.

Medium-sized forest trees 40–75 ft. in height and 12–30″ in trunk diameter, with opposite, elliptic-lanceolate leaves which have slender, long-decurrent, petiolate bases and shortly obtuse-acuminate tips. The small, white, multibracteolate flowers are produced from late January through most of February in short, dense paniculate cymes, rendering the trees very conspicuous from a distance. The fruits have not been seen, but are described as being white, compressed drupes about 1″ in length. The hard, pale-pink wood is not used locally, but would appear to be suitable for tool handles, carved objects, and articles of turnery as a substitute for South American Boxwood (*Casearia praecox*).—Frequent in the forested hills near Golfito, particularly across the bay from the townsite, *Allen 5857*.

LIPPIA OXYPHYLLARIA (Donn. Sm.) Standl.—Verbenaceae—*Caragra* (Costa Rica).

Shrubs or small trees, with rather thick, opposite, oblong-elliptic, acute or acuminate, short-pilose leaves which have crenate margins and average about 3″ in length. The flowers are produced in paired axillary, subglo-

bose, bracteate heads which are about 1″ in diameter and followed by dry, drupaceous fruits which separate at maturity into 2 nutlets. The genus is usually typical of dry highland savannas and thickets.—Known in our area only from the vicinity of Buenos Aires and Terraba, *Tonduz 3951*—Ujarras de Buenos Aires, *Pittier 10632*.

Lirio—See *Alibertia edulis* (Honduras).

Llayo—See *Licaria Cervantesii* (Local). Fruits acorn-like.

Llayo—See *Luehea Seemannii* (Honduras). Fruits not acorn-like.

Llora sangre—See *Croton xalapensis* (Guatemala).

LONCHOCARPUS—KEY

1. Leaves minutely puberulent on the lower surface. Sap turns blood red when exposed to the air...*Lonchocarpus sericeus* var. *glabrescens*
1. Leaves completely glabrous. Sap not red.
   2. Leaflets usually less than 2″ long...*Lonchocarpus minimiflorus*
   2. Leaflets usually more than 2″ long.
      3. Fruits conspicuously thickened along one margin...*Lonchocarpus guatemalensis*
      3. Margins of the fruits not thickened...................................*Lonchocarpus latifolius*

LONCHOCARPUS GUATEMALENSIS Benth.—Leguminosae; Papilionatae—*Chaperno* (Costa Rica)—*Chapel* (Honduras)—*Chapelno hediondo, Cincho,* or *Chaperno prieto* (Salvador)—*Habin, Ixec-subin, Palo de gusano, Sibicte,* or *Xaxmujin* (Guatemala).

Medium-sized or tall forest trees, to about 90 ft. in height, with alternate, pinnate leaves, the 5–9 ovate to elliptic-oblong, shortly obtuse-acuminate leaflets 2–4½″ in length. The pale purple flowers are produced in axillary racemes, probably in February, and are followed in late March and April by the 1- to 3-seeded, flat, straw-colored pods which are conspicuously thickened on the upper margin.—Occasional in forests near the Tinoco Station, and probably elsewhere, *Allen 6040*.

LONCHOCARPUS LATIFOLIUS (Willd.) HBK—Leguminosae; Papilionatae—*Dogwood* (Bocas del Toro)—*Cincho* (Honduras)—*Almendro* or *Mataboy* (Guatemala).

Medium-sized or sometimes large forest trees, to about 75 ft. in height, with alternate, long-petiolate, odd-pinnate leaves, the 5–9 oblong or obovate, subobtuse to acuminate leaflets 3–6″ in length. The reddish-purple or greenish-purple flowers are produced in axillary racemes, near the ends of the branches, and are followed by thin, pale-brown, linear or lanceolate, 1- to 5-seeded pods and are produced in our area in late March and early April. The wood is yellowish when first cut, deepening to russet brown on exposure, and is hard, heavy, tough, and strong and is highly resistant to decay. It is used in Honduras for many kinds of heavy, durable con-

struction.—Occasional in the hillside forests near Palmar and Farm 2, *Allen 6023.*

LONCHOCARPUS MINIMIFLORUS Donn. Sm.—Leguminosae; Papilionatae —*Chaperno* (Costa Rica)—*Chapelno, Chapelno negro, Chaperno,* or *Chapuno* (Salvador)—*Chaperno* (Guatemala).

Medium-sized trees, 30–60 ft. in height, with small, alternate, pinnate leaves, the 7–13 thin, ovate or obliquely lanceolate, acute or acuminate leaflets 1–1½″ in length. The small, rose-purple flowers are produced in short axillary racemes and are followed by the grayish-brown, flat, papery, 1-seeded pods, which are oblanceolate and apiculate in outline. We have seen only fruiting specimens in our area, which seem to differ somewhat from typical material, and may ultimately prove to be referable to some other species. The wood of *Lonchocarpus minimiflorus* is reported to be used for charcoal and fence posts in Guatemala, and is valued for fuel, since it burns for a long time.—Very common along small streams in pastures near Palmar Norte, 100 ft., *Allen 5737.*

LONCHOCARPUS SERICEUS var. GLABRESCENS Benth.—Leguminosae; Papilionatae—*Chaperno* (Local)—*Chaperno* or *Dogwood* (Bocas del Toro)—*Siete cueros* (Panama)—*Malvecino* (Panama, Chepo).

Medium-sized trees, 40–75 ft. in height and up to about 18″ in trunk diameter, with alternate, pinnate leaves, the 7–13 leathery, ovate or oblanceolate, shortly acute, glabrous leaflets from 2–5″ in length. The pinkish-lavender or purple flowers are borne in showy terminal or axillary racemes in early February and are followed by flat, 1- to 4-seeded pods. The white or pale-yellow wood is not durable, and has no local uses. The sap of the trunk is at first pink, drying blood-red on exposure to the air. —Santo Domingo de Osa, *Tonduz 9950*—Lowland forest near Kilometer 42, *Allen 5850.*

Loro—See *Dipterodendron costaricense* (Costa Rica).

LOZANIA PEDICELLATA (Standl.) L. B. Smith—Flacourtiaceae.

Forest trees, 15–20 ft. in height, with alternate, narrowly oblong or lanceolate, rather abruptly acuminate leaves about 4½″ long, the blades usually with minutely serrate margins. The minute green flowers are produced in December on very slender axillary spikes which about equal the leaves in length.—Esquinas Forest, 200 ft., *Skutch 5396.*

Lucuma chiricana—See *Pouteria chiricana.*

LUEHEA SEEMANII Triana & Planch.—Tiliaceae—*Guacimo colorado* or *Guacimo macho* (Local)—*Guacimo* (Costa Rica)—*Guacimo* or *Guacimo*

molenillo (Panama)—*Guacimo colorado, Guacimo macho, Guacimo molinero,* or *Molinillo* (Nicaragua)—*Caulote, Cotonron, Guacimo, Guacimo colorado, Llayo, Yayo,* or *Tapasquit* (Honduras)—*Cotonron, Yayo,* or *Tapasquit* (Guatemala). (Fig. 16)

Medium-sized or very large trees, often 90–120 ft. in height, the irregular trunks 2–4 ft. in diameter above the very strongly developed buttresses. The alternate, short-petiolate leaves are elliptic-oblong, with acute or acuminate apices and serrate margins, the lower surface covered with a fine brown tomentum. The small white flowers are produced in short terminal or axillary panicles in January and are the source of the delightful fragrance so noticeable in the woodlands at that season. The fruits are dark-brown or nearly black 5-angled capsules, which are star-shaped in cross section. They mature in March and April, rendering the trees quite conspicuous. Young plants in pastures and on cutover land are often little more than shrubs and differ from the mature trees in the very large juvenile leaves which may be nearly 1 ft. in length. Small, round-topped specimens in pastures often have the aspect, when seen from a distance, of a tree in flower, when the wind exposes the brownish undersurfaces of the foliage. The light, soft, pale-brown wood is straight-grained and easy to work, but is weak and not durable. It has no known uses. One of the most common and characteristic species of the lowland forests.—Buenos Aires, *Tonduz 6636 & 6672*—Forests near Jalaca, sea level, *Allen 5212*—Pastures near Palmar Norte, 100 ft., *Allen 5798*—Pastures near Palmar, sea level, *Allen 6700.*

LYSILOMA GUANACASTENSE Standl. & L. Wms.—Leguminosae; Mimosoideae—*Quiebrahacha* (Tenorio, Guanacaste).

Tall forest trees, to 90 ft. or more in height and 3 ft. in trunk diameter, with very conspicuous, shaggy bark and bipinnate leaves. The flowers have been seen from a distance only with binoculars, but appear to be white, in slender axillary spikes. The flat, dark-brown pods are 5½–6½″ long and about 1½″ wide. The species is occasional in the forests throughout the area from sea level to about 1,500 ft. Identification must be regarded as provisional, since actual specimens were not obtained.

LYTHRACEAE—One genus, *Lafoensia.*

Macano—See *Chlorophora tinctoria* (Panama). Leaves simple.
Macano—See *Diphysa robinioides* (Panama). Leaves pinnate.
Macetero—See *Hedyosmum mexicanum* (Guatemala).

Fig. 16. LUEHEA SEEMANNII. 1 — Flowering branch. 2 — Profile of individual fruit. 3 — Cross section of individual fruit. 4 — Individual flower. 5 — Dissection of flower, showing stamen fascicles and style.

MACROCNEMUM GLABRESCENS (Benth.) Wedd.—Rubiaceae—*Palo cuadrado* (Costa Rica and Panama). (Pl. 32)

Medium-sized forest trees, 30–75 ft. in height and up to about 24″ in diameter, old specimens usually having irregular, deeply furrowed trunks. The opposite, elliptic-lanceolate or oblanceolate, nearly glabrous leaves are obtuse or acute at the apex and rounded or acute at the base, with short, slender petioles. The vivid rose-pink, attractive flowers are produced in dense, long-stalked axillary panicles and remain in bloom from December through most of April. Frequent in hillside forests throughout the area.—Boruca, 1,000 ft., *Pittier 12077*—Palmar Norte, *Allen 5246, 5500, & 5783.*

Macuelizo—See *Tabebuia pentaphylla* (Nicaragua and Honduras).

Madera negra—See *Gliricidia sepium* (Costa Rica and Panama).

Madero negro—See *Gliricidia sepium* (Local).

Madre de cacao—See *Gliricidia sepium* (Local and general).

Madriado—See *Gliricidia sepium* (Honduras).

Madrial—See *Gliricidia sepium* (Honduras).

Madroño—See *Alibertia edulis* (Costa Rica and Panama). Plants without yellow latex. Flowers without conspicuous white bracts.

Madroño—See *Calycophyllum candidissimum* (Local and general). Plants without yellow latex. Flowers with showy white bracts.

Madroño—See *Rheedia madruno* (Local and Panama). Plants with sticky yellow latex.

Magnolia—See *Vochysia hondurensis* (Costa Rica).

Mahoe—See *Hibiscus tiliaceus* (Bocas del Toro).

Majagua—See *Apeiba Tibourbou* (Local). Leaf margins serrate.

Majagua—See *Hibiscus tiliaceus* (Costa Rica, Panama, and Honduras). Leaf margins entire, the blades heart-shaped in outline.

Majagua—See *Xylopia frutescens* (Costa Rica). Leaf blades slender, with entire margins.

Majaguillo—See *Muntingia calabura* (Chiriquí).

Majao—See *Heliocarpus appendiculatus* (Honduras). Flowers not Hibiscus-like.

Majao—See *Hibiscus tiliaceus* (Honduras). Trees confined to sea beaches.

Majauha—See *Heliocarpus appendiculatus* (Guatemala).

Malady—See *Aspidosperma megalocarpon* (British Honduras).

Mala gente—See *Oreopanax xalapense* (Costa Rica).

Malagueta—See *Xylopia frutescens* (Costa Rica). Outer petals valvate in bud.

Malagueto—See *Guatteria aeruginosa* (Panama). Outer petals imbricate in bud.

Malagueto—See *Guatteria chiriquiensis* (Chiriquí). Outer petals imbricate in bud.

Malagueto—See *Xylopia sericophylla* (Local). Outer petals valvate in bud.

Malagueto hembra—See *Xylopia frutescens* (Panama).

Malcajaco—See *Curatella americana* (Salvador).

MALPIGHIACEAE—KEY

1. Flowers yellow. Styles sharply pointed at the apex. Fruits yellow............... *Byrsonima*
1. Flowers pink or white. Styles obtuse, or truncate at the apex. Fruits red... *Malpighia*

MALPIGHIA TENUIFOLIA Standl. & L. Wms.—Malpighiaceae.

Small trees, about 18 ft. in height, with opposite, nearly sessile, oblong-lanceolate, acuminate or sometimes abruptly and falcately acuminate leaves 2½–5″ long and 1½–2″ wide. The small white flowers are produced in December in short axillary racemes and are almost immediately followed by the bright-red fruits.—Esquinas Forest, sea level, *Skutch 5375*—Forested hills near Palmar Norte, 1,000 ft., *Allen 6333*.

MALVACEAE—KEY

1. Leaves conspicuously paler on the lower surface. Flowers yellow, turning reddish brown with age. Trees found only along sea beaches.. *Hibiscus*
1. Leaves not paler on the lower surface. Flowers dark pink. Plants never found along sea beaches.. *Malvaviscus*

MALVAVISCUS ARBOREUS var. PENDULIFLORUS (M. & S.) Schery—Malvaceae—*Amapola* (Local)—*Amapolilla, Mapola,* or *Quesito* (Costa Rica)—*Mapola* (Chiriquí)—*Papito de monte* (Panama)—*Arito, Manzana, Manzanita, Manzanilla, Flor de arito,* or *Quesillo* (Salvador)—*Amapola, Manzanita, Monacillo, Pico de gorrion, Polvo de monte, Poro, Sobon, Tamanchich,* or *Tulipancillo* (Guatemala).

Shrubs or small trees to about 15 ft. in height, with broad, alternate, long-petiolate leaves, the nearly glabrous blades ovate to lanceolate in outline, with acuminate apices and coarsely dentate margins. The relatively large, showy, dark-pink flowers are produced in short, few-flowered terminal fascicles in December, and are soon followed by the round, flattened, translucent white fruits. Common in lowland forests near Palmar Norte and throughout the area.—Santo Domingo de Osa, *Tonduz 10010*—Palmar Norte, 100 ft., *Allen 5734 & 6641*.

Malvecino—See *Lonchocarpus sericeus* var. *glabrescens* (Panama). Flowers purple.

Malvecino—See *Swartzia panamensis* (Panama). Flowers yellow.

Malvecino—See *Sweetia panamensis* (Panama). Flowers white.

Mameycillo—Large trees in the Esquinas Forest. Probably *Calocarpum borucanum.*

Mameyito—See *Rheedia edulis* (Guatemala).

Manaca—See *Scheelea rostrata* (Local and Panama).

Manax—See *Pseudolmedia spuria* (Peten, Guatemala).

Manazana—See *Malvaviscus arboreus* (Salvador).

Mancha-mancha—See *Miconia argentea* (Panama).

Mang—See *Mangifera indica* (Guatemala).

Mangabe—See *Didymopanax Morototoni* (Panama). Leaves digitately compound.

Mangabe—See *Pourouma aspera* (Panama). Leaves deeply lobed, but not digitately compound.

MANGIFERA INDICA L.—Anacardiaceae—*Mango* (Local and general)—*Mang* (Guatemala).

The common mango, native of the Asiatic tropics, is planted throughout the area and occasionally persistent near old house sites and along trails.

Mangle—See *Avicennia nitida* (Panama). Trees without stilt roots.

Mangle—See *Rhizophora mangle* (Local and general). Trees with conspicuous stilt roots.

Mangle blanco—See *Bravaisia integerrima* (Costa Rica, Panama, and Nicaragua). Trees with stilt roots.

Mangle blanco—See *Laguncularia racemosa* (Panama to Honduras). Trees without stilt roots.

Mangle caballero—See *Rhizophora mangle* (Local).

Mangle colorado—See *Rhizophora mangle* (Local and general).

Mangle de agua—See *Bravaisia integerrima* (Local).

Mangle marequita—See *Conocarpus erecta* (Local). Leaves alternate.

Mangle marequita—See *Laguncularia racemosa* (Local). Leaves opposite.

Mangle negro—See *Conocarpus erecta* (Costa Rica).

Mangle piñuela—See *Pelliciera rhizophorae* (Local). Trunks buttressed. Fruits not conelike.

Mangle piñuelo—See *Conocarpus erecta* (Panama). Trunks not buttressed. Fruits conelike.

Mangle salado—See *Avicennia nitida* (Local and Panama). Trees without stilt roots.

Mangle salado—See *Rhizophora mangle* (Panama). Trees with conspicuous stilt roots.

Mangle torcido—See *Conocarpus erecta* (Panama).

Mango—See *Mangifera indica* (Local and general).

Mangrove—See *Rhizophora mangle* (Local and general).

Mano de leon—See *Oreopanax xalapense* (Guatemala). Leaves digitately compound.

Mano de leon—See *Pourouma aspera* (Honduras). Leaves deeply lobed, but not digitately compound.

Mano de tigre—See *Oreopanax xalapense* (Guatemala).

Mano tigre—Reported from the area, but not identified.

Manteco—See *Cupania guatemalensis* (Costa Rica).

Manu—See *Caryocar costaricense* (Costa Rica). Leaves trifoliate.

Manu—See *Guarea Guara* (Local). Leaves pinnate.

Manu—See *Guarea Hoffmanniana* (Local). Leaves pinnate.

Manu—See *Minquartia guianensis* (Costa Rica). Leaves simple.

Manu negro—See *Minquartia guianensis* (Costa Rica).

Manwood—See *Minquartia guianensis* (Bocas del Toro and Limón).

Manzana de montaña—See *Bellucia costaricensis* (Guatemala).

Manzana de playa—See *Crataeva tapia* (Nicaragua).

Manzanilla—See *Casearia arguta* (Guatemala). Leaf margins serrate. Flowers white, in axillary fascicles.

Manzanilla—See *Malvaviscus arboreus* var. *penduliflorus* (Salvador). Flowers dark pink.

Manzanilla—See *Ximenia americana* (Guatemala). Bark red. Branches spiny. White flowers in subterminal clusters.

Manzanillo—See *Hippomane mancinella* (Local and general). Milky, poisonous latex.

Manzanillo de la playa—See *Hippomane mancinella* (Local and general).

Manzanita—See *Malvaviscus arboreus* var. *penduliflorus* (Salvador and Guatemala).

Mapaguite—See *Trichilia acutanthera* (Guatemala).

Mapahuite—See *Trichilia acutanthera* (Guatemala).

Mapahuito—See *Trichilia acutanthera* (Guatemala).

Mapola—See *Malvaviscus arboreus* var. *penduliflorus* (Costa Rica and Chiriquí).

Maquenque—See *Oenocarpus panamanus* (Local and Panama). Trees without stilt roots.

Maquenque—See *Socratea durissima* (Costa Rica). Trees with very conspicuous stilt roots.

Marañon—See *Anacardium occidentale* (Local and general).

Marchucha—See *Tecoma stans* (Salvador).

Marequito—See *Conocarpus erecta* (Costa Rica).

Margarita—See *Ardisia revoluta* (Panama).

Maria—See *Calophyllum braziliense* var. *Rekoi* (Local and general). Leaves with a single longitudinal nerve.

Maria—See *Miconia argentea* (Costa Rica). Leaves with 5 longitudinal nerves.

Maria colorado—See *Calophyllum braziliense* var. *Rekoi* (Costa Rica).

MARILA PLURICOSTATA Standl. & L. Wms.—Guttiferae.

Slender forest trees, 25–50 ft. in height, with large, opposite, narrowly lanceolate, acuminate leaves 12–18″ long and about 4½″ wide, the lower surface of the blades with about 30–40 pairs of lateral nerves. The fragrant white flowers are about ½″ in diameter and are borne on slender pendulous racemes which are often more than 1 ft. in length. The species is thus far known only from the Esquinas Forest, at about 100–250 ft. in elevation.—*Skutch 5420* & *Allen 6538.*

Marillo—See *Symphonia globulifera* (Local).

Mario—See *Calophyllum braziliense* var. *Rekoi* (Salvador and Guatemala).

Mariquito—See *Conocarpus erecta* (Costa Rica).

Masica—See *Brosimum terrabanum* (Nicaragua and Honduras).

Masicaran—See *Astronium graveolens* (Honduras). Leaves pinnate.

Masicaran—See *Brosimum terrabanum* (British Honduras). Leaves simple.

Masicaran—See *Tabebuia chrysantha* (Honduras). Leaves digitately compound.

Masico—See *Brosimum alicastrum* (Guatemala). Lateral veins of the lower leaf surface not prominent.

Masico—See *Brosimum terrabanum* (Honduras). Lateral veins of the lower leaf surface very prominent.

Mastate—See *Brosimum utile* (Costa Rica).

Mastate blanco—Probably *Castilla fallax* (Boruca).

Mastate colorado—See *Brosimum utile* (Boruca).

Mataboy—See *Lonchocarpus latifolius* (Guatemala).

Mata cansada—See *Eschweilera calyculata* (Chiriquí).

Matacartago—See *Casearia aculeata* (Costa Rica). Leaves usually less than 3″ long.

Mata Cartago—See *Urera alceifolia* (Costa Rica). Leaves more than 3″ long.

Mata gente—See *Oreopanax xalapense* (Guatemala).

Matapalo—Applied generally to epiphytic species of *Ficus.*

Mata piojo—See *Trichilia acutanthera* (Nicaragua, Guatemala, and Honduras).

Mata raton—See *Gliricidia sepium* (Panama).

Matasanillo—See *Crataeva tapia* (Guatemala).

Matasano de mico—See *Diospyros ebenaster* (Guatemala).

MATAYBA INGAEFOLIA Standl.—Sapindaceae.

Trees about 20–25 ft. in height, with alternate, pinnate leaves, the 4 lanceolate or oblong-lanceolate, acuminate leaflets alternate on the terete rachis and unequal in size, the upper 2 about 4″ long and about twice as big as the lower pair. The small white flowers are produced in slender axillary racemes, and are followed by small 2-celled capsular fruits.—Esquinas Forest, 50 ft., *Skutch 5372.*

Matuhua—See *Plumeria rubra* forma *acutifolia* (Guatemala).

MAURIA GLAUCA Donn. Sm.—Anacardiaceae—*Cirri amarillo* (Costa Rica)—*Dantisco de montaña* (Central America).

Small trees, 12–25 ft. in height, with alternate, pinnate leaves, the 5–7 glabrous leaflets oblong-lanceolate in outline, with acute or acuminate apices. The small whitish flowers are produced in terminal or axillary panicles and are followed by small, somewhat compressed, drupaceous fruits.—Cañas Gordas, 3,300 ft., *Pittier 11207.*

Mauro—See *Casearia myriantha* (Panama).

Mayamaya—See *Pithecolobium longifolium* (Honduras).

MAYNA ECHINATA Spruce—Flacourtiaceae.

Shrubs or small trees, to about 18 ft. in height, with large, alternate, lanceolate or oblanceolate, caudate-acuminate leaves 10–18″ long and up to 5½″ wide, with coarsely sinuate-serrate margins on the upper half of the blade. The small but attractive white flowers are usually solitary and nearly sessile in the leaf axils and are followed by spiny, bright-yellow, globose fruits 1–1½″ in diameter, each containing several bright-red seeds. Our specimens have larger flowers and fruits than those described for this species, and may prove to be referable to *Mayna*

grandiflora Karsten of Colombia, or to be an undescribed entity.—Esquinas Forest, 200 ft., *Skutch 5376*—Forested hills above Palmar Norte, 250–1,000 ft., *Allen 6554*.

Mayo—See *Vochysia* sp. var. (Local).
Mayo blanco—See *Vochysia* sp. (Local).
Mayo colorado—See *Vochysia* sp. (Local).

MAYTENUS PALLIDIFOLIUS Standl. & L. Wms.—Celastraceae.

Trees about 75 ft. in height, with deeply furrowed and often somewhat twisted trunks, mature specimens averaging about 24″ in diameter. The rather leathery, lanceolate or oblong-lanceolate leaves are alternate on the slender branches, the blades 3–5″ long and up to 2″ wide, usually with obtuse or subacute apices and cuneate bases. The minute white flowers are produced in very small sessile fascicles from the nodes or internodes.—Locally very common in climax forest in the hills above Palmar Norte at 1,000–1,500 ft., *Allen 6327*.

Mazorco—See *Hedyosmum mexicanum* (Guatemala). Leaves simple.
Mazorco—See *Oreopanax xalapense* (Guatemala). Leaves digitately compound.
Mecate de agua—See *Heliocarpus appendiculatus* (Honduras).
Mecri—See *Vochysia ferruginea* (Panama).

MELASTOMACEAE—KEY

1. Leaves with a single longitudinal nerve .... *Mouriria*
1. Leaves with 3–7 longitudinal nerves.
   2. Inflorescences terminal .... *Miconia*
   2. Inflorescences axillary.
      3. Leaves broadly ovate. Petals obtuse .... *Bellucia*
      3. Leaves narrowly lanceolate. Petals acute .... *Henriettella*

MELIACEAE—KEY

1. Filaments of the stamens joined together to form a tube or cup. Seeds not winged.
   2. Fruits very large, woody, more than 4″ in diameter .... *Carapa*
   2. Fruits less than 2″ in diameter.
      3. Capsule 2- to 3-valved, or indehiscent and tuberculate. Anthers borne on the apex of the stamen tube .... *Trichilia*
      3. Capsule 4-valved. Anthers borne inside the apex of the stamen tube .... *Guarea*
1. Filaments of the stamens not joined or forming a cup or tube. Seeds winged .... *Cedrela*

MELIOSMA—KEY

1. Leaf blade long-decurrent at the base, the petiole about ½″ in length .... *Meliosma anisophylla*
1. Leaf blade cuneate at the base, but not long-decurrent, the petiole more than 2″ long.

2. Leaf blades very large, usually more than 8″ wide........................*Meliosma Allenii*
2. Leaf blades usually less than 3″ wide..*Meliosma longipetiola*

MELIOSMA ALLENII Standl. & L. Wms.—Sabiaceae. (Pl. 33)

Forest trees 35–75 ft. in height and up to 24″ in trunk diameter, old specimens characterized by their twisted and deeply furrowed boles. The elliptic-oblong, acute or shortly acuminate blades are about 20″ long, with cuneate bases and slender petioles up to 3½″ in length. The small orange flowers are produced in short pyramidal panicles, and are followed by the globose fruits, which are about 1″ in diameter. These are at first a pure waxy white, but become black at maturity.—A common species of the hillside forests near Palmar Norte, *Allen 5648.*

MELIOSMA ANISOPHYLLA Standl. & L. Wms.—Sabiaceae.

Small trees, averaging about 30 ft. in height, with very large alternate leaves which are 2 ft. or more in length and about 6″ in width. The oblanceolate blades have abruptly acuminate tips and long-decurrent bases, with stout petioles less than 1″ long. The obovoid, black fruits are produced in late January and early February on short axillary panicles and average about 1″ in diameter.—The species is known only from the Esquinas Forest, where it is not common, *Allen 5829.*

MELIOSMA LONGIPETIOLATA Standl. & L. Wms.—Sabiaceae.

Medium-sized trees, about 40 ft. in height, with alternate, linear-lanceolate, acute or acuminate leaves which average 9–12″ in length, including the 2 or 3″ petiole. The minute greenish-yellow flowers are produced in slender axillary panicles in April, and are followed by the nearly globose, black fruits which are about 1″ in diameter.—Occasional in lowland forests near Palmar Norte, *Allen 5248.*

Membrillo—See *Cespedesia macrophylla* (Panama). Flowers bright yellow, in showy terminal panicles.

Membrillo—See *Grias Fendleri* (Chiriquí). Flowers white, in short, dense clusters on the trunk and lower branches.

Membrillo—See *Terminalia amazonia* (Honduras). Flowers greenish tan, in slender axillary spikes.

Membrillo macho—See *Grias Fendleri* (Panama).

Membrillo de monte—See *Ximenia americana* (Guatemala).

Miacaguite—See *Cupania guatemalensis* (Salvador).

MICONIA—KEY

1. Leaves completely sessile, the base of the blade auriculate and clasping the stem...*Miconia impetiolaris*

1. Leaves with a short or elongate petiole, the base of the blade never auriculate or clasping the stem.
   2. Leafy branches and inflorescence densely covered with conspicuous rusty red hairs .... *Miconia Mathaei*
   2. Leafy branches and inflorescence not as above.
      3. Lower surfaces of the leaves reddish brown or pale tan, densely and uniformly covered with a minute, stellate pubescence which contrasts strongly with the dark-green upper surface.
         4. Leaf blades 2½–4″ long, the blades with 3–5 longitudinal nerves .... *Miconia rubiginosa*
         4. Leaf blades 4–12″ long, the blades with 5–9 longitudinal nerves.
            5. Flowers relatively large, the buds enveloped in conspicuous rounded, whitish bracts. Calyx about ¼″ long .... *Miconia dodecandra*
            5. Flowers small, the buds not enveloped in conspicuous bracts. Calyx about ⅛″ long.
               6. Calyx with short, distinct, pointed teeth .... *Miconia* sp.—*Allen 5814*
               6. Calyx obscurely dentate.
                  7. Leaves obovate-oblong, the blades acute at the base .... *Miconia elata*
                  7. Leaves ovate or oblong-ovate, the blades rounded or very obtuse at the base .... *Miconia argentea*
      3. Lower surfaces of the leaves green, the veins often minutely stellate-pubescent, but never with a dense, uniform covering over the entire blade.
         4. Lateral longitudinal nerves attached to the central nerve somewhat above the base of the blade.
            5. Petioles of the largest leaves less than ½″ long. Inflorescences very slender, with very short lateral branches, the scapes equaling or exceeding the largest leaves in length .... *Miconia scorpioides*
            5. Petioles of the largest leaves more than ½″ long. Inflorescences not as above, usually shorter than the largest leaves.
               6. Leaf blades long-decurrent at the base, so that most of the petiole appears to be narrowly winged .... *Miconia pteropoda*
               6. Leaf blades wedge-shaped, but never long-decurrent at the base.
                  7. Leaves narrowly lanceolate, tapering gradually to an acuminate tip. Margins serrate .... *Miconia Schlimii*
                  7. Leaves broadly oblong, abruptly and shortly acute at the apex. Margins not serrate .... *Miconia hondurensis*
         4. Lateral longitudinal nerves arising from the base of the blade.
            5. Leaves with very long petioles and broadly elliptic, caudate-acuminate blades. Flowers about ½″ long .... *Miconia caudata*
            5. Leaves with very short petioles and narrowly lanceolate or elliptic-lanceolate blades. Flowers about ¼″ long .... *Miconia laevigata*

MICONIA ARGENTEA (Swartz) DC.—Melastomaceae—*Santa Maria* (Local)—*Maria* (Costa Rica)—*Canillo, Cainillo, Cainillo de cerro, Dos caras, Friega platos, Gorgojillo, Gorgojo, Mancha-mancha, Oreja de mula, Palo negro,* or *Papelillo* (Panama)—*Capirote* or *Capirote blanco* (Nica-

ragua)—*Cenizo, Sirin,* or *Sirinon* (Honduras)—*Sirin macho* or *Sirinon* (Salvador)—*Jolte, Tolte, Sirin cacal, Sirinon, Sirin,* or *Siril* (Guatemala). (Pl. 29)

Small, attractive trees, 25–40 ft. in height, with opposite, long-petiolate leaves, the broadly oblong-ovate, acute or shortly acuminate blades white or pale tan on the lower surface and with 5 conspicuous longitudinal nerves. The small white flowers are produced during the dry season, usually in January or February in dense, pyramidal terminal panicles. The brown wood is moderately hard, heavy, fine-textured, straight-grained, and easy to work, but is not durable. It is seldom used, but has been reported as a source of second-grade railroad ties in southern Mexico. A common species, found in second-growth thickets and on the margins of forest near Palmar Norte and elsewhere throughout the drier parts of the area.—Buenos Aires, 800 ft., *Tonduz 3750*—Terraba, 800 ft., *Tonduz 3775*—Boruca, 1,400 ft., *Tonduz 3799 & 6870*—Santo Domingo de Osa, *Tonduz 9965*—Palmar Norte, 50 ft., *Allen 5790 & 5941.*

MICONIA CAUDATA (Bonpl.) DC.—Melastomaceae.

Shrubs or small trees, to about 30 ft. in height, with opposite, long-petiolate, elliptic or broadly ovate-elliptic leaves which are characterized by their 5 prominent longitudinal nerves and conspicuously caudate-acuminate tips. The attractive flowers are produced in large terminal panicles, the pinkish or white petals contrasting handsomely with the purple and yellow anthers.—Boruca, *Tonduz 6873.*

MICONIA DODECANDRA (Desr.) Cogn.—Melastomaceae—*Santa Maria* (Costa Rica).

Shrubs or small trees, to about 20 ft. in height, with opposite, long-petiolate leaves, the ovate to ovate-oblong blades acuminate at the apex and with 5–9 longitudinal nerves. The relatively small pink or white flowers are produced in narrow panicles 5–12″ long, the individual blooms enveloped in bud by a pair of conspicuous whitish bracteoles.—Cañas Gordas, *Pittier 10959.*

MICONIA ELATA (Swartz) DC.—Melastomaceae.

Shrubs or trees, to about 40 ft. in height, with opposite, elliptical to obovate-oblong leaves, the 5-nerved blades abruptly acuminate at the apex. The small flowers are produced in dense pyramidal panicles.—Buenos Aires, 1,000 ft., *Tonduz 4941.*

MICONIA HONDURENSIS Donn. Sm.—Melastomaceae.

Shrubs or small trees, to about 18–20 ft. in height, with opposite, broadly oblong, acute or abruptly acuminate leaves 6–10″ long, the blades with 3 conspicuous longitudinal nerves. The small white flowers are pro-

duced in terminal panicles.—Vicinity Puerto Jiménez, sea level, *Brenes 12168.*

MICONIA IMPETIOLARIS (Swartz) D. Don—Melastomaceae—*Hoja de pasmo* (Costa Rica)—*Dos caras* or *Oreja de mula* (Panama)—*Sirin* (Honduras).

Shrubs or small trees, to about 25 ft. in height, with large, opposite, oblong or obovate, acute or abruptly acuminate leaves 8–15″ long, the more or less auriculate bases of the blades completely sessile and clasping the stems. The lower leaf surfaces are characteristically pale tan, or occasionally reddish in color. The small white flowers are produced in terminal racemes and are followed by small blue fruits.—Boruca, 1,400 ft., *Tonduz 3776, 3777, & 6866*—Buenos Aires, 700 ft., *Tonduz 3786 & 4962*—Santo Domingo de Osa, sea level, *Tonduz 10022*—Río Tigre de Osa, *Brenes 12192*—Golfito, sea level, *Brenes 12322*—Second-growth thickets near Palmar Norte, 50 ft., *Allen 5940.*

MICONIA LAEVIGATA (L.) DC.—Melastomaceae—*Joint bush* (Bocas del Toro)—*Sirin* (Salvador)—*Tinajito* (Guatemala).

Shrubs or small trees, to about 15 ft. in height, with opposite, short-petiolate leaves, the oblong-elliptic, acuminate blades green on both surfaces and mostly 7–8″ long. The small white flowers are produced in February in short, terminal panicles. They have a rather unpleasant odor. —Occasional on the margins of forest near Palmar Norte, 50 ft., *Allen 5881.*

MICONIA MATTHAEI Naud.—Melastomaceae—*Canillito* (Costa Rica).

Small trees, to about 35 ft. in height, with narrow, opposite, short-petiolate leaves, the often somewhat bullate blades lanceolate or oblong-lanceolate, up to about 8″ in length, with acuminate apices. The small white flowers are produced in large terminal panicles in February and March, and are followed by small black globose fruits. The species is easily recognized by the abundant long, soft, rusty hairs that cover the branches, leaf petioles, and inflorescences.—Buenos Aires, *Tonduz 4970*—Cañas Gordas, 3,400 ft., *Pittier 11064*—Diquis, 1,000 ft., *Pittier 12107.*

MICONIA PTEROPODA Benth.—Melastomaceae.

Shrubs or small trees, to about 20 ft. in height, with opposite, ovate-oblong or oblong-lanceolate, acuminate leaves 6–8″ in length, the blades with 5 longitudinal nerves and long-decurrent at the base, so that the petiole is winged for most of its length. The flowers are small and white and are produced in a terminal panicle, as in the genus.—Buenos Aires, *Tonduz 6519*—Diquis, *Pittier 12143.*

MICONIA RUBIGINOSA (Bonpl.) DC.—Melastomaceae.

Shrubs or small trees, with opposite, ovate to ovate-oblong leaves, the 3—5-nerved, shortly acuminate blades 2½–4″ in length and rough to the touch. The small white flowers are produced in dense pyramidal panicles and are followed by black fruits.—Buenos Aires, *Tonduz 4965.*

MICONIA SCHLIMII Triana—Melastomaceae—*Sirin* or *Sirin blanco* (Honduras)—*Quina blanca* (Salvador)—*Siril de shara* (Guatemala).

Shrubs or small trees, to about 25 ft. in height, with opposite, oblong-lanceolate, acuminate leaves and short terminal cymes of rather attractive white flowers with showy yellow stamens, which are produced from March until about May.—Santo Domingo de Osa, *Tonduz 10017*—Diquis, 800 ft., *Pittier 12113*—Río Terraba, near La Presa, 100 ft., *Allen 5286*—Forested hills near the Golfito airfield, sea level, *Allen 5989.*

MICONIA SCORPIOIDES (S. & C.) Naud.—Melastomaceae.

Small or medium-sized trees, to about 40 ft. in height, with large, opposite, short-petiolate leaves, the elliptic-oblong or oblanceolate blades with obtuse or acute apices and decurrent bases, and averaging about 9 or 10″ in length. The small flowers are produced in late February and early March in very long, slender, dark-pink panicles, which have short, lateral scorpioid branches.—Esquinas Forest, 100 ft., *Allen 6005*—Diquis, 1,500 ft., *Tonduz 6521*—Hills near the Golfito dairy, 100 ft., *Allen 6620.*

MICONIA sp.—Melastomaceae.

Slender forest trees, 40–60 ft. in height, with opposite, elliptic-lanceolate leaves 4–6″ long, the blades pale tan in color on the lower surface, and with 5 longitudinal nerves. The relatively small white flowers appear in January and February in short, few-flowered terminal panicles.—Esquinas Forest, 250 ft., *Allen 5814.*

Mierda de gallina—See *Casearia myriantha* (Guatemala).

Milady blanco—See *Aspidosperma megalocarpon* (Guatemala).

MINQUARTIA GUIANENSIS Aubl.—Olacaceae—*Cuajada negro, Manu,* or *Black manu* (Local)—*Manu negro* or *Palo de piedra* (Costa Rica)—*Criollo, Cuajada negro,* or *Nispero negro* (Chiriquí)—*Palo criollo* (Panama)—*Manwood* or *Black manwood* (Bocas del Toro and Limón).

Tall forest trees, 75–100 ft. in height and up to 3 ft. in diameter above the moderately developed buttresses, with alternate, oblong or oblong-ovate, acute or shortly acuminate leaves and short axillary paniculate racemes of small flowers which are produced from January until May and are followed by small, 1-celled drupaceous fruits. The dark grayish-brown, hard, heavy wood is tough and strong and extremely durable in contact with the ground, lasting more than 30 years in some instances

under extremely difficult climatic conditions. It has been used by the United Fruit Company in Bocas del Toro and Limón for railroad ties, and for house and fence posts. The species has acquired a poor reputation in our area due to confusion with *Guarea Hoffmanniana,* a nondurable timber which is also known as "Manu." Occasional in hillside forests, but apparently not common. A few logs have been seen in Golfito, and the species is listed from Golfo Dulce by Standley, but without citation of a definite locality or specimen. The nearest authentic material would appear to be *Cooper & Slater 312,* from Progreso, in Chiriquí Province, Panama.

Mirto—See *Myrcia Oerstediana* (Costa Rica).
Molinillo—See *Luehea Seemannii* (Nicaragua).
Monacillo—See *Malvaviscus arboreus* (Guatemala).

MONIMIACEAE—One genus, *Siparuna.*
Monkey apple—See *Dialium guianense* (Bocas del Toro). Leaves pinnate.
Monkey apple—See *Posoqueria latifolia* (Bocas del Toro). Leaves simple.
Monkey comb—See *Apeiba aspera* (Canal Zone, Bocas del Toro, and Limón). Spines of fruits short and hard.
Monkey comb—See *Apeiba Tibourbou* (Canal Zone, Bocas del Toro, and Limón). Spines of fruits long and pliant.
Monkey tambrin—See *Inga spectabilis* (Panama).
Mora—See *Ardisia revoluta* (Guatemala). Leaves leathery.
Mora—See *Chlorophora tinctoria* (Local and general). Leaves not leathery.
Mora amarilla—See *Chlorophora tinctoria* (U.S. trade name).

MORACEAE—KEY

1. Leaves conspicuously palmately lobed.
  2. Leaves peltate. Branches hollow, usually inhabited by colonies of stinging ants........ *Cecropia*
  2. Leaves not peltate. Branches not as above........ *Pourouma aspera*
1. Leaves not conspicuously palmately lobed. (Sometimes irregularly lobed in *Chlorophora.*)
  2. Fruiting receptacles sessile, or nearly sessile in the leaf axils.
    3. Fruiting receptacles globose; solitary or in pairs.
      4. Fruits hollow, with many tiny seeds........ *Ficus*
      4. Fruits not hollow, usually with 1 or 2 relatively large seeds.
        5. Fruits more than 1" in diameter, like a miniature Osage orange *(Ma-*

clura) in superficial appearance. Branches and fruits with pale orange latex .... *Batocarpus*

5. Fruits less than 1″ in diameter, never with pale orange latex.
   6. Pistillate flowers and fruits solitary.
      7. Fruits enveloped in a fleshy, accrescent perianth.
         8. Leaves with coarsely serrate margins.......... *Olmedia falcifolia*
         8. Leaves with entire margins.......... *Pseudolmedia spuria*
      7. Fruits not enveloped in an accrescent perianth.
         8. Branches often spiny. Plants without latex. Wood yellow, hard, and heavy. Male and female flowers on separate trees.......... *Chlorophora tinctoria*
         8. Branches never spiny. Plants with abundant creamy latex. Wood white, soft. Male and female flowers produced on the same receptacle.......... *Brosimum*
   6. Pistillate flowers and fruits in pairs.
      7. Leaves with coarsely serrate margins.......... *Olmedia falcifolia*
      7. Leaves with entire margins.......... *Clarisia mexicana*

3. Fruiting receptacles flattened; concave or disklike.
   4. Staminate flowers without a perianth.......... *Castilla*
   4. Staminate flowers with a perianth.......... *Perebea*

2. Fruits produced in slender racemes or on the branches of a slender, diffuse panicle.
   3. Fruits in short, slender racemes.......... *Trophis racemosa*
   3. Fruits on the branches of a slender, diffuse panicle.
      4. Leaves caudate-acuminate at the apex, conspicuously reticulate-veined beneath. Plants never epiphytic.......... *Trophis chorizantha*
      4. Leaves shortly and abruptly acute at the apex, never reticulate-veined beneath. Plants often epiphytic.......... *Coussapoa parviceps*

Morado—See *Peltogyne purpurea* (Panama).

Mora megistosperma—See *Mora oleifera.*

MORA OLEIFERA (Triana) Ducke—Leguminosae; Caesalpinoideae [*Mora megistosperma* (Pittier) Britt. & Rose]—*Alcornoque* (Costa Rica and Panama). (Pl. 16)

Medium-sized or large trees, 50–120 ft. in height and up to about 3 ft. in trunk diameter, with rather large, pinnate leaves, the 2 pairs of smooth, leathery leaflets elliptic-oblong in outline, with acute or acuminate apices, averaging about 6″ in length. The small white flowers are produced in April in dense, terminal or axillary spikes and are followed in August to about October by the very large, 1- or 2-seeded pods. The huge kidney-shaped seeds are over 7″ long and are among the largest known for any dicotyledonous plant. The hard, heavy, reddish-brown wood is sometimes used for crossties, bridge boards, fence posts, and general heavy construction but is reported to check and warp badly. The trees often form near-

ly pure stands back of the mangrove swamps, notable examples in our area being found near Farms 19 and 20 and in the lower valley of the Río Esquinas. The leaves begin to turn yellow in September and October during the heavy rains, and the trees are usually completely leafless during November and December, rendering the stands very conspicuous when seen from the air.—Río Terraba, near the Boca Brava, sea level, *Allen 5603*.

Morillo—See *Brosimum sapiifolium* (Local). Very large, unarmed trees, with milky latex.

Morillo—See *Chlorophora tinctoria* (Local). Medium-sized or large trees, often with spiny branches, without milky latex.

Morillo—See *Trophis racemosa* (Panama). Unarmed shrubs or small trees.

Morita—See *Ardisia revoluta* (Guatemala).

Morito de rio—See *Enallagma latifolia* (British Honduras).

Moro—See *Quararibea guatemalteca* (Guatemala).

Morro—See *Crescentia cujete* (Honduras).

MORTONIODENDRON—Key

1. Leaves broadly elliptic-lanceolate or elliptic-oblong, usually more than 7″ long. Capsules 5-celled. Tall trees, to about 100 ft., with strongy buttressed trunks.......... *Mortoniodendron anisophyllum*
1. Leaves lanceolate, usually less than 6″ long. Capsules 3-celled. Shrubs or small trees, to about 18 ft. in height.......... *Mortoniodendron guatemalense*

MORTONIODENDRON ANISOPHYLLUM (Standl.) Standl. & Steyerm.—Tiliaceae. (Pl. 33)

Trees, 35–100 ft. in height, the larger specimens with strongly buttressed trunks. Leaves oblong or broadly elliptic, acute or acuminate at the apex and usually rather obliquely cuneate and 3-nerved at the base, the blades 4–10″ long and 1¾–4½″ wide. The flowers are produced in terminal panicles and are followed in August and September by the pale-yellow, 5-valved capsular fruits which average about 1″ in diameter. —Very common in the forested hills near Golfito at low elevations, *Allen 6561*.

MORTONIODENDRON GUATEMALENSE var. AUSTRALE Morton & Steyermark —Tiliaceae.

Small trees, to about 18 ft. in height, with alternate, oblong-lanceolate, acuminate leaves 4¾–6¼″ long and 1½–2″ wide. The small pale-yellow flowers are produced in December in very short, slender, few-flowered ter-

minal or subterminal cymes, followed by small 3-celled fruits.—Esquinas Forest, 50 ft., *Skutch 5389*—Vicinity of Jalaca, 100 ft., *Allen 5304.*

Mostrenco—See *Randia armata* (Costa Rica).

Mountain guava—See *Posoqueria latifolia* (British Honduras).

MOURIRIA PARVIFOLIA Benth.—Melastomaceae—*Cierito* (Chiriquí)—*Arracheche, Kenna,* or *Solacra* (Panama)—*Camaron* or *Capulin verde* (Salvador)—*Isna* (Honduras)—*Cacho de venado, Cuerno de venado,* or *Chicharillo* (Guatemala)—*Half crown* (British Honduras).

Small or medium-sized trees 12–45 ft. in height, with small, opposite, ovate-lanceolate, acuminate, nearly sessile, 1-nerved, glabrous leaves which are very unlike those of other genera of the Melastomaceae, being superficially rather reminiscent of those of the common Surinam cherry *(Eugenia uniflora)* of the Myrtaceae. The small, solitary white flowers are produced in late January and early February from the leaf axils, and are soon followed by the attractive, bright-red, globose, juicy fruits which are about ½″ in diameter. The brown or reddish-brown wood is very hard, heavy, tough, and strong, and has a reputation for being durable in contact with the ground. A closely allied species has been found to be very satisfactory for fence posts in Panama. Fairly frequent in hillside forests above about 1,000 ft.—Hills above Palmar Norte, 1,800 ft., *Allen 5926.*

Mox—See *Ceiba pentandra* (Guatemala).

M'shal—See *Oreopanax xalapense* (Guatemala).

Much—See *Diphysa robinioides* (Guatemala).

Mucut—See *Cassia grandis* (Guatemala).

Mulato—See *Bursera simaruba* (Honduras). Leaves pinnate.

Mulato—See *Triplaris melaenodendron* (Honduras and Guatemala). Leaves simple.

MUNTINGIA CALABURA L.—Tiliaceae—*Capulin* (Local)—*Capulin blanco* (Boruca)—*Majaguillo* (Chiriquí)—*Pacito* (Panama)—*Capulin* (Nicaragua)—*Capulin* or *Capulin blanco* (Guatemala). (Pl. 28)

Small trees, 15 to about 45 ft. in height, with alternate, oblong-lanceolate short-petiolate leaves which are paler in color on the lower surface, the blades with acuminate tips and strongly oblique, rounded bases. The rather attractive white flowers are about 1″ in diameter, and are produced in November and December in 2- to 3-flowered pedunculate fascicles from the internodes. The globose red fruits are about ⅝″ in diameter and are sweet and edible. The bark contains a tough fiber that is sometimes used

for cordage, and the manufacture of bark cloth. The light, firm, pale-brown wood works easily and finishes smoothly, but is not durable and is seldom used. A common species of very rapid growth, found on the margins of forest and on cutover land in association with Balsa and Guarumo.—Palmar Sur, 50 ft., *Allen 6337*.

Muñeca—See *Hasseltia floribunda* (Nicaragua). Leaves simple.
Muñeco—See *Crataeva tapia* (Costa Rica). Leaves 3-parted.
Murcielago—See *Cornutia grandifolia* (Chiriquí).
Murta—See *Myrcia Oerstediana* (Costa Rica).
My Lady—See *Aspidosperma megalocarpon* (British Honduras).

MYRCIA OERSTEDIANA Berg—Myrtaceae—*Mirto, Murta,* or *Turro* (Costa Rica)—*Pimento* (Panama)—*Agal* (Guatemala)—*Pigeon plum* (British Honduras).

Shrubs or small trees, to about 30 ft. in height, with slender hairy branches and opposite, very shortly petiolate, conspicuously veined leaves, the oblong-lanceolate, acuminate blades 2½–4″ in length. The plants have very much the aspect of a small-flowered *Eugenia*. The small white flowers are produced in large, lax, terminal or axillary panicles and are followed by purple or blackish edible fruits which have a spicy flavor.—Santo Domingo de Osa, sea level, *Tonduz 10016*.

MYRIOCARPA LONGIPES Liebm.—Urticaceae—*Estrella* (Local and Boruca)—*Chichicaste colorado* or *Picapica* (Salvador)—*Chichicastillo* or *Tapon* (Honduras)—*Chichicaste* or *Chichicaste manso* (Guatemala).

Shrubs or small trees, to about 18 ft. in height, with large, alternate, elliptic-ovate leaves, the blades acute at the apex, and with minutely serrate margins. The minute whitish flowers are produced in December in very long, pendulous, thread-like axillary spikes. A common species found in forests and on cutover land throughout the area.—Lagarto, *Tonduz 4826*—Esquinas Forest, 200 ft., *Skutch 5379*—Palmar Norte, 50 ft., *Allen 5739*.

MYRISTICACEAE—KEY

1. Fleshy aril enveloping the seeds deeply and conspicuously divided, usually for more than ½ its length.
   2. Leaf petioles and flushes of new growth with stellate hairs (Lens!) .... *Virola*
   2. Leaf petioles and flushes of new growth with simple or branching hairs .... *Dialyanthera*
1. Fleshy aril not deeply divided, usually only very briefly lobed at the apex .... *Compsoneura*

MYROXYLON BALSAMUM (L.) Harms var. PEREIRAE (Royle) Harms—Leguminosae; Papilionatae—*Chirraca, Sandalo,* or *Balsamo* (Local)—*Chilaca* (Local)—*Balsam of Peru* (Honduras and U. S. Pharmacopoeia).

Very large forest trees 90–120 ft. in height and 2–4 ft. in trunk diameter. The odd-pinnate, alternate leaves have 7–11 ovate or oblong leaflets which are usually acute or shortly acuminate at the apex and rounded and very shortly petiolate at the base. The small whitish flowers are produced in terminal or axillary racemes and are followed in January and February by the fragrant winged fruits which superficially resemble a single unit of the familiar paired Maple seeds of temperate woodlands. The trees are the source of the famous, though misnamed, Balsam of Peru, a viscid, dark reddish-brown, fragrant aromatic liquid obtained from the bark, most of the industry being centered in El Salvador. It has long been an official drug in the U. S. Pharmacopoeia and has stomachic and expectorant properties. In preconquest times it formed one of the regular items of tribute to the Aztec court, and has been much appreciated in Europe for the manufacture of perfumes. It is reported still to be used in Central American churches for incense and the preparation of the chrism. The seeds are added to aguardiente in Guatemala and are said to impart a delightful flavor. The hard, heavy, reddish-brown wood is handsomely figured and takes a high polish. It has an agreeable odor that it retains for a long time, and is suitable for furniture, millwork, shipbuilding, interior trim and paneling, veneer, and railroad ties.—Fairly frequent in hillside forests near Palmar Norte, *Allen 5751.*

MYRSINACEAE—One genus, *Ardisia.*

MYRTACEAE—KEY

1. Calyx undivided in bud, splitting irregularly on flowering.................. *Psidium*
1. Calyx with 4 or 5 distinct lobes in bud.
   2. Calyx lobes and petals 4.................. *Eugenia*
   2. Calyx lobes and petals 5.................. *Myrcia*

Nacaspiro—See *Inga multijuga* (Guatemala). Leaflets 5–10 pairs.

Nacaspiro—See *Inga punctata* (Guatemala). Leaflets 2–4 pairs.

Nance—See *Byrsonima* sp. (Local and general). Fruit a yellow, edible drupe.

Nance—See *Clethra lanata* (Costa Rica). Fruit a 3-valved capsule.

Nance macho—See *Clethra lanata* (Costa Rica).

Nancite—See *Byrsonima crassifolia* (Nicaragua).

Nancito—See *Byrsonima crassifolia* (Honduras).

Napahuite—See *Trichilia acutanthera* (Guatemala).

Naranjillo—See *Heisteria concinna* (Costa Rica). Leaves simple.

Naranjillo—See *Swartzia simplex* (Panama). Leaves superficially simple. Flowers yellow.

Naranjillo—See *Zanthoxylum procerum* (Guatemala). Leaves pinnate.

Naranjo—See *Terminalia amazonia* (Honduras and Guatemala).

Nargusta—See *Terminalia amazonia* (U. S. trade name).

Nazareno—See *Peltogyne purpurea* (Local and Panama).

NECTANDRA—KEY

1. Axils of the veins of the lower leaf surface with small but conspicuous tufts of hairs.......... *Nectandra perdubia*
1. Axils of the veins of the lower leaf surface without tufts of hairs.
   2. Fruits nearly globose, subtended by a very shallow cupule.......... *Nectandra latifolia*
   2. Fruits ellipsoid or conic, subtended by a deep, acorn-like cupule.
      3. Leaf blades mostly narrowly lanceolate, tapering gradually to an acuminate tip. Flower buds glabrescent. Flowering season in December.......... *Nectandra salicifolia*
      3. Leaf blades mostly elliptic-lanceolate, abruptly acuminate. Flower buds densely pilose. Flowering season in April.......... *Nectandra nervosa*

Nectandra glabrescens—See *Nectandra salicifolia.*

NECTANDRA LATIFOLIA (HBK) Mez.—Lauraceae—*Sigua negro* (Panama)—*Sigua blanca* or *blanco* (Panama to Honduras)—*Sacalante* or *Coyokiche* (Guatemala).

Trees 40–60 ft. in height, with alternate, short-petiolate, oblong-lanceolate, acuminate, glabrous leaves and slender axillary corymbose panicles of small white flowers, followed in late August and early September by the small, black, ovoid fruits. The yellowish or pale olive-brown wood burns easily and resembles that of *Nectandra salicifolia,* and should be suitable for the same uses.—Terraba, *Tonduz 3952*—Santo Domingo de Osa, *Tonduz 10047*—Rocky beaches near the delta of the Río Esquinas, *Allen 5629.*

NECTANDRA NERVOSA Mez & Pittier—Lauraceae.

Small trees 30–40 ft. in height, with alternate, elliptic-lanceolate, acuminate, glabrous leaves and short subterminal panicles of small white flowers that appear in April.—Delta of the Río Diquis (Terraba), *Tonduz 6758*—Margins of forest near Palmar Norte, 50 ft., *Allen 5251.*

NECTANDRA PERDUBIA Lundell—Lauraceae—*Aguacatillo* (Mexico)—*Bastard timbersweet* (British Honduras).

Almost identical with *Nectandra latifolia,* at least on the basis of herbarium material, but differing in the presence of small tufts of barbate hairs in the axils of the veins of the lower leaf surface and in the apiculate fruits. It seems probable that this will eventually be regarded as a variety

of the former species or reduced to synonomy.—Golfito, *Brenes 12314*—Puerto Jiménez, *Brenes 12163*.

NECTANDRA SALICIFOLIA (HBK) Nees—Lauraceae—*Canelo* or *Sigua blanca* (Local)—*Aguacatillo, Quizarra, Quina, Quizarra quina*, or *Sigua amarillo* (Costa Rica)—*Sigua* (Bocas del Toro)—*Pimiento* (Salvador)—*Aguacatillo* or *Pubabac* (Guatemala)—*Laurel* or *Timbersweet* (British Honduras).

Common, round-topped trees, 40–75 ft. in height and up to about 18″ in trunk diameter, with alternate, leathery, linear or elliptic-lanceolate, long-acuminate leaves which are from 4–8″ in length. The small fragrant white flowers are produced in slender terminal or axillary corymbose panicles in November and December, often rendering the trees rather conspicuous from a distance. The black, oblong-ovoid fruits have a well-developed cupule and mature in late March and early April. The yellowish or olive-brown wood is fine-textured and straight-grained, working easily and finishing smoothly with a waxy feel and high polish. It is used in Costa Rica for furniture, interior trim, and general construction. The species is listed in *Flora of Costa Rica* under the synonymous name of *Nectandra glabrescens* Benth. One of the most frequent and characteristic species of trailsides and pastures near Palmar Norte.—Cañas Gordas, 3,500 ft., *Pittier 11102*—Palmar Norte, 50 ft., *Allen 5449, 5518, & 5224*.

NEEA—KEY

1. Branches and veins of the lower leaf surface densely ferrugineous-tomentulose........ .... *Neea elegans*
1. Branches and veins of the lower leaf surface glabrous.......... *Neea Popenoei*

NEEA ELEGANS P. H. Allen, sp. nov.—Nyctaginaceae. (Pl. 28)

Small trees, 15–20 ft. in height, with alternate, opposite or verticillate leaves, the narrowly lanceolate, acuminate blades 7–10″ long and 2–3″ wide; the 12–16 pairs of lateral nerves of the lower surface, petioles and branchlets densely ferrugineous-tomentulose. The small flowers are produced in relatively dense pedunculate terminal cymes and are followed in December and January by very showy pendulous clusters of blood-red, narrowly ellipsoidal fruits which average about ½″ in length. A particularly attractive species which would be well worth cultivating for ornament. Infrequent in wet lowland forests at sea level.—Tinoco Station, *Allen 6683* TYPE (Herbarium Escuela Agrícola Panamericana).

NEEA POPENOEI P. H. Allen, sp. nov.—Nyctaginaceae.

A spreading tree, 60 ft. in height, with alternate or paired leaves, the

thin-textured elliptic or linear-lanceolate, acute or acuminate, glabrous blades 4–8″ in length, and with 6–10 pairs of lateral nerves. The small, pale-yellow flowers are produced in rather open, long-pedunculate, terminal or axillary cymes during late March and early April. Fruits have not been seen. The species is dedicated to Dr. Wilson Popenoe, Director of the Escuela Agrícola Panamericana in Honduras, who did so much to make completion of our project possible.—Margins of forest near Palmar Norte, 50 ft., *Allen 5225* TYPE (Herbarium Escuela Agrícola Panamericana).

Negrito—See *Simaruba glauca* (Honduras).

NEONICHOLSONIA WATSONII Dammer—Palmaceae.

Dwarf, single-stemmed, unarmed palms with about 8 pinnate fronds, the plants averaging about 8–9 ft. in height to the tips of the terminal leaflets. The erect or arching undivided spicate inflorescences about equal the fronds in length. Very common in climax forests throughout the area.—Coto Junction, 50 ft., *Allen 6659*.

Nicte de monte—See *Plumeria rubra* forma *acutifolia* (Guatemala).
Nispero—See *Pouteria* sp. (Local). Plants with milky latex.
Nispero—See *Sacoglottis excelsa* (Local). Plants without milky latex. Leaf margins serrate.
Nispero—See *Vantanea Barbourii* (Costa Rica). Plants without latex. Leaf margins not serrate.
Nispero colorado—See *Pouteria heterodoxa* (Local).
Nispero negro—See *Minquartia guianensis* (Chiriquí).
Nispero zapote—See *Pouteria heterodoxa* (Local).
Nune—See *Hura crepitans* (Chiriquí).
Nuo—See *Ceiba pentandra* (Guatemala).

NYCTAGINACEAE—One genus, *Neea*.

O—See *Persea americana* (Guatemala).

OCHNACEAE—KEY

1. Leaves conspicuously clustered at the ends of the branches. Blades broadest above the middle, with coarsely sinuate-serrate margins .... *Cespedesia*
1. Leaves not conspicuously clustered at the ends of the branches. Blades broadest at about the middle, with entire margins .... *Ouratea*

OCHROMA LAGOPUS Swartz var. BICOLOR (Rowlee) Standl. & Steyermark —Bombacaceae—*Balsa* (Local and general)—*Balso* (Costa Rica)—*Balsa, Cotton tree,* or *Lana* (Panama)—*Balsa* or *Gatillo* (Nicaragua)—

Balsa, Guano, or *Tambor* (Honduras)—*Algodon* or *Balsa* (Salvador)—*Balsa, Cajeto, Corcho, Jujul, Lana, Lanilla,* or *Puj* (Guatemala). (Pl. 12)

Very common medium-sized or occasionally large trees, 30 to about 90 ft. in height and up to 36″ in trunk diameter, with smooth grayish bark and large, alternate, long-petiolate leaves. The broadly ovate, palmately 5- to 7-veined blades are usually more or less lobed, and with rounded or subacute tips and deeply cordate bases, the lower surface with a contrasting, pale greenish-brown tomentum. In juvenile specimens the leaves are very large, often 24″ or more in diameter, while in mature trees they are much smaller, averaging about 8″ across. The large white, rather fleshy, tubular-campanulate flowers are solitary on short, stout peduncles in the upper leaf axils and are produced from October through December and are followed by the long, slender woody 5-valvate capsular fruits which mature on the trees in mid-February, the ground for some distance being covered with the pale-brown, fluffy, kapok-like fiber. The white, extremely light wood is used for model airplanes, life rafts, and as a substitute for cork in refrigerators, the best quality material being obtained from young, fast-growing specimens. The kapok-like fiber is suitable for filling pillows and mattresses. Several attempts have been made in the American tropics to establish commercial plantations of this very useful species, but many have failed because of the difficulty in transplanting the young trees from the nursery beds. This can be overcome by sowing the seeds at stake, in their permanent positions, after which a light brush fire should be lighted to assure uniform germination. The trees are of very rapid growth, and are common to the margins of forest and on cutover land, where they often form nearly pure stands.—Buenos Aires, *Tonduz 6619*—Palmar Norte, 50 ft., *Allen 5799.*

OCOTEA—Key

1. Leaves very large, usually more than 1 ft. in length.
  2. Veins of the lower leaf surface conspicuously ferrugineous-tomentose .... *Ocotea mollifolia*
  2. Veins of the lower leaf surface glabrous.
    3. Leaf blades broadest above the middle, the base strongly decurrent, with recurved margins .... *Ocotea rivularis*
    3. Leaf blades broadest at about the middle, the base rounded and never decurrent .... *Ocotea nicaraguensis*
1. Leaves less than 1 ft. in length.
  2. Leaf blades strongly decurrent at the base.
    3. Leaves oblanceolate, more than 7″ long .... *Ocotea Ira*
    3. Leaves lanceolate, less than 6″ long .... *Ocotea Williamsii*
  2. Leaf blades rounded or cuneate at the base, never strongly decurrent.

3. Leaf blades narrowly linear-lanceolate, with parallel margins, about 4 times as long as broad .... *Ocotea pergamentacea*
3. Leaf blades never with parallel margins, much less than 4 times as long as broad .... *Ocotea veraguensis*

OCOTEA IRA Mez & Pittier ex Mez—Lauraceae—*Aguacaton* or *Pocora* (Local)—*Ira* (Costa Rica)—*Aguacaton* (Panama).

Medium-sized forest trees, 40–75 ft. in height, with relatively large, alternate, short-petiolate, leathery leaves, the oblong-elliptic or obovate, glabrous blades shortly and obtusely acuminate at the apex and long-decurrent at the base. The very small pale-yellow flowers are usually produced in late August and early September in large, dense subterminal panicles, rendering the trees fairly conspicuous from a distance. The soft white wood is reported to be used for nondurable construction. Very common in lowland forests throughout the area.—Tinoco Station, sea level, *Allen 5645*—Puerto Jiménez, *Brenes 12157 & 12278.*

OCOTEA MOLLIFOLIA Mez & Pittier—Lauraceae—*Quizarra* (Local).

Small trees, about 35 ft. in height, with very large, alternate, short-petiolate leaves, the oblanceolate, acuminate blades soft-pilose on the lower surface and varying from 9–15″ in length. The oblong-ovoid fruits have large cupules with dentate margins, and are produced in late January and early February in very long, slender axillary panicles.—Esquinas Forest, 250 ft., *Allen 5824.*

OCOTEA NICARAGUENSIS Mez—Lauraceae.

Slender forest trees, 50–60 ft. in height, with very large, leathery, alternate, short-petiolate leaves, the broadly oblong-elliptic, glabrous blades obtuse or abruptly obtuse-acuminate at the apex and rounded at the base. The black, ellipsoidal fruits are produced on very strongly angled, subalate, axillary panicles in late March, the margins of the cupules irregularly lobate.—Frequent in forests near Palmar and Palmar Norte, *Allen 5503.*

OCOTEA PERGAMENTACEA Standl. & L. Wms.—Lauraceae.

Forest trees, about 60 ft. in height, with short-petiolate, alternate leaves, the narrowly oblong, thin blades with acuminate apices and decurrent bases, and averaging about 9″ in length. The very small yellow flowers are produced in late February in slender axillary panicles. The fruits have not been seen. Thus far known only from the type collection.—Forested hills near Palmar Norte, 1,500 ft., *Allen 5950.*

OCOTEA RIVULARIS Standl. & L. Wms.—Lauraceae.

Medium-sized, rather bushy trees, 30–50 ft. in height, with large, alternate, short-petiolate, glabrous leaves, the broadly oblanceolate,

rather leathery blades shortly obtuse-acuminate at the apex and long decurrent at the base, averaging about 1 ft. long and 6″ wide. The small green flowers are produced in late July and early August in slender axillary panicles, the strongly angled rachis being dark red in color. The ellipsoidal fruits are about ½″ long, the cupules with spreading marginal teeth.—Frequent in pastures and on the margins of forest near Esquinas, *Allen 5590.*

OCOTEA VERAGUENSIS (Meissn.) Mez—Lauraceae—*Aguacatillo, Canelo, Canelilla, Quizarra,* or *Sigua canela* (Costa Rica)—*Bambito* or *Sigua canelo* (Panama)—*Palo colorado* (Nicaragua)—*Canelo, Canelito,* or *Pimiento* (Salvador)—*Aguacatillo* (Honduras)—*Pimiento, Pimienton,* or *Pububuc* (Guatemala).

Medium-sized or large forest trees, 30–80 ft. in height, with alternate or subopposite, grayish-green, short-petiolate leaves, the blades oblong or elliptic-lanceolate in outline, usually with acute or rounded apices and cuneate bases. The bark is reported to have a cinnamon-like odor. The small, grayish-white flowers are produced from November until February in slender, subterminal panicles and are followed from February to June by ellipsoidal, apiculate fruits which average about ¾″ in length, the shallow cupule having a double margin. The dark walnut-brown wood is easy to work and is reported to be durable. It is suitable for furniture, paneling, interior trim, veneer, and general construction. Reported to be very common on the Atlantic slope near the Panama border.—Buenos Aires, 800 ft., *Tonduz 4032 & 6637.*

OCOTEA WILLIAMSII P. H. Allen, sp. nov.—Lauraceae—*Canelo* (Local). (Pl. 27)

Forest trees, to about 80 ft. in height. Leaves alternate, the lanceolate, elliptic-lanceolate, or oblanceolate blades 2¾–5¼″ long and ⅞–1¾″ wide with acute or very shortly acuminate apices and long-decurrent, very shortly petiolate bases. The upper leaf surface is glabrous, the lower surface glabrous, but with small but distinct tufts of tawny hairs in the axils of the secondary and sometimes tertiary veins. The small, pale-yellow flowers are produced in May in rather diffuse subterminal panicles 4½–6″ in length and are followed by black, large-seeded ellipsoidal fruits about 2″ long, which are subtended by a very shallow red cupule which is without marginal thickenings and conspicuously less than the fruit in diameter. Dr. Louis O. Williams, for whom the species is named, believes it to be allied to *O. Endresiana* and *O. Tonduzii,* but it will be found to be amply distinct in the curious axillary tufts of hairs on the lower leaf surfaces and in the unusually large fruits.—Margins of forest near Golfito, 100

ft., *Allen 5983* TYPE & *6246* (Herbarium Escuela Agrícola Panamericana).

Ocze—See *Hedyosmum mexicanum* (Guatemala).

OENOCARPUS PANAMANUS Bailey—Palmaceae—*Maquenque* (Local and Panama). (Pl. 6)

Handsome, multiple-trunked, unarmed palms, 20–65 ft. in height, with 8–10 arching pinnate fronds, the well-developed, often somewhat bulging crownshaft dark blackish purple in color. The smooth, almost bamboo-like individual canes are conspicuously ringed and average 4–6″ in diameter. The 2–4 inflorescences are borne well below the crownshaft and are at first covered by 2 cylindric spathes which are deciduous. The short, woody, nearly horizontal spadix has a great number of slender, pendulous, wine-red strands which bear the small white flowers and hard black or dark-purple, plumlike fruits. Frequent in hillside forests throughout the area.—Hills near Palmar, 100 ft., *Moore 6541* & *Allen 6680.*

Oj—See *Persea americana* (Guatemala).

Ojoche—See *Brosimum alicastrum* (Local and general). Fruits not enveloped in a perianth. Lateral veins of the lower leaf surface not prominent.

Ojoche—See *Brosimum terrabanum* (Local and general). Fruits not enveloped in a perianth. Lateral veins of the lower leaf surface prominent.

Ojoche—See *Pseudolmedia spuria* (Costa Rica). Fruits enveloped in a perianth.

Ojoche macho—See *Batocarpus costaricense* (Local). Leaves obtuse or shortly acute.

Ojoche macho—See *Brosimum sapiifolium* (Local). Leaves long-acuminate with entire margins.

Ojoche macho—See *Trophis racemosa* (Chiriquí). Leaves long-acuminate, with serrate margins.

Ojushte—See *Trophis racemosa* (Salvador).

OLACACEAE—KEY

1. Leaves rounded at the apex. Branches armed with short, sharp spines............*Ximenia*
1. Leaves acute or acuminate at the apex.
   2. Fruits subtended by a showy, red, accrescent calyx. Small trees, to about 30 ft.............*Heisteria*
   2. Fruits not subtended by a red, accrescent calyx. Tall trees, to about 100 ft.............*Minquartia*

OLEACEAE—One genus, *Linociera.*

OLIGANTHES FERRUGINEA Gleason—Compositae.

Small trees, 30–40 ft. in height, with broad, alternate, acuminate leaves 4–6″ in length and dense terminal white corymbiform panicles of 2-flowered heads. The species seems to be very closely allied to *O. discolor,* and may eventually prove to be identical.—Forests of Mano de Tigre, Diquis, 2,300 ft., *Pittier 12138*—Buenos Aires, *Tonduz 4910.*

Olivo—See *Sapium jamaicense* (Chiriquí). Plants with milky latex.
Olivo—See *Simarouba glauca* (Costa Rica). No latex.
Olivo macho—See *Sapium thelocarpum* (Panama).
Ollito—See *Eschweilera calyculata* (Chiriquí).

OLMEDIA FALCIFOLIA Pittier—Moraceae.

Small trees, to about 30 ft. in height with tomentulose branches and alternate, short-petiolate leaves, the linear or elliptic-oblong blades about 9″ in length, with abruptly long-acuminate apices and sinuate-serrate margins. The small dioecious flowers are produced in late January and early February from the leaf axils, the staminate form in small heads, the pistillate solitary or in pairs, surrounded by bracts, and followed by small drupaceous fruits. The species is very closely allied to *O. aspera* R. & P., if not actually identical.—Terraba, 450 ft., *Tonduz 12101*—Golfito, *Brenes 12139*—Esquinas Forest, 250 ft., *Skutch 5294* & *Allen 5823.*

On—See *Persea americana* (Guatemala).
Onc—See *Hedyosmum mexicanum* (Guatemala).
Onion cordia—See *Cordia alliodora* (Costa Rica).
Onj—See *Hedyosmum mexicanum* (Guatemala).
Oox—See *Bixa Orellana* (Guatemala).
Oreja de mula—See *Miconia argentea* (Panama).
Oreja de mula—See *Miconia impetiolaris* (Panama).

OREOPANAX—KEY

1. Leaves simple .... *Oreopanax capitatus*
1. Leaves digitately compound .... *Oreopanax xalapense*

OREOPANAX CAPITATUS (Jacq.) Dcne. & Planch.—Araliaceae—*C'ab-choh, Chuh, Cohetillo, Palo de cohetillo, Pata de leon, Tapacopal,* or *Tronador* (Guatemala).

Shrubs or small trees to about 45 ft. in height, often of epiphytic habit. The leaves are alternate and clustered at the ends of the branches and have very long, slender petioles, the glabrous blades variable in size and shape but usually broadly ovate in outline, with obtuse or short acuminate apices and cuneate, rounded, or subcordate bases. The small white flowers

are produced in attractive terminal panicles of capitate heads in January and February and are followed by small, globose, juicy fruits. A common, variable species of wide distribution in the American tropics.—Cañas Gordas, 3,300 ft., *Pittier 11163.*

OREOPANAX XALAPENSE (HBK) Dcne. & Planch.—Araliaceae—*Higuera* or *Mala gente* (Costa Rica)—*Aticuej, Chilil, Chilil mazorco, Mano de leon, Mano de tigre, Mata gente, M'shal,* or *Mazorco* (Guatemala)—*Brasil* (Salvador).

Small or medium-sized trees, to about 60 ft. in height, with alternate, long-petiolate, digitately compound leaves which are mostly clustered near the ends of the thick branches. The 5–9 nearly glabrous leaflets vary from 4 to about 12″ in length and are elliptic-lanceolate to oblanceolate in outline, with acute or acuminate apices and entire or somewhat serrate margins. The small white capitate flowers are produced in January and February in very long, slender terminal paniculate racemes. A striking and attractive highland species, easily recognized by the digitately compound leaves.—Cañas Gordas, 3,300 ft., *Pittier 11165.*

ORMOSIA PANAMENSIS Benth.—Leguminosae; Papilionatae.

Tall forest trees, to about 90 ft. in height, with alternate, pinnate leaves, the 7–9 narrowly lanceolate leaflets about 2–3″ long and up to 1¼″ wide. The attractive lavender flowers are produced in terminal panicles and are followed by dehiscent pods with several very pretty red and black seeds. Known only from a single collection of a quantity of fallen seeds, and hence open to some question.—Esquinas Forest, 250 ft., *Allen 6351.*

OURATEA VALERII Standl.—Ochnaceae.

Small trees, to about 30 ft. in height, with very large, alternate, shortly-petiolate, leathery leaves, the oblong-lanceolate acute blades 12–20″ long and 3½–5″ wide. The rich-yellow, attractive flowers are produced in October in large terminal, spreading panicles which about equal the leaves in length. Locally very common in wet lowland forest.—Playa Blanca de Golfo Dulce, *Valerio 435*—Area between the Estero Azul and the Río Sierpe, sea level, *Allen & Chittenden 6596.*

Ortiga—See *Urera caracasana* (Costa Rica).
Ortiga blanca—See *Urera caracasana* (Costa Rica).
Ox—See *Bixa Orellana* (Guatemala). No milky latex.
Ox—See *Brosimum alicastrum* (Guatemala). Milky latex.
Pacaya—See *Chamaedorea* sp. (Local and general).

PACHIRA AQUATICA Aubl.—Bombacaceae—*Jelinjoche, Quirihuillo,* or *Quiriguillo* (Costa Rica)—*Provision tree* (Bocas del Toro and Limón)—*Pumpunjuche, Saba nut, Cacao cimmaron,* or *Cacao de playa* (Nicaragua)—*Pumpunjuche* or *Zapoton* (Honduras)—*Shila blanca* (Salvador)—*Pumpunjuche, Uacoot, Zapote bobo,* or *Zapoton* (Guatemala). (Pl. 13)

Small or medium-sized trees, to about 65 ft. in height, with alternate, long-petiolate, digitately compound leaves, the 5–7 broadly oblanceolate, obtuse or apiculate leaflets often paler in color on the lower surface. The very large fragrant flowers are produced in short, few-flowered terminal fascicles in July and have very long, linear, creamy-white, rather fleshy petals which soon fall, the densely branched stamen tube deep wine red in color. The large russet-brown, ovoid, woody fruits are about 9″ long and 7″ in diameter and mature in December and January in our area. They contain many edible, chestnut-like seeds, which are often cooked and offered for sale in the Central American markets. Occasional in wet lowland forests and open freshwater swamps, the trees in such situations often having a few stilt roots.—Esquinas Forest, 50 ft., *Allen 6270.*

Pacito—See *Muntingia calabura* (Panama).
Pacon—See *Sapindus saponaria* (Nicaragua and Honduras).
Pacun—See *Sapindus saponaria* (Guatemala and Salvador).
Pacxoc—See *Theobroma Cacao* (Guatemala).
Paipute—See *Casearia arguta* (Costa Rica). Trunks unarmed.
Paiputo—See *Xylosma excelsum* (Local). Trunks armed with dense fascicles of long spines.
Palal—See *Inga laurina* (Guatemala).
Paleta—See *Dialium guianense* (Honduras and Guatemala).
Paleto—See *Dialium guianense* (Honduras).
Paleto negro—See *Dialium guianense* (Honduras).

PALMACEAE—KEY

1. Plants armed with spines.
   2. Leaves fan-shaped .... *Cryosophila guagara*
   2. Leaves pinnate, or entire.
      3. Individual leaflets conspicuously wedge-shaped, broadest at the apex. Lower trunk with conspicuous prickly stilt roots.
         4. Stilt roots in mature specimens as high as a man. Stamens more than 50. Seeds with an apical embryo .... *Socratea durissima*
         4. Stilt roots in mature specimens about 2–3 ft. high. Stamens about 14. Seeds with a lateral embryo .... *Iriartea gigantea*
      3. Individual leaflets never wedge-shaped or broadest at the apex.

4. Trunks solitary.
 5. Trunks very short and massive, often reclining. Spines confined to the margins of the base of the frond........*Corozo oleifera*
 5. Trunks of varying diameter but never short, massive, or reclining. Spines abundant on the trunk and other parts.
  6. Trunks more than 8″ in diameter. Fruits globose, greenish yellow........*Acrocomia vinifera*
  6. Trunks about 6″ or less in diameter. Fruits not globose or greenish yellow.
   7. Fruits intensely spiny, in short compact clusters. Pinnae often in broad, confluent blocks........*Astrocaryum alatum*
   7. Fruits never spiny, in long pendulous clusters. Pinnae never in broad confluent blocks........*Astrocaryum Standleyanum*
4. Trunks multiple.
 5. Fronds entire, with a bifid apex.
  6. Fronds more than 6 ft. long. Common palms of swampy forests at sea level........*Bactris militaris*
  6. Fronds less than 3 ft. long. Rare palms of forested ridges at 1,500–2,000 ft........*Bactris* sp.—*Allen 6765*
 5. Fronds pinnatisect.
  6. Spines conspicuously pale tan in color, winged or flattened throughout most of their length........*Bactris divisicupula*
  6. Spines not pale tan or conspicuously flattened.
   7. Fruits about ½″ in diameter, red in color at maturity. Spathes densely woolly but not armed with spines on the expanded portion........*Bactris Baileyana*
   7. Fruits about 1″ in diameter, deep purple at maturity. Spathes intensely spiny on the expanded portion........*Bactris balanoidea*

1. Plants not armed with spines.
 2. Fronds entire, bifid at the apex........*Asterogyne Martiana*
 2. Fronds pinnatisect.
  3. Individual leaflets wedge-shaped, broadest at the apex. Lower trunk with stilt roots.
   4. Stilt roots in mature specimens usually more than 6 ft. in height. Stamens more than 50. Seeds with an apical embryo........*Socratea durissima*
   4. Stilt roots in mature specimens about 2–3 ft. high, often absent in young plants. Stamens about 14. Seeds with a lateral embryo........*Iriartea gigantea*
  3. Individual leaflets never wedge-shaped or broadest at the apex.
   4. Trunks multiple.
    5. Fruits covered with overlapping scales. A species usually found in nearly pure stands in coastal swamps........*Raphia taedigera*
    5. Fruits not scaly. Plants not found in swamps nor in pure stands.
     6. Staminate and pistillate flowers produced on separate scapes........*Chamaedorea Woodsoniana*
     6. Staminate and pistillate flowers produced on the same scape.
      7. Spadix broomlike, the slender, rodlike basal part much longer than the terminal cluster of flowering or fruiting strands........*Synechanthus angustifolius*

7. Spadix not broomlike, the basal part much shorter than the flowering or fruiting strands.
 8. Flowers produced in deep pits in the rachis.
 9. Canes about 2″ in diameter. Fruits about ½″ in diameter..*Geonoma congesta*
 9. Canes less than ½″ in diameter. Fruits about ¼″ in diameter...*Geonoma* sp.—*Allen 6750*
 8. Flowers not immersed in pits in the rachis.
 9. Trunks 4–6″ in diameter. Scapes large with many pendulous strands 1 ft. or more in length..........................*Oenocarpus panamanus*
 9. Trunks less than 2″ in diameter. Scapes small, the strands less than 1 ft. long..*Hyospathe Lehmannii*

4. Trunks solitary.
 5. Trunks very short and massive, often reclining. Fruits produced in large, very compact clusters which are deeply seated in the axils of the fronds..*Corozo oleifera*
 5. Trunks not massive or reclining. Fruits not in compact clusters in the frond axils.
 6. Plants large, usually 30–65 ft. or more in height.
 7. Fruits usually more than 6″ in diameter. Cultivated or naturalized on sea beaches..*Cocos nucifera*
 7. Fruits less than 2″ in diameter.
 8. Strands of the rachis about 1″ in diameter and octagonal in cross section. Flowers produced in deep pits. Fruits almond-shaped..*Welfia Georgii*
 8. Strands of the rachis much less than 1″ in diameter and never octagonal in cross section. Fruits not almond-shaped.
 9. Trunks slender, usually less than 6″ in diameter. Fruits globose, about ½″ in diameter...*Euterpe panamensis*
 9. Trunks massive, to about 12–18″ in diameter. Fruits ellipsoidal, conspicuously beaked at the apex.......................................*Scheelea rostrata*
 6. Plants usually less than 20 ft. in height.
 7. Plants dwarf, to about 8 ft. to the tips of the fronds, the trunk usually less than 1 ft. in height.
 8. Spadix spicate, undivided. Staminate and pistillate flowers both present on the same scape......................................*Neonicholsonia Watsonii*
 8. Spadix branched. Staminate and pistillate flowers produced on separate scapes.
 7. Plants with trunks more than 3 ft. in height.
 8. Staminate and pistillate flowers produced on separate scapes.
 9. Scapes erect. Plants found in wet forests at sea level...*Chamaedorea* sp.—*Allen 6262*
 9. Scapes pendulous. Plants found on forested ridges at 1,800–2,000 ft.
 10. Fruits obovoid, less than ½″ long...*Chamaedorea* sp.—*Moore 6527*
 10. Fruits oblong, more than ⅝″ long...*Chamaedorea Wendlandiana*
 8. Staminate and pistillate flowers produced on the same scape.

9. Flowers borne in pits on the rachis. Fruits dark purple or black...... *Geonoma binervia*

9. Flowers not borne in pits. Fruits yellow, orange, or red................. *Synechanthus angustifolius*

Palma conga—See *Welfia Georgii* (Local and Panama).

Palma real—See *Raphia taedigera* (Local). Fruits scaly.

Palma real—See *Scheelea rostrata* (Local and Panama). Fruits prominently beaked at the apex, not scaly.

Palma real—See *Welfia Georgii* (Local and Panama). Fruits not scaly or beaked.

Palmiche—See *Corozo oleifera* (Nicoya).

Palmilera—See *Socratea durissima* (Boruca).

Palmito—See *Socratea durissima* (Costa Rica). Plants with conspicuous stilt roots.

Palmito—See *Welfia Georgii* (Local). Plants without stilt roots.

Palo amarillo—See *Diphysa robinioides* (Guatemala).

Palo bayo—See *Vochysia hondurensis* (Guatemala).

Palo blanco—See *Godmania aesculifolia* (Guatemala). Leaves digitately compound.

Palo blanco—See *Vernonia patens* (Chiriquí). Leaves simple.

Palo camaron—See *Hamelia patens* (Costa Rica).

Palo chancho—See *Vochysia hondurensis* (Nicaragua).

Palo chino—See *Bursera simaruba* (Honduras and Guatemala).

Palo colorado—See *Ocotea veraguensis* (Nicaragua).

Palo criollo—See *Minquartia guianensis* (Panama).

Palo cuadrado—See *Cornutia grandifolia* (Panama). Flowers blue.

Palo cuadrado—See *Macrocnemum glabrescens* (Costa Rica and Panama). Flowers pink.

Palo de agua—See *Bravaisia integerrima* (Costa Rica, Panama, and Nicaragua). Trees with conspicuous stilt roots.

Palo de agua—See *Godmania aesculifolia* (Guatemala). Leaves digitately compound.

Palo de agua—See *Hedyosmum mexicanum* (Guatemala). Leaves simple, the inflated bases joined and clasping the stem.

Palo de agua—See *Psychotria chiapensis* (Guatemala). Leaves simple, the bases not inflated.

Palo de barril—See *Jacaratia dolichaula* (Costa Rica).

Palo de buba—See *Jacaranda copaia* (Panama).

Palo de chancho—See *Vochysia hondurensis* (Costa Rica).

Palo de cohetillo—See *Oreopanax capitatus* (Guatemala).

Palo de cruz—See *Plumeria rubra* forma *acutifolia* (Guatemala). Plants with milky latex.

Palo de cruz—See *Randia armata* (Guatemala). Plants without milky latex.

Palo de culebra—See *Lafoensia punicifolia* (Guatemala).

Palo de danto—See *Cornutia grandifolia* (Local).

Palo de escoba—See *Zexmenia frutescens* (Guatemala).

Palo de guaco—See *Crataeva tapia* (Panama).

Palo de gusano—See *Lonchocarpus guatemalensis* (Guatemala).

Palo de Lacandon—See *Dialium guianense* (Guatemala).

Palo de la cruz—See *Casearia guianensis* (Panama).

Palo de lagarto—See *Zanthoxylum procerum* (Costa Rica).

Palo de leche—See *Brosimum utile* (Panama). Leaves without glands at the base of the blade.

Palo de leche—See *Sapium thelocarpum* (Local). Leaves with 2 glands at the base of the blade.

Palo de Maria—See *Calophyllum braziliense* var. *Rekoi* (Honduras).

Palo de Mayo—See *Vochysia* sp. (Local and Panama).

Palo de peine—See *Calycophyllum candidissimum* (Guatemala).

Palo de piedra—See *Minquartia guianensis* (Costa Rica).

Palo de sal—See *Avicennia nitida* (Local and general). Leaves petiolate, without glands.

Palo de sal—See *Laguncularia racemosa* (Costa Rica). Leaves petiolate, with 2 glands at the base of the blade.

Palo de sal—See *Pelliciera rhizophorae* (Panama). Leaves sessile.

Palo de sangre—See *Acalypha diversifolia* (Guatemala).

Palo de sebo—See *Virola guatemalensis* (Guatemala).

Palo de vaca—See *Brosimum utile* (Local).

Palo de zope—See *Phyllanthus acuminatus* (Salvador).

Palo jiote—See *Bursera simaruba* (Honduras and Guatemala).

Palo malin—See *Vochysia ferruginea* (Panama, Madden area).

Palo Maria—See *Casearia arguta* (Costa Rica).

Palomo—See *Lafoensia punicifolia* (Local).

Palo morillo—See *Trophis chorizantha* (Guatemala).

Palo mulato—See *Bursera simaruba* (Honduras and Guatemala). Leaves pinnate.

Palo mulato—See *Triplaris melaenodendron* (Guatemala). Leaves simple.

Palo negro—See *Miconia argentea* (Panama).

Palo obero—See *Astronium graveolens* (Honduras and Guatemala).
Palo santo—See *Triplaris melaenodendron* (Panama).
Panama—See *Sterculia apetala* (Local, Panama, and Nicaragua).
Papaturo—See *Coccoloba Standleyana* (Local).
Papaturrillo—See *Coccoloba acuminata* (Nicaragua).
Papaturro agrio—See *Bellucia costaricensis* (Costa Rica).
Papaya—See *Carica papaya* (Local and general). Leaves simple, but deeply lobed.
Papaya—See *Jacaratia dolichaula* (Panama). Leaves digitately compound.
Papaya de monte—See *Jacaratia dolichaula* (Costa Rica).
Papayo—See *Terminalia amazonia* (Local).
Papayo de montaña—See *Carica pennata* (Guatemala).
Papayo de monte—See *Carica pennata* (Guatemala).
Papayillo—See *Jacaratia dolichaula* (Costa Rica).
Papayillo de venado—See *Carica pennata* (Local). Leaves simple, but deeply lobed.
Papayillo de venado—See *Jacaratia dolichaula* (Local). Leaves digitately compound.
Papelillo—See *Miconia argentea* (Panama).
Papito de monte—See *Malvaviscus arboreus* var. *penduliflorus* (Panama).

PARMENTIERA MACROPHYLLA Standl. (?)—Bignoniaceae—*Wild calabash* (Bocas del Toro).

A single fallen fruit, possibly of this species has been found in the forested hills above Palmar Norte at about 1,800 ft.

Pasac—See *Simarouba glauca* (Guatemala).
Pasta—See *Hirtella americana* (Honduras).
Pastora de montaña—See *Warscewiczia coccinea* (Local).
Pata de leon—See *Oreopanax capitatus* (Guatemala).
Pata de vaca—See *Bauhinia ungulata* (Guatemala).
Pata de venado—See *Bauhinia ungulata* (Salvador).
Pataiste—See *Theobroma bicolor* (Costa Rica).
Pataste—See *Theobroma bicolor* (Costa Rica).
Paternillo—See *Inga laurina* (Salvador).
Paterno—See *Swartzia panamensis* (Honduras).
Paujil—See *Croton glabellus* (Guatemala).
Pava—See *Didymopanax Morototoni* (Costa Rica and Panama).

Pavilla—See *Didymopanax Morototoni* (Costa Rica). Leaves digitately compound.

Pavilla—See *Cornutia grandifolia* (Costa Rica). Leaves simple.

Pavo—See *Didymopanax Morototoni* (Costa Rica).

Pegle—See *Vochysia ferruginea* (Panama).

Peinecillo—See *Apeiba Tibourbou* (Chiriquí).

Peine de mico—See *Apeiba aspera* (Local and Guatemala). Spines of fruits short and hard.

Peine de mico—See *Apeiba Tibourbou* (Local and Panama). Spines of fruits long and pliant.

Pejibaye—See *Astrocaryum Standleyanum* (Local). Wild species. Pulp of fruits very thin.

Pejibaye—See *Guilielma utilis* (Costa Rica). Cultivated species. Pulp of fruits thick.

Pela nariz—See *Croton xalapensis* (Honduras).

Pellejo de toro—See *Tachigalia versicolor* (Local).

PELLICIERA RHIZOPHORAE Triana & Planch.—Theaceae—*Mangle piñuela* (Local)—*Palo de sal* (Panama). (Pls. 5 & 20)

Small trees, usually about 15–30 ft. in height but sometimes as much as 60 ft., with a very straight, unbranched trunk and narrow crown. The trees have a very strongly developed, deeply fluted pyramidal buttress at the base, even in young specimens, and are unique in this character in the Mangrove association. The alternate, leathery leaves are clustered at the ends of the branches and are oblanceolate in outline with acute apices and sessile bases, the margins often with many small brown glands which give them a serrate appearance. The large and attractive pink or white flowers are sessile and subterminal, and are subtended by 2 stipule-like bracts. The large, woody, reddish-brown, deeply corrugated fruits are broadly ovate or subcordate in profile, with the apical portion abruptly contracted into a pointed beak. The species is very common in the local Mangrove swamps, particularly back of shingle beaches near the mouths of small streams on Golfo Dulce, where they seem to flower and fruit at irregular intervals throughout the year, but with a particularly heavy fruiting season from August to about October.—Diquis, *Tonduz 6723*—Delta of the Río Esquinas, Golfo Dulce, *Allen 5628*.

Pelo de indio—See *Hirtella americana* (Nicaragua).

PELTOGYNE PURPUREA Pittier—Leguminosae; Caesalpinoideae—*Naza-*

reno (Local)—*Morado, Nazareno,* or *Purple heart* (Panama and general). (Fig. 17)

Very tall, handsome forest trees, 120–150 ft. in height and up to about 3 ½ ft. in diameter above the moderately buttressed base. The alternate, pinnate leaves are composed of a single pair of elliptic-lanceolate, long-acuminate leathery leaflets which are from 2–3″ long. The trees are briefly deciduous in late December or early January, the brown leaves being somewhat conspicuous before they fall. The new foliage appears almost immediately, and the trees can be distinguished for miles when in this condition. The small white fragrant flowers are produced in subterminal panicles, individual specimens flowering several times, from early August until about mid-September at intervals of about 2 weeks, each crop of flowers lasting about 3 days. The brown, flat, obliquely obovoid-oblong 1-seeded pods are about 2″ long and mature in great abundance in early February. The moderately hard, very heavy wood is rich purple in color when freshly cut, deepening somewhat upon exposure. It is used for dock fenders in Golfito and makes beautiful paneling, furniture, inlaid work, or flooring. Some of the doors in the National Museum in San José are made of this wood. Fairly common in the rainy hills near Golfo Dulce and in the Esquinas Forest, but always on high, well-drained land.—Esquinas Forest, 200 ft., *Allen 5595, 5608, & 5854.*

PENTAGONIA GYMNOPODA Standl.—Rubiaceae.

Slender trees 15–20 ft. in height with very large, opposite, pinnately lobed leaves which are about 2 ft. in length, with linear acuminate lateral segments. The rather large tubular, pale-pink flowers are produced in February in dense sessile terminal clusters enveloped by 2 very broad, ovate, acuminate bracts. The russet-brown fruits are nearly globose and average about 1″ in diameter. Common in hillside forests throughout the area, to about 1,800 ft. in elevation.—Hills near Palmar Norte, *Allen 5925.*

PENTAPLARIS DOROTEAE L. Wms. & Standl.—Tiliaceae. (Fig. 18)

Handsome gray-barked trees 120 ft. in height and about 30″ in trunk diameter with open, spreading, rather flat-topped crowns. Leaves alternate, with slender petioles, the blades usually narrowly ovate-lanceolate, acute or shortly acuminate at the apex and cordate at the base. The small flowers are produced during early December within a conspicuous pink or rich reddish-brown 5-lobed calyx in very showy terminal racemes or panicles, the color effect when seen from a distance being reminiscent

Fig. 17. Peltogyne purpurea. 1 — Fruiting branch. 2 — Cross section of seed pod. 3 — Ovary and style. 4 — Open flower. 5 — Partially opened flower bud.

Fig. 18. Pentaplaris Doroteae. 1 — Flowering branch. 2 — Calyx, with 2 lobes removed, showing ovary and corolla.

of *Triplaris.* The sapwood is not differentiated from the heartwood, both being hard and light brown in color. A very remarkable genus with no close relatives in the family to which it has been assigned. The species is dedicated to Dorothy O. Allen, who has prepared the line drawings which accompany the text. Locally common on the precipitous slopes of deep ravines near Palmar Norte at about 1,000–1,500 ft. Thus far known only from the type collection.—*Allen 6326.*

Pepenance—See *Ximenia americana* (Costa Rica and Salvador).
Pepeto—See *Inga punctata* (Salvador).
Pepinillo—See *Cyphomandra costaricensis* (Costa Rica).

PEREBEA—KEY

1. Fruits glabrous. Leaves shortly acute or acuminate, with entire margins .... *Perebea trophophylla*
1. Fruits velvety. Leaves abruptly caudate-acuminate, often with sinuate-serrate margins .... *Perebea molliflora*

PEREBEA MOLLIFLORA Standl. & L. Wms.—Moraceae.

Medium-sized forest trees, 30–50 ft. in height, with alternate, short-petiolate, nearly glabrous leaves 7–10″ long, the oblong-elliptic blades with abruptly caudate-acuminate apices and rounded bases. The minute white flowers are produced on round, flattened axillary receptacles about ¼″ in diameter in mid-February and are soon followed by sessile clusters of small, obovoid blood-red velvety fruits.—Frequent in the forested hills above Palmar Norte at 1,500–1,800 ft., *Allen 5933 & 5952.*

PEREBEA TROPHOPHYLLA Standl. & L. Wms.—Moraceae. (Pl. 32)

Slender trees, about 65 ft. in height, with alternate, oblong-lanceolate, abruptly acute or acuminate, short-petiolate, leathery leaves 6½–9″ long and 2½–3½″ wide, all parts of the plant with sticky, milky latex. The small flowers are produced on a broad, disklike, multibracteate axillary receptacle and are followed by sessile clusters of bright-red, glabrous fruits.—Forested hills near Golfito, 100 ft., *Allen 6348.*

Perescuch—See *Croton glabellus* (Guatemala).

PERSEA—KEY

1. Floral pedicels with a conspicuous pair of linear bracts .... *Persea pallida*
1. Floral pedicels without bracts.
   2. Fruits about ¼″ in diameter .... *Persea Skutchii*
   2. Fruits 3″ or more in diameter .... *Persea americana*

PERSEA AMERICANA Mill.—Lauraceae—*Aguacate* or *Avocado* (Local and general)—*Aguacate, Ju, O, Oj, On, Um,* or *Un* (Guatemala). (Pl. 27)

The familiar long-fruited West Indian form of this species is frequently cultivated in townsites throughout the area. In addition to these there is, however, a wild tree which often reaches 100 ft. in height and up to 2 ft. in trunk diameter, which is found in the hillside climax forests in association with things like *Brosimum utile, Anacardium excelsum,* and *Apeiba aspera,* but always on clay and above the level of the swamps. The alternate leaves have slender petioles, the glabrous blades elliptic-oblong or oblanceolate in outline with shortly acute apices and cuneate bases and paler in color on the lower surface. The *globose* green fruits mature in late May and June and average about 3″ in diameter and have an edible flesh about 3/8″ thick, which is sweet and much relished by the local bands of spider monkeys. What would appear to be this same wild form is occasionally grown in the small valleys near the delta of the Río Esquinas and by the Boruca Indians near Palmar Norte. The freshly cut seeds of this and the more familiar forms leave an indelible stain on cloth or skin and can be used for printing monograms or decorative designs, which can be cut into the surface. Since the cultivated varieties of the avocado do not thrive on heavy clay or in areas of high rainfall it would seem that this forest type might have great promise as a rootstock for some of the better kinds, since it grows under conditions that would quickly eliminate any of its known relatives.—Esquinas Forest, 250 ft., *Allen 5552.*

PERSEA PALLIDA Mez & Pittier—Lauraceae—*Timber sweetwood* (Bocas del Toro)—*Pizarra* (Panama).

Medium-sized forest trees, 30–75 ft. in height, with alternate, elliptic-oblong, shortly acuminate leaves and slender axillary racemes of relatively large white flowers.—Valle de Coto, Río Diquis, 4,800 ft., *Pittier 11111*—Hills above Palmar Norte, 1,800–2,000 ft., *Allen 5885.*

PERSEA SKUTCHII Allen—Lauraceae—*Aguacatillo* (Costa Rica).

Small or medium-sized trees 15–45 ft. in height, with alternate, elliptic to oblong, obtuse, acute or acuminate leaves and attractive yellowish flowers followed by small globose fruits.—Buenos Aires, 900 ft., *Tonduz 6680*—Boruca, 1,400 ft., *Tonduz 6861*—Vicinity of Palmar Norte, 100 ft., *Allen 6722.*

Pestaño de mula—See *Heliocarpus appendiculatus* (Nicaragua).

PHOEBE COSTARICANA Mez & Pittier ex Mez—Lauraceae—*Sigua* or *Sigua blanco* (Panama).

Medium-sized trees, up to about 60 ft. in height, with alternate, leathery, glabrous leaves, the blades narrowly elliptic in outline, with acute or

subacuminate apices and cuneate bases. The small yellowish- or greenish-white flowers are produced in slender axillary or subterminal panicles, which exceed the leaves in length, and are followed by ellipsoidal, apiculate fruits, the subtending cupule with thickened marginal lobes. A fairly frequent highland species, known in our area only from Cañas Gordas.—*Pittier 11107.*

PHYLLANTHUS ACUMINATUS Vahl—Euphorbiaceae—*Chile* (Local)—*Chilillo* or *Gallina* (Costa Rica)—*Jobotillo* (Panama)—*Pimientillo* or *Palo de zope* (Salvador)—*Ciruelillo* (British Honduras).

Shrubs or small trees, 6 to about 20 ft. in height, with small, short-petiolate, oblong-ovate, acute or acuminate leaves which average about 1″ in length. The green primary branches and their slender distichous secondary branchlets produce the effect of a large, bipinnate, fernlike frond. The minute, greenish-yellow flowers are produced in dense axillary fascicles in June and July and are soon followed by the small green, nearly globose, 3-celled capsular fruits. The species is usually listed in floras and hand lists as *P. brasiliensis* (Aubl.) Poir., a South American plant whose leaves and stems are crushed and used as a barbasco for catching fish. Very common along trails and in brushy places near Palmar and Palmar Norte at low elevations.—*Tonduz 6727* & *Allen 5568.*

Phyllanthus brasiliensis—See *Phyllanthus acuminatus.*

PHYLLOCARPUS SEPTENTRIONALIS Donn. Sm.—Leguminosae; Caesalpinoideae—*Flor de mico, Guacamayo,* or *Guatemalan flame tree* (Honduras, Guatemala, and general). (Pl. 13)

Tall, very handsome forest trees, 90–120 ft. in height and up to 36″ in trunk diameter, with alternate, pinnate leaves which have 4–6 pairs of elliptic-oblong, obtuse or acute, glabrous leaflets 1½-2″ in length. The trees are leafless from late November until and during the flowering season in late December and early January, the small blood-red flowers being produced in dense clusters of short racemes from the defoliate nodes, rendering the trees very conspicuous and attractive when seen against the dark hillsides. The fruits have not been seen, but are reported to be thin, oblong, compressed, indehiscent 1- or 2-seeded pods 4½–7″ long. The nearly white wood is moderately hard and heavy, rather coarse-textured, works easily and finishes smoothly but is probably not durable. It has no known uses. The species is locally frequent in the low, forested hills between Farm 18 and Tinoco and occasional near Palmar and Palmar Norte. Maximum flowering seems to be limited to years in which there

is an early onset of the dry season, with little or no rain in December. In wet years a few blooms are set, but the trees do not produce the magnificent burst of color for which they are so justly famous. Young specimens are common in cultivation in both Palmar and Golfito, having been imported from the Canal Zone Experiment Station at Summit before it was realized that the species was native in the area, but few of these have reached blooming size.—Forested hills near Tinoco, 100 ft., *Allen 5698.*

Picapica—See *Myriocarpa longipes* (Salvador).
Pico de gorrion—See *Acacia spadicigera* (Guatemala). Leaves pinnate. Plants thorny.
Pico de gorrion—See *Malvaviscus arboreus* (Guatemala). Leaves simple. Plants not thorny.
Pico de gurrion—See *Acacia spadicigera* (Guatemala).

PICRAMNIA LATIFOLIA Tul.—Simaroubaceae—*Coralillo* (Local).

Small forest trees, 15–30 ft. in height, with large, alternate, pinnate leaves, the 5–8 broadly elliptic or ovate, shortly acuminate, almost glabrous leaflets alternate on the rachis and 2½–4″ long and 1½–2″ wide. The small greenish-white or pinkish-green flowers are produced on slender pendulous catkin-like spikes from the nodes, and may be from about 6″–1 ft. in length. The small, ellipsoidal scarlet fruits are borne in late March and early April and are fairly conspicuous. All parts of the plant, but particularly the bark and wood, are intensely bitter. Frequent throughout the area.—Woodlands near Jalaca Station, 25 ft., *Allen 5206*—Forested hills near Palmar Norte, 200 ft., *Allen 5511 & 6671.*

Pie de cabro—See *Bauhinia ungulata* (Salvador).
Pie de vaca—See *Bauhinia ungulata* (Honduras).
Pie de venado—See *Bauhinia ungulata* (Guatemala).
Pigeon plum—See *Hirtella americana* (British Honduras). Leaves alternate.
Pigeon plum—See *Myrcia Oerstediana* (British Honduras). Leaves opposite.
Pijibay—See *Guilielma utilis* (Nicaragua).
Pilijushte—See *Trophis racemosa* (Salvador).
Pilon—See *Andira inermis* (Panama). Leaves pinnate.
Pilon—See *Hieronyma tectissima* (Local). Leaves simple.
Pimento—See *Myrcia Oerstediana* (Panama).

Pimientillo—See *Phyllanthus acuminatus* (Salvador). Plants without yellow sap.

Pimientillo—See *Symphonia globulifera* (Honduras and Guatemala). Abundant sticky yellow sap.

Pimiento—See *Nectandra salicifolia* (Salvador). Leaves mostly acute to caudate acuminate at the apex.

Pimiento—See *Ocotea veraguensis* (Salvador and Guatemala). Leaves mostly obtuse or obtusely acute at the apex.

Pimienton—See *Ocotea veraguensis* (Guatemala).

Piñon—See *Jatropha curcas* (Honduras and Panama).

Pipa—See *Cocos nucifera* (Immature fruits) (Local and general).

PIPER—KEY

1. Leaves with 5–7 longitudinal nerves........ *Piper reticulatum*
1. Leaves with a single longitudinal nerve........ *Piper arboreum*

PIPER ARBOREUM Aubl.—Piperaceae.

Stout woody herbs, sometimes arborescent and up to 15–18 ft. in height, with alternate, elliptic-oblong, acuminate leaves about 7–8″ long, the bases short-petiolate and very strongly asymmetrical. The minute flowers are produced in slender axillary spikes, as in the genus.—Esquinas Forest, 50 ft., *Skutch 5296*.

PIPER RETICULATUM L.—Piperaceae—*Cordoncillo* (Local).

Shrubs or small trees, to about 18 ft. in height, with very large, alternate, short-petiolate, nearly glabrous leaves, the broadly elliptic-ovate blades with abruptly acuminate tips and rounded bases, averaging about 1 ft. in length and 6–7″ in width. The minute greenish-white flowers are borne on slender, erect pseudoterminal spikes. Common along trails and in brushy places throughout the area.—Woods near Río Nuevo de Osa, *Brenes 12212*—Esquinas Forest, *Skutch 5404* & *Allen 5727*.

PIPERACEAE—One genus, *Piper*.

Pisabed—See *Cassia spectabilis* (British Honduras).

Pisma—See *Brosimum terrabanum* (Honduras).

Pissi—See *Hamelia patens* (Costa Rica).

Pit—See *Enterolobium cyclocarpum* (Guatemala).

PITHECOLOBIUM—KEY

1. Leaves once pinnate........ *Pithecolobium glanduligerum*
1. Leaves bipinnate.
    2. Leaflets 3 or 4 in number........ *Pithecolobium longifolium*
    2. Leaflets more than 4.

3. Lower leaf rachis with a conspicuous, vase-shaped gland .... *Pithecolobium macradenium*
3. Lower leaf rachis without a conspicuous gland .... *Pithecolobium austrinum*

PITHECOLOBIUM AUSTRINUM Standl. & L. Wms.—Leguminosae; Mimosoideae—*Algarrobo* or *Tamarindo* (Local). (Pl. 13)

Medium-sized to very large forest trees, 75–120 ft. in height and up to about 3 ft. in trunk diameter. Old specimens often have spreading, flattened crowns and rough bark, which scales off in large, loose plates. The alternate, fernlike leaves are bipinnate, with about 8–12 pairs of pinnae, the terete rachis with a very small gland at the base of each pair. The small, linear-oblong, rather oblique, obtuse, glabrous leaflets are about ½″ in length, occurring as 18–24 sessile pairs, each pair with 2 small but conspicuous glands at their base. The small, white, fragrant flowers are produced in dense, long-pedunculate, globose heads about 1″ in diameter in late February or early March and are followed in about 2 weeks by the conspicuous red, pendulous, many-seeded fleshy pods. The trees are briefly deciduous in early December, the new flushes of bright lettuce-green foliage rendering them very conspicuous in the forest from late December until early January. The species is closely allied to *Pithecolobium arboreum* (L.) Urban of northern Central America and the West Indies, but differs in the much larger habit, and in small but well-marked flowering and fruiting characters. The dark, hard, heavy, reddish-brown wood may be found to be suitable for flooring, general construction, cabinetwork, paneling, and furniture, but is not used at present. The species seems to lend itself well to reforestation projects and has grown well in trial plots at Esquinas. A very common and characteristic local tree, found throughout the area from sea level to about 2,500 ft. in elevation.—Hills near Palmar, 40 ft., *Allen 5230, 5468, & 5520.*

PITHECOLOBIUM GLANDULIGERUM Standl. & L. Wms.—Leguminosae; Mimosoideae.

Small trees, about 25–30 ft. in height, with alternate, bipinnate leaves (sometimes appearing superficially to be once-pinnate), the asymmetrical individual leaflets about 2½″ long and 1″ in width. The white flowers are produced in showy globose heads about 2″ in diameter at the ends of the slender, pendulous axillary scapes which are up to about 6″ in length.—Forests near Palmar Norte, 100 ft., *Allen 6340.*

PITHECOLOBIUM LONGIFOLIUM (H. & B.) Standl.—Leguminosae; Mimosoideae—*Azote caballo* (Local)—*Sotacaballo* (Costa Rica)—*Azote caballo* or *Swampwood* (Nicaragua)—*Mayamaya* (Honduras).

Small trees, 12–35 ft. in height, often with gnarled, multiple trunks, usually confined to rocky riverbanks, where they form nearly pure stands overhanging the water. The rather large, alternate, bipinnate leaves consist of a single, often unequal pair of pinnae, the mostly 2–3 short-petiolate, obliquely lanceolate, glabrous leaflets 1½–3″ long, with obtuse or acute apices and cuneate bases. The small, pale-pink, very fragrant flowers are produced in several successive flowerings from about October to December in short, sessile fascicles of spikes from the older, woody branches, and are soon followed by the long, slender, flattened, somewhat curved, many-seeded pods which mature in May. The wood is hard and heavy, but is seldom used because of the difficulty in securing large sizes. Very common along the Río Terraba above the influence of the tide.—Vicinity of La Presa, Río Terraba, 100 ft., *Allen 5285*—Banks of the Río Terraba, near Palmar, *Allen 6604.*

PITHECOLOBIUM MACRADENIUM Pittier—Leguminosae; Mimosoideae—*Guavo* or *Guavo de montaña* (Local).

Very large forest trees, 90–120 ft. in height and 3–4 ft. in trunk diameter, with rather large, alternate, bipinnate leaves, which have 3 or 4 pairs of pinnae, one of the lowermost pair usually being replaced by a very large, unique, funnel- or vase-shaped gland up to ⅝″ in length. The 5–8 pairs of obliquely oblong, nearly glabrous leaflets are broadly rounded at the apex, with inequilateral, truncate, nearly sessile bases, the uppermost 3 or 4 pairs having a minute, stipitate, cupular gland just below the point of insertion. The small white flowers are produced in late March in dense globose, subterminal pedunculate heads about 1½″ in diameter, rendering the trees fairly conspicuous from a distance. The fruits are narrow, thick, laterally compressed, strongly curved, many-seeded pods which average about 4″ in length. The very hard, white, fine-grained wood has no known uses. The species is immediately separable from all others in the genus by the enormous petiolar glands.—Occasional in hillside forests near Palmar Norte at about 1,500 ft., *Allen 6042.*

Pixoy—See *Guazuma ulmifolia* (Guatemala).
Pizarra—See *Persea pallida* (Panama).
Platano—See *Hieronyma tectissima* (Local).

PLATYMISCIUM PINNATUM (Jacq.) Dugand—Leguminosae; Papilionatae—*Cachimbo, Cristobal,* or *Quira* (Local and Costa Rica generally)—*Sangrillo* (Chiriquí)—*Swamp kaway* (Bocas del Toro)—*Coyote* (Nicaragua)—*Granadillo* or *Hormigo* (Honduras).

Medium-sized or large forest trees, 65 to about 100 ft. in height and

up to 3 ft. in trunk diameter, with opposite, odd-pinnate leaves, the 3–5 short-petiolate, broadly ovate, acute or acuminate, glabrous leaflets from 2–8″ in length. The trees are usually deciduous during the dry season. The small, pale-orange flowers are produced from December to about April, in clusters of slender racemes from the subterminal or axillary defoliate nodes, rendering the trees very conspicuous and attractive from a distance. The fruits are flat, rather leathery, elliptic or nearly oblong, 1-seeded, indehiscent pods about 3″ in length and mature within approximately 30 days after the individual trees flower. The hard, strong, reddish-brown wood has an irregular, handsome figure and is much used for high-quality flooring, paneling, cabinetwork, and furniture. It should make beautiful veneer. It is reported to be durable in contact with the ground and has been used for railroad ties in Colombia. Apparently not very common locally, but fairly frequent near Quepos and in Chiriquí.—Curres de Diquis, *Pittier 11954.*

Plomillo—See *Caryocar costaricense* (Costa Rica).
Plomo—See *Belotia macrantha* (Local). Leaves simple.
Plomo—See *Tachigalia versicolor* (Local). Leaves pinnate.
Plumajillo—See *Schizolobium parahybum* (Honduras and Guatemala).

PLUMERIA RUBRA L. forma ACUTIFOLIA (Poir.) Woodson—Apocynaceae—*Cacalojoche, Juche, Flor blanca,* or *Torito* (Costa Rica)—*Caracucha* or *Caracucha blanca* (Panama)—*Frangipani* (General)—*Sacuanjoche* (Nicaragua)—*Flor blanca* or *Flor de ensarta* (Salvador)—*Cumpap, Flor de la cruz, Flor de Mayo, Matuhua, Nicte de monte,* or *Palo de cruz* (Guatemala).

Small or medium-sized trees, 15 to about 40 ft. in height, with abundant milky latex, the large, rather thick, alternate, elliptic-oblong, glabrous leaves clustered at the ends of the stout branches. The trees are usually completely deciduous during the dry season. The large, white, intensely fragrant flowers are produced from the leafless branches in handsome terminal cymes, and are at their best from February until about April. The fruits are pendulous, and consist of divergent pairs of thick follicles about 6″ in length, which contain the winged seeds. The wood is not used locally, but is described as being compact, fine-textured and hard, usually of a yellowish-brown color often more or less marked with faint purplish streaks. It takes a high polish, and is sometimes used for articles of turnery in Guatemala. The species is one of the typical elements found in dry gorges and on rocky hillsides in Central America. They are not

common in our area, and appear to be confined to stony outcrops along the Río Terraba between Palmar and Potrero Grande. It seems probable that the specimen collected by Tonduz in the "potreros" at Santo Domingo de Osa (his 9919) may have come from a cultivated plant.

Plumillo—See *Schizolobium parahybum* (Guatemala).

Poc—See *Spondias mombin* (Guatemala).

Pochote—See *Ceiba pentandra* (Central America generally). Large trees, with buttressed trunks.

Pochote—See *Cochlospermum vitifolium* (Guatemala and British Honduras). Small trees.

Pocora—See *Ocotea Ira* (Local).

Polvo de monte—See *Malvaviscus arboreus* (Guatemala).

POLYGONACEAE—KEY

1. Calyx lobes conspicuously elongate in fruit, deep pink or bright purplish red in color, very showy from a distance. Staminate and pistillate flowers produced on separate trees .... *Triplaris*
1. Calyx lobes not as above. Flowers perfect .... *Coccoloba*

Pomp—See *Cochlospermum vitifolium* (Guatemala).

Poponjoche—See *Huberodendron Allenii* (Local).

Poro—See *Erythrina costaricensis* (Local). Leaves pinnate.

Poro—See *Malvaviscus arboreus* (Guatemala). Leaves simple.

Poro colorado—See *Erythrina costaricensis* (Costa Rica).

Poro espinas—See *Erythrina costaricensis* (Local).

Poroporo—See *Cochlospermum vitifolium* (Costa Rica, Panama, and Nicaragua).

Poro trinidad—See *Erythrina costaricensis* (Local).

POSOQUERIA LATIFOLIA (Rudge) Roem. & Schult.—Rubiaceae—*Fruta de mono* or *Guayabo de mico* (Costa Rica)—*Wild coffee* or *Monkey apple* (Bocas del Toro)—*Boca vieja, Boca de vieja, Fruta de murcielago,* or *Huevo de mono* (Panama)—*Jasmin de arbol* or *Jicarillo* (Nicaragua)—*Cachito* (Honduras)—*Chintorol* (Guatemala)—*Chintonrol, Mountain guava,* or *Snake seed* (British Honduras).

Small trees, 12–30 ft. in height, with opposite, short-petiolate, glabrous leaves, the elliptic-oblong blades with acute or shortly acuminate tips and rounded bases and varying from about 4–10″ in length and 2–5″ in width. The attractive white flowers are produced in lax terminal corymbs at irregular intervals throughout the year, and are conspicuous because of the very long, slender corolla tubes. They have a strong,

pleasant, carnation-like fragrance. The yellow, nearly globose fruits have a number of pale-orange, angular seeds surrounded by a scanty, sweet pulp which is much appreciated by birds and animals. The hard, heavy, white, fine-grained wood has no local uses. The species is sometimes planted for ornament.—Occasional in pastures and on the margins of forest near Palmar Norte, *Allen 5243 & 6652.*

Possumwood—See *Hura crepitans* (U. S. trade name).

POUROUMA ASPERA Trecul—Moraceae—*Guarumo* or *Guarumo de montaña* (Costa Rica)—*Mangabe* (Panama)—*Guarumo macho* (Nicaragua)—*Guarumo de montaña* or *Mano de leon* (Honduras)—*Guarumo de montaña* (Guatemala).

Tall forest trees, 65–90 ft. in height, with large, alternate, long-petiolate leaves, which are mostly clustered at the ends of the stout branches. The blades are deeply 3- to 5-lobed, the individual lobes elliptic-oblong in outline, with shortly acuminate tips, the undersurfaces strongly contrasting and whitish in color. All parts of the plant have a characteristic black, watery sap when cut. The small dioecious flowers are produced in subterminal paniculate cymes, and are followed by ovoid, purplish-black, pubescent, juicy fruits which are sweet and edible. Occasional in hillside forests to about 3,000 ft. in elevation.—Valley of the Río Diquis (Terraba), *Pittier 12105.*

POUTERIA—KEY

1. Leaves usually more than 8″ long.
   2. Leaves conspicuously clustered at the ends of the branches. Fruits globose, densely covered with thick, brownish-orange, woolly processes.................. *Pouteria neglecta*
   2. Leaves not clustered at the ends of the branches. Fruits densely covered with minute, brown, scurfy scales.................. *Pouteria triplarifolia*
1. Leaves usually less than 6″ long.
   2. Flowers in sessile, few-flowered fascicles.
      3. Leaves elliptic-oblong or often oblanceolate, broadly rounded or rather obscurely acute at the apex, the petioles usually about 1″ long.................. *Pouteria subrotata*
      3. Leaves lanceolate, acute or acuminate at the apex, the petioles usually about ½″ long.................. *Pouteria chiricana*
   2. Flowers in dense heads on slender, often branching scapes, up to 1″ or more in length.................. *Pouteria heterodoxa*

POUTERIA CHIRICANA (Standl.) Baehni—Sapotaceae—*Nispero* (Local).

Trees, to about 80 ft. in height, with alternate, elliptic-lanceolate leaves. Older specimens have a somewhat irregular trunk up to about 2 ft. in diameter, rather reminiscent of *Vitex Cooperi* in superficial appearance, though a nick with a machete will reveal the characteristic dark-pink

or red inner bark and white sticky latex. The small, pale-green flowers have a yellow calyx, and are followed by dark-purple, plum-like fruits which become woody and last long on the branches. A fairly frequent species found in hillside forests throughout the area.—Golfito, 100 ft., *Allen 6317.*

POUTERIA HETERODOXA Standl. & L. Wms.—Sapotaceae—*Nispero, Nispero colorado, Nispero zapote,* or *Zapotillo* (Local). (Pl. 30)

Tall forest trees, 90–140 ft. in height and up to 4 ft. in trunk diameter, with alternate, short-petiolate, glabrous leaves, the elliptic-lanceolate, shortly acuminate blades with rounded tips and cuneate bases, averaging 3–4″ in length. The minute tan flowers are produced in late January and early February on very short, dense spikes from the defoliate nodes just below the leaves, and are followed by the brownish, ellipsoidal, 1-seeded fruits which mature in March and April. All parts of the tree have a white, milky latex. The reddish-brown, very hard, heavy, durable wood is reported to be used locally for fence posts, bridge stringers, piles, and dragline mats. Railroad ties are reported to last for 4 or 5 years, but are seldom used because of their tendency to crack.—A common species of hillside forests, from about 1,000 to 2,500 ft. in elevation, *Allen 5905.*

POUTERIA NEGLECTA Cronquist—Sapotaceae—*Zapote del monte* or *Zapotillo* (Local). (Pl. 30)

Medium-sized, or sometimes very large forest trees, 60–100 ft. in height and up to 36″ in diameter, old specimens often having rather irregular, deeply fluted trunks. The alternate, short-petiolate leaves are densely clustered at the ends of the branches, the lanceolate or oblanceolate blades acute or shortly acuminate at the apex and cuneate at the base, either nearly glabrous (in local specimens) or with brown-tomentulose petioles and veins, particularly on the lower surface (in specimens from Honduras), individual examples varying from about 9 to 15″ in length. The relatively large, pale-green flowers are produced in dense, sessile clusters in April from the defoliate nodes just below the leaves, and are followed in August by the large, nearly globose fruits which are 2 or 3″ in diameter and densely covered with soft, velvety-orange, spinelike protuberances. The soft, translucent pulp surrounding the seeds has a sweet, pleasant taste like that of some of the cultivated Amazonean *Lucumas.* There is a slight milky exudate from the broken fruits and leaf petiole, but none from the trunk. The hard, heavy wood does not seem to be used, and the trees are sometimes left standing in pastures and clearings, presumably because they are considered too hard to cut. Occasional in

forests throughout the area and extending to Honduras where it is fairly common in the Lancetilla Valley.—Palmar Norte, 50 ft., *Allen 5607*.

POUTERIA SUBROTATA Cronquist—Sapotaceae.

Small, bushy trees, about 45 ft. in height, with alternate, slender-petiolate leaves, the elliptic-oblong, nearly glabrous blades mostly shortly acute at the apex and rounded or cuneate at the base, varying from about 4 to 8″ in length and 1½–3″ in width. The very small white flowers are produced in dense sessile fascicles from the defoliate nodes below the leaves in late March and early April. The fruits have not been seen.—Occasional along small streams in pastures near Palmar Norte, *Allen 5523*.

POUTERIA TRIPLARIFOLIA Standl. & L. Wms.—Sapotaceae—*Sapote* or *Zapote* (Local). (Pl. 30)

Medium-sized forest trees, to about 65 ft. in height, with lustrous dark-green, shortly petiolate leaves, the elliptic-oblong, glabrous blades acute or abruptly and shortly acuminate at the apex, averaging about 9″ in length in the specimens seen. The flowers are not known. The pale-brown, minutely furfuraceous, ovoid fruits average about 2″ in length, and are produced from the defoliate nodes below the foliage, maturing in late February and early March. They are reported to be edible when fully ripe. All parts of the tree have a fairly abundant, chicle-like latex. The hard, pink, heavy wood is apparently not used locally.—Occasional in the forested hills above Palmar Norte at about 1,800 ft., *Allen 5907*.

Pozolillo—See *Cupania guatemalensis* (Costa Rica).

PRIORIA COPAIFERA Griseb.—Leguminosae; Caesalpinoideae—*Cativo* or *Camibar* (Costa Rica and Nicaragua)—*Cativo* or *Amansa mujer* (Panama). (Fig. 19)

Tall trees, 80–120 ft. in height and up to about 60″ in trunk diameter, with alternate, pinnate leaves composed of 2 pairs of ovate or elliptic-oblong, acute, glabrous leaflets which vary from about 4 to 6″ in length. The very small, white flowers turn yellowish or brownish with age and are produced in large terminal paniculate racemes, rendering the trees fairly conspicuous from a distance. The species flowers and fruits at irregular intervals throughout the year, judging by the rather extensive series of specimens examined at the National Museum in Washington, but at least one conspicuous flowering was seen near Limón and in the upper Estrella Valley October 18, 1951. The fruits are flat, brown, 1-seeded, obovate, dehiscent pods which average about 3½–4″ in length. The trunks exude a copious black gum when cut. The heartwood is light brown, with abundant short, black longitudinal streaks from the gum ducts, the often very thick sapwood being nearly white when freshly

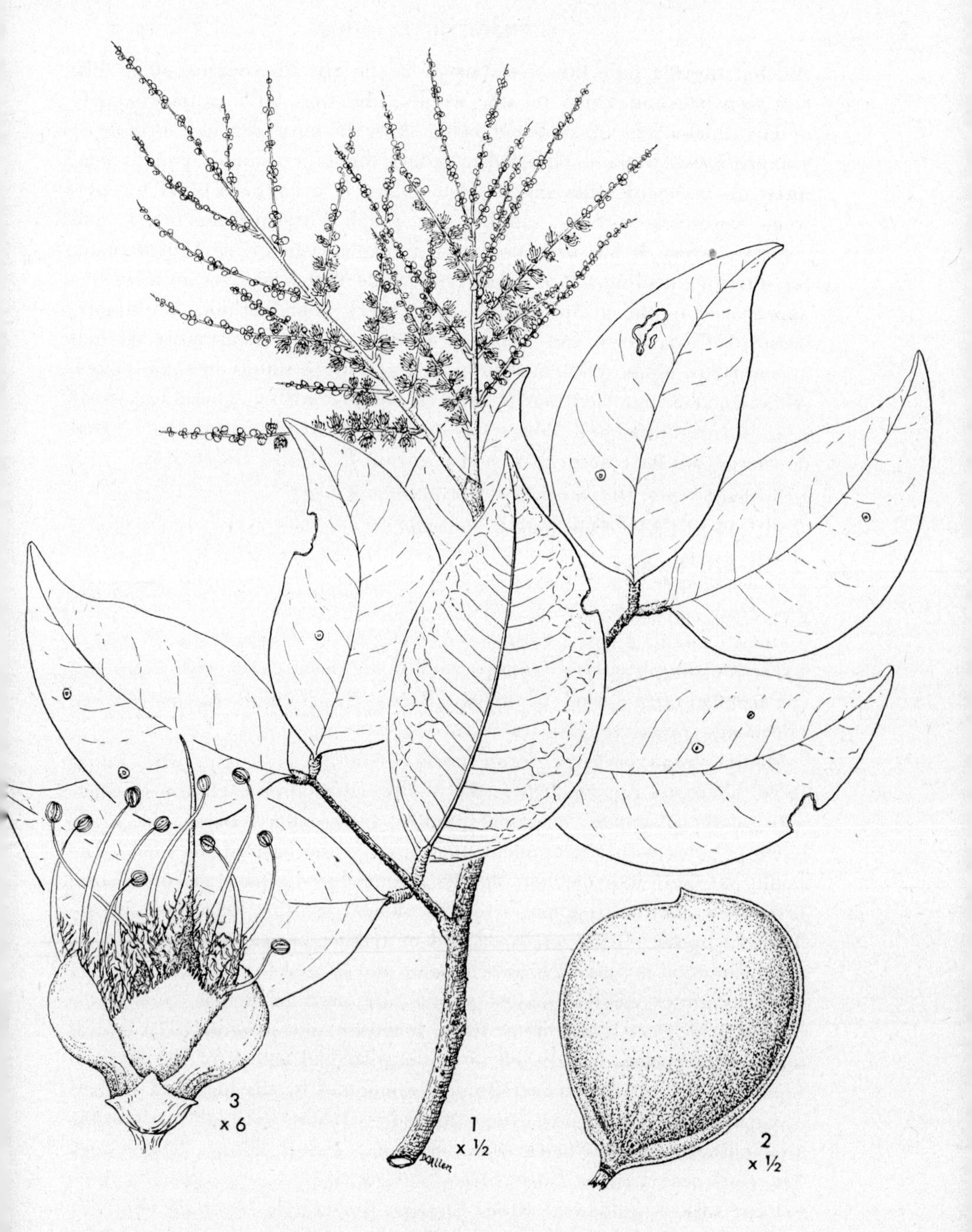

Fig. 19. Prioria copaifera. 1 — Flowering branch. 2 — Seed pod. 3 — Individual flower.

cut, but turning pale tan on exposure to the air. The heartwood is difficult to work and tends to saw woolly, the logs with proportionately greater thicknesses of sapwood being those in most demand. Prolonged soaking in sea water is reported to reduce this objectionable gummy condition to a considerable extent. The lumber is principally used for plywood corestock, but is suitable for rough construction, boxes, and concrete forms. It has been used to a limited extent in the United States for interior paneling, since it has been found that wax gives an attractive appearance to the grain. The trees are very abundant on the Atlantic coast of Costa Rica and adjacent Panama, along tidal estuaries and ascending to about 1,000 ft., forming nearly pure stands in some places. Although it is reported that a stand of some 2 million board feet exists near Burica Point, and Pittier has cited its occurrence near the Laguna de Sierpe, we have seen no specimens from the Golfo Dulce area.

Probado—See *Didymopanax Morototoni* (Local).

PROTEACEAE—One genus, *Roupala*.

PROTIUM—KEY

1. Flowers sessile............ *Protium neglectum* var. *sessiliflorum*
1. Flowers pedicellate.
   2. Leaves broadly elliptic, usually more than 4½" long...... *Protium copal* var. *glabrum*
   2. Leaves narrowly lanceolate, usually less than 4½" long............ *Protium costaricense*

PROTIUM COPAL var. GLABRUM (Rose) Swart—Burseraceae—*Caraña* (Boruca)—*Fosforito* (Nicaragua).

Medium-sized or large forest trees, 40–80 ft. in height, with rather large, alternate, pinnate leaves, the 5–7 broadly elliptic, glabrous leaflets with acute or shortly acuminate apices and rounded bases, the short, terete petioles with a prominent thickening at the base of the blade. The small, pale-yellow pedicellate flowers are produced during the dry season from December until about April in slender axillary panicles and are followed by the elliptic or ovoid, red or reddish-brown apiculate drupes. The heartwood is pinkish brown in color and is used for fuel and kindling since it ignites easily. Related species are used in British Guiana for interior trim, door and window stock, furniture, and plywood. The typical form of the species is believed to be the principal source of the fragrant copal so frequently used in religious ceremonies by the highland Indians of Guatemala. Our variety is occasional in hillside forests throughout the area.—Buenos Aires, *Tonduz 6682*—Esquinas Forest, 200 ft., *Skutch 5340*—Forests near Palmar Norte, *Allen 5502 & 5731*.

PROTIUM COSTARICENSE (Rose) Engler—Burseraceae—*Caraña* (Boruca)—*Copal* (Costa Rica).

Medium-sized or tall trees, 40–80 ft. in height, with alternate, pinnate leaves, the 5 lanceolate leaflets acuminate at the apex and cuneate or somewhat decurrent at the base. The very small, pale-yellow, pedicellate flowers are produced in late August and early September in short, subterminal, minutely puberulent panicles, and are followed by drupaceous fruits. Our specimens differ from descriptions of *P. costaricense* in their glabrous leaves and minutely puberulent rather than pubescent panicles, but match material exactly in the Chicago Natural History Museum which has been identified by Swart.—Diquis, *Pittier 12106*—Sea beaches near the delta of the Río Esquinas, *Allen 5641 & 5644.*

PROTIUM NEGLECTUM var. SESSILIFLORUM (Rose) Swart—Burseraceae—*Caraña* (Boruca)—*Chutras* or *Comida de mono* (Chiriquí)—*Anime* (Panama)—*Tontol* (Guatemala)—*Copal* or *Copal macho* (British Honduras).

A medium-sized or large glabrous tree, with alternate, pinnate leaves, the 5–7 oblong or elliptic leaflets short-acuminate at the apex and acute at the base. The small, sessile flowers are produced in short, somewhat pubescent panicles.—Santo Domingo de Osa, *Tonduz 6989 & 9952.*

Provision tree—See *Pachira aquatica* (Bocas del Toro and Limón).

PSEUDIMA COSTARICENSIS L. Wms.—Sapindaceae.

Shrubs or small, multiple-trunked trees, the erect, usually unbranched canes about 10–18 ft. in height, with alternate, pinnate leaves. The narrowly lanceolate, acuminate, rather rigid leaflets are alternate on the rachis, and average about 4–6″ long and 1–1¼″ wide. The flowers are produced in elongate terminal panicles, and are followed by bright-red, obovoid fruits (sometimes fused into pairs) about ¾″ in diameter, each with a solitary, jet-black spherical seed.—Locally common along the banks of small streams in pastures near Palmar Norte, at about 50 ft. elevation, *Allen 5525 & 6651.*

PSEUDOLMEDIA SPURIA (Sw.) Griseb.—Moraceae—*Ojoche* (Costa Rica)—*Manax* (Peten, Guatemala)—*Cherry* (British Honduras).

Tall forest trees, 75–90 ft. in height, with alternate, very short-petiolate, glabrous leaves, the lanceolate, acuminate blades with cuneate bases and averaging about 3″ in length. The minute white dioecious flowers are produced in late February and early March from the leaf axils, the staminate form solitary or in small sessile clusters, and the pistillate subtended by numerous imbricating bracts. The fruit is a small, edible drupe. All parts of the tree have an abundant creamy-white latex. Standley

reports that the branches are cut during the dry season in Guanacaste as forage for oxen. The latex is sometimes used to adulterate chicle in Yucatan. The hard white wood is apparently not used.—Occasional in the forested hills above Palmar Norte at about 1,500 ft., *Allen 5961.*

PSIDIUM—KEY

1. Lower part of the flower bud conspicuously contracted into a tube. Veins of the lower leaf surface elevated. Fruits sweet.......................................*Psidium Guajava*
1. Lower part of the flower bud not contracted into a tube. Veins of the lower leaf surface inconspicuous. Fruits sour..................................*Psidium Friedrichsthalianum*

PSIDIUM FRIEDRICHSTHALIANUM (Berg) Niedenzu—Myrtaceae—*Cas* or *Cas acido* (Costa Rica and general).

Small trees, to about 35 ft. in height, with small white flowers and acid fruits. Frequently seen in cultivation in townsites and sometimes persistent near old house sites and along trails.

PSIDIUM GUAJAVA L.—Myrtaceae—*Guayaba* or *Guayabo* (Local and general)—*Guayaba perulera* or *Guayaba de gusano* (Nicaragua).

Small trees, to about 30 ft. in height and up to 6–10″ in trunk diameter, with opposite, elliptic-oblong, obtuse or acute leaves and small white flowers about 1″ in diameter, which are produced on slender peduncles from the leaf axils, and followed by the variously globose, ovoid or pyriform, usually yellow fruits. The sweet flesh is commonly deep pink in color and filled with many small, very hard seeds. It is eaten out of hand by children and used for the preparation of the familiar guava paste and guava jelly. The bark is rich in tannin, and is sometimes used in Central America for curing skins. The hard wood is apparently durable, but is seldom used because of the difficulty in securing large sizes. The species is probably not native in our area, but is common in pastures and on gravel bars near the Río Terraba, often forming nearly pure stands of considerable extent.

PSYCHOTRIA CHIAPENSIS Standl.—Rubiaceae—*Cocobolito* (Chiriquí)—*Palo de agua* (Guatemala)—*Casada* or *Whitewood* (British Honduras).

Shrubs or small trees, to about 18 ft. in height, with opposite, short-petiolate, lanceolate, acuminate, glabrous leaves which vary from about 5 to 8″ in length. The long, white, tubular flowers are produced in dense terminal cymes in late May and early June. As has been noted by Standley, the species is very unlike most *Psychotrias,* and has rather the aspect of a *Cephaelis.*—Fairly frequent in the forested hills near the Golfito airport, *Allen 5291.*

PTEROCARPUS—KEY

1. Outer surface of the floral calyx densely brown-tomentose. Fruits thin, the margins broadly winged.......... *Pterocarpus Hayesii*
1. Outer surface of the floral calyx nearly glabrous. Fruits conspicuously thickened, the margins narrowly winged.......... *Pterocarpus officinalis*

PTEROCARPUS HAYESII Hemsl.—Leguminosae; Papilionatae—*Cheja* (Guatemala). (Pl. 15)

Medium-sized or tall trees, 60–90 ft. in height, with alternate, pinnate leaves, the 5–7 short-petiolate leaflets alternate or subopposite on the rachis. The thin, glabrous blades are elliptic-oblong or oblanceolate in outline, with abruptly acuminate apices which are rounded or obscurely emarginate at the tip. The small yellow or orange, somewhat fragrant flowers are produced in late February and early March in slender terminal or axillary fasciculate racemes, either from the nearly bare branches or sometimes while the trees are in full leaf, and are followed in July by the nearly circular flat, pale-brown, papery, 1-seeded indehiscent fruits which average about 1½″ in diameter. The species is apparently somewhat variable, but is often very showy when in flower in the hills near Palmar Norte.—Pastures near the Golfito airport, 25 ft., *Allen 5992 & 6281*—Hills near Palmar Norte, 100 ft., *Allen 5980.*

PTEROCARPUS OFFICINALIS Jacq.—Leguminosae; Papilionatae—*Sangrillo* or *Sangregado* (Local)—*Chajada amarilla* (Costa Rica)—*Bloodwood, Huevos de gato,* or *Sangre de drago* (Panama)—*Sangregado* or *Sangre de drago* (Nicaragua)—*Sangre* or *Cowee* (Honduras)—*Sangre de drago* or *Sangregado* (Guatemala). (Pl. 15)

Medium-sized or tall trees, 50–90 ft. in height and up to 30″ in diameter above the strongly developed buttresses. The bark is thin, gray, and rather smooth, with blood-red sap. The relatively large, alternate leaves are pinnate, each with 6–9 alternate, short-petiolate, glabrous leaflets, the individual elliptic-oblong blades having acute or acuminate apices and rounded bases, varying from about 3 to 4″ in length. The trees are usually briefly deciduous in late December or early January. The yellow or pale-orange flowers are produced in May or June in our area (in September in Panama), the large subterminal or axillary panicles rendering the trees conspicuous and attractive from a distance. The nearly circular, narrowly winged, 1-seeded, indehiscent fruits average about 2½″ in diameter, and mature in late August. The white or yellowish wood is light in weight, firm, easy to work, but not durable. There is a report from eastern Nicaragua that it burns when freshly cut, but it does not seem

to be used locally, although it is considered suitable for cheap interior construction, boxes, and concrete forms. The species is common throughout the area on the margins of mangrove swamps and in wet lowland forest, where it forms one of the dominant elements in association with *Mora oleifera* or *Carapa Slateri.*—Esquinas Forest, in the low, wet area below the residence, *Allen 5530*—Margins of mangrove swamp near the Golfito dairy, *Allen 5616.*

Pubabac—See *Nectandra salicifolia* (Guatemala). Leaves long-acuminate at the apex.

Pububuc—See *Ocotea veraguensis* (Guatemala). Leaves acute or rounded at the apex.

Pucte—See *Terminalia amazonia* (Guatemala).

Puj—See *Ochroma lagopus* (Guatemala).

Pullun—See *Casearia aculeata* (Guatemala).

Pumpo—See *Cochlospermum vitifolium* (Guatemala).

Pumpumjuche—See *Cochlospermum vitifolium* (Guatemala).

Pumpunjuche—See *Cochlospermum vitifolium* (Guatemala). Leaves simple, but deeply lobed.

Pumpunjuche—See *Pachira aquatica* (Nicaragua, Honduras, and Guatemala). Leaves digitately compound.

Purple heart—See *Peltogyne purpurea* (Panama and general).

Putzil—See *Ximenia americana* (Guatemala).

QUARARIBEA GUATEMALTECA (D. Sm.) Standl.—Bombacaceae—*Garroche* (Local)—*Moro* (Guatemala). (Pl. 28)

Medium-sized trees, to about 75 ft. in height, usually with a narrow, pyramidal crown and conifer-like branching habit. The often angular trunks average about 1 ft. in diameter and have thin greenish-gray bark which often flakes off in long irregular patches. The leaves are dark green, alternate, and characteristically 2-ranked and more or less pendulous on the branches, and have a sweet and pleasantly distinctive odor which they seem to retain indefinitely after they have been dried. The small white flowers are produced from the axils of the current leafy growth, apparently in June and again in November, probably coinciding with the heavy rains. A common and very distinctive species found throughout the area from the river floodplains to about 2,500 ft. in elevation.—Pastures near Palmar Norte, 100 ft., *Allen 6635, 6653, & 6719.*

Quaruba—See *Vochysia hondurensis* (U. S. trade name).

QUASSIA AMARA L.—Simaroubaceae—*Hombre grande, Hombron,* or *Guavo* (Costa Rica).

Shrubs or small trees, to about 25 ft. in height, with alternate, pinnate leaves, the conspicuously winged rachis with 5 lanceolate or oblanceolate, acute or acuminate leaflets. The attractive pink or red flowers are more than 1″ long and are produced in elongate pendulous racemes which are followed by fruiting receptacles bearing 5 black drupes. The bark and wood are intensely bitter and are used locally for the treatment of fevers. Known in our area only from the region of Boruca and Potrero Grande.

Quebracho—See *Tabebuia chrysantha* (Honduras). Leaves digitately compound.

Quebracho—See *Terminalia amazonia* (Guatemala). Leaves simple.

Quebracho blanco—See *Hasseltia quinquenervia* (Local).

Quebracho de cerro—See *Diphysa robinioides* (Honduras).

QUERCUS sp.—Fagaceae—*Encino* or *Roble* (Costa Rica).

Various species of true oaks occur in the highlands near Cañas Gordas and Agua Buena but have not been collected.

Quesillo—See *Astronium graveolens* (Guatemala). Leaves pinnate.

Quesillo—See *Malvaviscus arboreus* (Salvador). Leaves simple.

Quesito—See *Malvaviscus arboreus* var. *penduliflorus* (Costa Rica). Leaves simple.

Quiebrahacha—See *Lysiloma guanacastense* (Tenorio, Guanacaste).

Quina—See *Ladenbergia Brenesii* (Costa Rica). Leaves opposite.

Quina—See *Nectandra salicifolia* (Costa Rica). Leaves alternate.

Quina blanca—See *Miconia Schlimii* (Salvador).

Quina sylvestre—See *Sweetia panamensis* (Guatemala).

Quira—See *Andira inermis* (Chiriquí). Leaves pinnate, alternate.

Quira—See *Platymiscium pinnatum* (Local). Leaves pinnate, opposite.

Quiriguillo—See *Pachira aquatica* (Costa Rica).

Quirihuillo—See *Pachira aquatica* (Costa Rica).

Quiura—See *Terminalia lucida* (Local).

Quizarra—A generic term of wide application in the Lauraceae, used for various species of *Nectandra* and *Ocotea,* which see.

Quizarra copalchi—See *Croton glabellus* (Costa Rica).

Quizarra quina—See *Nectandra salicifolia* (Costa Rica).

Quon—See *Schizolobium parahybum* (Nicaragua).

Rajate bien—See *Vitex Cooperi* (U. S. trade name).

Ramon—See *Brosimum alicastrum* (Guatemala and British Honduras). Leaf margins entire.

Ramon—See *Trophis racemosa* (Panama to Honduras). Leaf margins serrate.

Ramon blanco—See *Brosimum alicastrum* (Guatemala and British Honduras).

Ramon colorado—See *Trophis racemosa* (Guatemala).

Ramoon—See *Trophis racemosa* (Bocas del Toro).

RANDIA—KEY

1. Leaves less than 6″ long. Lateral spur branches armed at the tip with 4 short spines...
.... *Randia armata*
1. Leaves usually more than 6″ long. Branches not armed with spines..........................
.... *Randia grandifolia*

RANDIA ARMATA (Sw.) DC.—Rubiaceae—*Limoncillo* (Local)—*Mostrenco* or *Crucillo* (Costa Rica)—*Cagalero, Crucetilla,* or *Jasmin cimarron* (Honduras)—*Caca de mico* or *Jicarillo* (Salvador)—*Flor de cruz, Crucito, Palo de cruz, Rosetillo,* or *Torolillo* (Guatemala).

Shrubs or small trees, to a maximum of about 40 ft. in height, with opposite leaves which are often clustered at the ends of the lateral, spurlike branches and frequently bearing 4 short sharp spines directly below the foliage. The small white dioecious flowers are produced in sessile fascicles at the ends of the branches and are followed by yellow, ellipsoidal fruits about 1–1¼″ in length which are reported to be edible, though they are also used in the West Indies as a barbasco for catching fish.—Pastures near Palmar Norte, 100 ft., *Allen 6324.*

RANDIA GRANDIFOLIA (Donn. Sm.) Standl.—Rubiaceae.

Shrubs or small trees to about 15 ft. in height, with large, opposite, oblong-lanceolate or elliptic-ovate leaves, the acute or acuminate blades 5 to about 16″ long and up to 6″ wide. These, as in the former species, are usually clustered at the ends of the branches. The small white dioecious flowers are produced in sessile clusters at the ends of the branches and are followed by yellow globose fruits about 1″ in diameter. —Santo Domingo de Golfo Dulce, *Tonduz 9878.*

RAPHIA TAEDIGERA Martius—Palmaceae—*Yolillo* or *Palma real* (Local) —*Holillo* or *Jolillo* (Bluefields, Nicaragua). (Pl. 8)

Robust palms, typically with 3–5 trunks, which may vary in height from about 30 to 65 ft., and covered by the overlapping bases of the 8–10 enormous pinnate fronds, which may be up to 40 ft. in length. Each side of the stout rachis bears some 95 pinnae, whose midribs and margins are armed with short, sharp spines. The small flowers are produced in 3–5 very large, arching, or commonly pendulous inflorescences which

originate in the axils of the upper fronds. These scapes may vary in length from 8 to about 16 ft., and lack a true spathe, but are enveloped throughout in a great number of persistent, spirally fused, papery bracts, each inverted cone enclosing a miniature cleft spathe and a distichous panicle of flowers. The handsome fruits are oblong or ellipsoidal, about 2–2½″ in length, with a short, sharp terminal beak, the whole entirely covered with closely overlapping brownish-orange glossy scales. Individual plants apparently flower and fruit throughout the year, the mature canes dying after they have exhausted their inflorescences, being replaced by basal suckers. The species forms vast unbroken stands thousands of acres in extent back of Jalaca Farm and near the Laguna de Sierpe, this however being the only known occurrence of the genus on the Pacific coast of Central America. These tremendous stands would undoubtedly be exploited in the Asiatic tropics for sugar by tapping the immature inflorescences, but nothing of the sort has ever been attempted in our hemisphere. The stout frond rachises have been used for banana prop poles in the Limón Division of the United Fruit Company.—Swamps near Jalaca, sea level, *Allen 6276.*

Raspa—See *Licania arborea* (Panama).

Raspa guacal—See *Curatella americana* (Costa Rica).

Raspa-lengua—See *Casearia arguta* (Panama to Guatemala). Leaves with a single longitudinal nerve, not rough to the touch.

Raspa lengua—See *Hasseltia floribunda* (Panama). Leaves with three longitudinal nerves.

Raspa lengua—See *Trophis racemosa* (Salvador). Leaves with a single longitudinal nerve, rough to the touch.

Raton—See *Roupala complicata* (Costa Rica).

Ratoncillo—See *Roupala complicata* (Costa Rica).

Red berry—See *Hamelia patens* (Panama).

Red mangrove—See *Rhizophora mangle* (General).

Reseco—See *Tachigalia versicolor* (Local).

RHEEDIA—KEY

1. Fruits very rough, bright yellow .... *Rheedia madruno*
1. Fruits smooth, orange .... *Rheedia edulis*

RHEEDIA EDULIS Triana & Planch.—Guttiferae—*Jorco* (Local)—*Cero* (Chiriquí)—*Sastra* (Panama)—*Jocomico* (Nicaragua)—*Chapparon* (Salvador)—*Caimito, Caimito de montaña,* or *Sastra* (Honduras)—*Arrayan* or *Mameyito* (Guatemala).

Medium-sized or tall forest trees, to about 80–100 ft. in height, with

opposite, linear-lanceolate, glabrous leaves which are acute or acuminate at the apex and decurrent at the base, the blades varying from about 2½ to 6″ in length. The small whitish flowers are produced in November and December in the leaf axils, and are followed by the smooth, ellipsoidal, orange or yellow, 1-seeded edible fruits which may vary from about 1″ in length to the size of a lemon and mature from late January to about March. All parts of the tree are characterized by an abundant sticky yellow latex. The wood is reported to be suitable for tool handles, temporary railroad ties, fence posts, and general construction. It has a reputation for being free from insect attack. Fairly frequent throughout the area.—Palmar Norte, 100 ft., *Allen 5766.*

RHEEDIA MADRUNO (HBK) Pl. & Tr.—Guttiferae—*Madroño, Cerillo,* or *Jorco* (Local)—*Madroño* or *Fruta de mono* (Panama).

Medium-sized forest trees, 30–65 ft. in height, with opposite, short-petiolate, oblong-elliptic, glabrous, leathery leaves which vary from 4–8″ in length, the blades with acute apices and cuneate bases. The small fragrant white flowers are produced in pedunculate fascicles from the defoliate nodes below the leaves from December to February, and are followed by the very rough, bright-yellow, acid, ellipsoidal fruits which mature from May until about August. All parts of the tree have an abundant sticky yellow latex. The rather hard, pinkish wood is apparently not used. One of our most common local trees, averaging 3–5 specimens per acre in some places in the hillside forests.—Esquinas Forest, 200 ft., *Allen 5553*—Forested hills above Palmar Norte, 500–2,000 ft., *Allen 5911.*

RHIZOPHORACEAE—KEY

1. Leaves leathery, obtuse or emarginate at the tip. Trees with conspicuous stilt roots, confined to coastal swamps .... *Rhizophora mangle*
1. Leaves not leathery; acute or acuminate at the tip. Trees of forested hillsides, without stilt roots .... *Cassipourea podantha*

RHIZOPHORA MANGLE L.—Rhizophoraceae—*Mangle colorado* or *Mangle caballero* (Local)—*Mangle, Mangrove,* or *Red mangrove* (General)—*Mangle, Mangle colorado,* or *Mangle salado* (Panama)—*Mangle* or *Mangle colorado* (Honduras)—*Red mangrove* or *Mangle colorado* (British Honduras). (Pl. 20)

Medium-sized, or sometimes very tall trees, 25–100 ft. in height, with very strongly developed stilt roots. Young plants, or those on the extreme limits of the species range, are often little more than shrubs, but old specimens in favorable situations develop a very straight trunk about

2 ft. in diameter, with clear lengths of 60 ft. or more. The very thick, opposite, leathery leaves have short petioles, the oblong-elliptic blades with obtuse or somewhat emarginate tips and broadly wedge-shaped bases, in mature specimens averaging about 6 or 7″ in length. The small, yellowish flowers are produced in late August and early September in few-flowered axillary cymes, and are followed by the oblong or ellipsoidal, leathery fruits which retain the conspicuous sepals at the base. The seeds germinate while the fruits are still attached to the branches, the long, stout pendulous primary roots protruding from the apex and often becoming 1 ft. or more in length before falling to the soft mud below. The reddish-brown, hard, heavy wood is very strong, tough, and fairly durable and is suitable for boat framing, house posts, piling, temporary railroad ties (which are reported to last one or two years in a wet climate), and fence posts, but is little used locally excepting for charcoal and fuel, the freshly cut wood being reported to burn green. The bark is rich in tannin, and when used alone yields a dark-red, rather brittle leather. A boiled concentrate of the bark is sometimes used for staining floors and furniture. A very common species, forming extensive, nearly pure stands from Punta Violines to the mouth of the Río Terraba, and in the deltas of the Río Esquinas and Río Coto.—Palmer, *Pittier 6772*—Santo Domingo de Osa, *Tonduz 10059*—Golfito, *Allen 5614.*

RICINUS COMMUNIS L.—Euphorbiaceae—*Higuerillo blanco* or *Higuerillo rojo* (Local)—*Higuerilla* or *Castor* (Costa Rica)—*Castor bean* or *Higuero* (Central America generally).

Stout herbs, often becoming woody and treelike, occasionally reaching as much as 18 ft. in height, with alternate, peltate, deeply lobed palmate leaves and terminal racemes of small flowers, followed by 3-celled, spiny fruits. The seeds are the source of castor oil, and are violently purgative or even poisonous, particularly to children, if eaten in quantity. Young plants are very frequently seen on disturbed soil or on fresh accumulations of railroad ballast throughout the area. A horticultural form with handsome coppery-red foliage is sometimes cultivated for ornament.

RINOREA PUBIPES Blake—Violaceae.

Small or medium-sized trees, 20–40 ft. in height, with opposite, elliptic-lanceolate, acuminate leaves which have obscurely serrulate margins, the upper surface of the blades glabrous, while the short petioles and veins of the lower surface are puberulent, and the axils of the flushes of new growth distinctly pilose. The very small, pale-yellow flowers are

produced in late March and early April in short, slender, terminal or axillary racemes, followed by the longitudinally dehiscent, 3-lobed capsular fruits.—Fairly frequent on stream banks in forested areas near Palmar Norte at low elevations, *Allen 5524.*

Robanchab—See *Vochysia hondurensis* (Guatemala).

Roble—See *Dialyanthera otoba* (Chiriquí). Fruits nutmeg-like.

Roble—See *Licania arborea* (Nicaragua to Guatemala). Fruits not nutmeg-like. Leaves whitish on the lower surface.

Roble—See *Quercus* sp. (General). Fruit an acorn.

Roble—See *Tabebuia pentaphylla* (Panama and Costa Rica). Leaves digitately compound.

Roble blanco—See *Licania arborea* (Costa Rica).

Roble de la sabana—See *Tabebuia pentaphylla* (Costa Rica and Panama).

ROLLINIA JIMENEZII Safford—Annonaceae—*Anonillo* (Local)—*Anona* or *Toreto* (Chiriquí)—*Torete* (Panama)—*Anona, Anonillo,* or *Chirimoya* (Guatemala). (Pl. 28)

Medium-sized trees, 40–60 ft. in height, with alternate elliptic-oblong, acute or acuminate leaves which vary from 5–8″ in length and are deciduous during the dry season. The tomentose flowers are produced in the leaf axils, and are followed from July to about February by the nearly globose yellow tuberculate fruits which average about 3″ in diameter. These are edible but are very seedy.—The species is frequent along trails and in areas of old second growth near Palmar Norte, *Allen 5316 & 6268.*

RONDELETIA UROPHYLLA Standl. & L. Wms.—Rubiaceae.

Shrubs or small trees, to about 30 ft. in height, with opposite, short-petiolate, lanceolate, long-acuminate leaves 7–9″ long and 2¼–2¾″ wide. The small white flowers are produced in July and August in narrow terminal panicles which nearly equal the leaves in length.—Esquinas Forest, 200 ft., *Skutch 5253* & *Allen 6291.*

Ronron—See *Astronium graveolens* (Local and general).

ROSACEAE—KEY

1. Petals conspicuously present .... *Hirtella*
1. Petals minute or absent .... *Licania*

Rosetillo—See *Randia armata* (Guatemala).

ROUPALA COMPLICATA HBK—Proteaceae—*Danto, Raton, Danto hediendo,* or *Ratoncillo* (Costa Rica).

Medium-sized trees, 25–60 ft. in height, with alternate, ovate, acuminate glabrous leaves which have serrate or entire margins and elongate petioles. The small, creamy-white flowers are produced in late January and early February in fascicles of slender axillary spikes, and are soon followed by the rather large, woody, obliquely bivalvate follicles which contain the winged seeds. The fresh wood and crushed leaves have a skunklike, offensive odor. The hard, heavy wood is apparently not used locally. Occasional in forests and along trails in the Terraba Valley.—Boruca, *Pittier 12075.*

Rubber tree—See *Castilla fallax* (Bocas del Toro and Limón).

RUBIACEAE—KEY

1. Leaves very large, usually 18″ or more in length.
    2. Blades deeply and conspicuously pinnatisect.......... *Pentagonia gymnopoda*
    2. Blades never pinnatisect.
        3. Flowers subtended by long, very showy blood-red bracts, the inflorescences superficially resembling those of the cultivated poinsettia... *Warscewiczia coccinea*
        3. Flowers not subtended by red bracts.......... *Condaminea corymbosa*
1. Leaves of moderate size or small, usually less than 1 ft. long.
    2. Flowers blue.
        3. Leaf blades with 3 longitudinal nerves.......... *Faramea suerrensis*
        3. Leaf blades with 1 longitudinal nerve.......... *Faramea sessifolia*
    2. Flowers not blue.
        3. Flowers subtended by a large, showy white petal-like calyx lobe.......... *Calycophyllum candidissimum*
        3. Flowers not subtended by a showy petal-like calyx lobe.
            4. Ends of the spurlike lateral branches armed with short, sharp spines.......... *Randia armata*
            4. Branches not armed with spines.
                5. Flowers about as long as broad, the lower corolla not contracted into a tube.
                    6. Leaf petioles about ¼″ long. Flowers usually more than ¾″ in diameter.......... *Genipa Caruto*
                    6. Leaf petioles usually more than ½″ long. Flowers less than ½″ in diameter.
                        7. Leaf petioles and veins densely covered with closely appressed silky hairs.......... *Sommera grandis*
                        7. Leaf petioles and veins completely glabrous.
                            8. Leaves of flowering branches very large, often 1 ft. in length. Flowers greenish yellow with purple stamens.......... *Sickingia Maxonii*
                            8. Leaves of flowering branches much less than 1 ft. long. Flowers white.
                                9. Inflorescences produced from the subterminal leaf axils. Petals free, or only slightly joined at the base.......... *Chimarrhis latifolia*

9. Inflorescences terminal. Petals joined for about ½ their length in a short but distinct tube..*Chione costaricensis*

5. Flowers longer than broad, the lower corolla contracted into a conspicuous slender tube.
 6. Floral tube more than 2½″ long.
  7. Flowers white..*Posoqueria latifolia*
  7. Flowers yellow..*Tocoyena Pittieri*
 6. Floral tube less than 2½″ long.
  7. Flowers solitary or in sessile, few-flowered fascicles at the ends of the leafy branches.
   8. Leaves large, 4½–16″ long. Outer surface of the flowers glabrous..*Randia grandifolia*
   8. Leaves 1½–8″ long. Outer surface of the flowers densely covered with fine, silky hairs.
    9. Floral tube up to 1½ times as long as the expanded lobes..*Alibertia edulis*
    9. Floral tube very slender, about 4 times as long as the expanded lobes..*Anisomeris Recordii*
  7. Flowers usually many, in branching cymes or panicles.
   8. Branches of the inflorescence terminating in very dense heads of closely overlapping bracts..*Psychotria chiapensis*
   8. Branches of the inflorescence not terminating in dense, bracteate heads.
    9. Individual flowers more than 1½″ long.
     10. Corolla tube about 1½ times as long as the lobes..*Ladenbergia Brenesii*
     10. Corolla tube about 3–4 times as long as the lobes..*Ladenbergia chariantha*
    9. Individual flowers less than 1″ long.
     10. Leaves narrowly lanceolate, tapering gradually to a caudate-acuminate tip. Flowers in slender, elongate panicles..*Rondeletia urophylla*
     10. Leaves elliptic-lanceolate, acute or shortly acuminate. Flowers in broad cymes.
      11. Floral tube with conspicuous longitudinal ribs, the lobes not conspicuously expanded. Flowers orange or yellow.
       12. Leaf blades usually less than 4″ long, the petioles and lower surface densely puberulent..*Hamelia patens*
       12. Leaf blades usually more than 6″ long, the petioles and lower surface glabrous..*Hamelia magnifolia*
      11. Floral tube without conspicuous longitudinal ribs.
       12. Calyx very long and slender, nearly as long as the corolla tube, the upper margin conspicuously lobed. Flowers dark pink..*Macrocnemum glabrescens*
       12. Calyx much shorter than the tube, the margins not lobed. Flowers white.
        13. Flower buds and corolla tube densely covered with

minute silky hairs, the tube about 3 times as long as the lobes. Leaves usually large.. *Guettarda macrosperma*

13. Flower buds and corolla tube glabrous, the tube about equaling the lobes in length. Leaves small........ ....*Ixora nicaraguensis*

Rum—See *Spondias purpurea* (Guatemala).

Run—See *Spondias purpurea* (Guatemala).

Saaxnic—See *Ximenia americana* (Guatemala).

Saba—See *Carapa Slateri* (Panama). Leaves pinnate.

Saba—See *Dialyanthera otoba* (Chiriquí). Leaves simple.

Saba nut—See *Pachira aquatica* (Nicaragua). Leaves digitately compound.

SABIACEAE—One genus, *Meliosma.*

Sacalante—See *Nectandra latifolia* (Guatemala).

Sacmuba—See *Casearia sylvestris* (Guatemala).

SACOGLOTTIS EXCELSA Ducke—Humiriaceae—*Nispero* (Local).

Very tall trees, to 140 ft. in height, with alternate, elliptic-lanceolate, glabrous leaves, the serrulate blades 1¾–2¼″ in length, with shortly acuminate apices and nearly sessile, cuneate bases. The very small greenish fragrant flowers are produced in terminal cymes in early January and are soon followed by the oblong or ellipsoidal, drupaceous fruits which are about 1″ long. The reddish-brown heartwood is very hard and heavy, and is reported to be durable in contact with the ground. It is used locally for fence posts, and would appear to be suitable for general heavy construction, such as bridge timbers, railroad ties, and piling. There is some indication that the wood of this species may sometimes be confused with that of *Vantanea Barbourii* and it seems possible that this may be the species represented by the sawmill sample from San Isidro del General forwarded to the Yale School of Forestry by John A. Scholten and William F. Barbour in 1944. The species is fairly frequent in the forested hills near Esquinas, and is otherwise known only from high land near Manaos and Belem do Pará in Brazil.—Esquinas Forest, 200 ft., *Allen 5812.*

Sacuanjoche—See *Plumeria rubra* forma *acutifolia* (Nicaragua).

Saha—See *Curatella americana* (Guatemala).

Salamo—See *Calycophyllum candidissimum* (Costa Rica to Guatemala).

Salvia—See *Vernonia patens* (Panama).

Sambogum—See *Symphonia globulifera* (Bocas del Toro and Limón).
Sambran prieto—See *Cassia reticulata* (Guatemala).
Sanalego—See *Vernonia patens* (Chiriquí).
San Andres—See *Tecoma stans* (Salvador and Honduras).
Sandalo—See *Cassia grandis* (Costa Rica). Fruit a terete pod. Sap not aromatic.
Sandalo—See *Myroxylon balsamum* var. *Pereirae* (Local). Fruit a winged seed. Sap pleasantly aromatic.
Sandbox—See *Hura crepitans* (Panama).
Sandio—See *Hedyosmum mexicanum* (Guatemala).
Sandpaper tree—See *Curatella americana* (Canal Zone).
Sangre—See *Compsoneura Sprucei* (Honduras and Guatemala). Leaves simple. Aril surrounding the seed only divided at the apex.
Sangre—See *Pterocarpus officinalis* (Honduras). Leaves pinnate.
Sangre—See *Virola Koschnyi* (Guatemala and Honduras). Leaves simple, brownish on the lower surface. Aril surrounding seed deeply divided.
Sangre blanca or blanco—See *Nectandra latifolia* (Panama to Honduras).
Sangre de drago—See *Gliricidia sepium* (Peninsula de Osa). Flowers pink.
Sangre de drago—See *Pterocarpus officinalis* (Panama to Guatemala). Flowers pale orange.
Sangre de perro—See *Vismia ferruginea* (Panama).
Sangre drago—See *Virola Koschnyi* (Nicaragua). Leaves with 20–30 pairs of lateral nerves.
Sangre drago—See *Virola sebifera* (Nicaragua). Leaves with 12–15 pairs of lateral nerves.
Sangregado—See *Pterocarpus officinalis* (Costa Rica to Guatemala).
Sangrillo—See *Platymiscium pinnatum* (Chiriquí). Leaves pinnate, opposite.
Sangrillo—See *Pterocarpus officinalis* (Local). Leaves pinnate, alternate.
Sangrillo—See *Vismia guianensis* (Local and Panama). Leaves simple.
Sangrillo negro—See *Dialium guianense* (Chiriquí).
San Juan—See *Vochysia hondurensis* (Honduras and Guatemala).
Sanpedrano—See *Vochysia hondurensis* (Guatemala).
San Ramon—See *Trophis racemosa* (Honduras).
Santa Maria—See *Calophyllum braziliense* var. *Rekoi* (Costa Rica and Honduras). Leaves with a single longitudinal nerve.

Santa Maria—See *Miconia argentea* (Local). Leaves with several longitudinal nerves. Flower buds not enveloped in conspicuous bracts.

Santa Maria—See *Miconia dodecandra* (Costa Rica). Leaves multinerved. Buds enveloped in large bracts.

Santa Rosa—See *Triplaris melaenodendron* (Nicaragua).

SAPINDACEAE—Key

1. Leaves bipinnate. Individual leaflets about 1″ long, with serrate margins........ *Dipterodendron costaricense*
1. Leaves once pinnate. Individual leaflets more than 2″ long, the margins entire.
  2. Pinnate leaves very large, more than 2 ft. long.
    3. Individual leaflets usually more than 8″ long. Fruits brown, nearly twice as long as broad, the apex acute........ *Talisia nervosa*
    3. Individual leaflets usually less than 6″ long. Fruits red, nearly globose, often with 2 fused together, the apex broadly rounded........ *Pseudima costaricensis*
  2. Pinnate leaves less than 2 ft. long.
    3. Fruits globose, waxy, 1-seeded........ *Sapindus saponaria*
    3. Fruits not waxy, 2–3 seeded.
      4. Capsules 2-celled........ *Matayba ingaefolia*
      4. Capsules 3-celled........ *Cupania*

Sapindus saponaria L.—Sapindaceae—*Jaboncillo* (Local)—*Limoncillo* (Panama)—*Jaboncillo, Cuyus,* or *Pacon* (Nicaragua)—*Pacon* (Honduras)—*Pacun* (Salvador)—*Guiril, Huiril, Jaboncillo,* or *Jaboncillal* (Guatemala)—*Soap berry* (Southern U. S.). (Pl. 34)

Medium-sized trees, 40–65 ft. in height, with alternate, pinnate leaves, the 6–12 large, opposite, elliptic-oblong, glabrous leaflets usually more or less asymmetrical in outline, with acuminate tips and wedge-shaped, short-petiolate bases, the blades varying from 3 to about 8″ in length. The small, greenish-white flowers are produced in November and early December in very large terminal panicles, rendering the trees conspicuous from a distance. The nearly globose, translucent, waxy-yellow fruits mature in late March, and are sometimes used in country districts of Central America as a soap substitute, since they lather well when mascerated in water. The seeds are reported to be used as a fish poison in Mexico. The yellowish or brownish, hard, heavy wood is coarse-textured and rather difficult to work, and is not durable. The species is frequent along trails and in the town of Palmar Norte, but has rather the aspect of an introduced plant.—Palmar Norte, 50 ft., *Allen 5515, 5722, 6011, & 6319.*

SAPIUM—KEY

1. Flower spikes in clusters. Tips of leaves not thickened, bifid or recurved.. *Sapium jamaicense*
1. Flower spikes solitary. Tips of leaves bifid, thickened and often recurved.. *Sapium thelocarpum*

SAPIUM JAMAICENSE Swartz—Euphorbiaceae—*Leche* (Local)—*Yos* (Costa Rica)—*Olivo* (Chiriquí)—*Wild fig* (Bocas del Toro).

Medium-sized to large forest trees, 65–90 ft. in height, with alternate, oblong-elliptic, leathery leaves which are obtuse or shortly acute at the apex and rounded at the base, the slender petioles with 2 small glands at the base of the blade. The small red flowers are produced in late February and early March in clusters of slender terminal spikes, and are followed by small capsular fruits. All parts of the tree have an abundant milky latex. The light, white wood is not used locally but should be suitable for boxes and crates. Frequent in forests throughout the area, and readily separable from the following species by the fasciculate rather than solitary spikes.—Golfito, *Pittier 9906* & *Allen 5991*—Esquinas Forest, *Allen 5773.*

SAPIUM THELOCARPUM Schum. & Pitt.—Euphorbiaceae—*Palo de leche* (Local)—*Olivo macho* (Panama).

Medium-sized trees, 40–60 ft. in height, with alternate elliptic-oblong leathery leaves 5–6″ long, the apex of the blade thickened, bifid and recurved. The margins are minutely serrulate, with distant, larger, irregularly spaced teeth, the slender petioles bearing a pair of small pointed glands at the base of the blade. The trunks are often characterized by low horizontal corky ridges placed at frequent intervals and sometimes having 4–5 short spines which are arranged in a definite symmetrical pattern. The small red flowers are produced in July in long slender solitary terminal spikes, and are followed in September by the 3-celled capsular fruits. All parts of the tree have an abundant milky latex. Young specimens up to 15–20 ft. in height develop very rapidly and have much larger leaves, up to about 15″ in length, with serrate, recurved margins. The white or pale-yellow wood is rather light and soft and is not used.—Frequent in pastures and fence rows near Palmar and Palmar Norte at low elevations, *Allen 5567 & 5660.*

Sapo—See *Grias Fendleri* (Chiriquí).

SAPOTACEAE—KEY

1. Leaves usually more than 8″ long.
   2. Leaves conspicuously clustered at the ends of the branches.

3. Fruits globose, densely covered with thick, brownish-orange processes which give them a woolly texture........*Pouteria neglecta*
3. Fruits ellipsoidal, glabrous, pale grayish white in color...*Calocarpum borucanum*
2. Leaves not clustered at the ends of the branches. Fruits densely covered with minute brown scurfy scales........*Pouteria triplarifolia*

1. Leaves usually less than 6″ long.
2. Leaves conspicuously reddish brown on the lower surface........*Chrysophyllum mexicanum*
2. Leaves not reddish brown on the lower surface.
3. Flowers in dense heads on slender, often branching scapes, up to 1″ or more in length........*Pouteria heterodoxa*
3. Flowers in sessile or nearly sessile fascicles.
4. Leaf petioles about 1″ long, the blades rounded or subacute at the apex........*Pouteria subrotata*
4. Leaf petioles about ½″ long, the blades acute or acuminate at the apex.
5. Flowers completely sessile........*Pouteria chiricana*
5. Flowers on slender peduncles about ⅛″ long........*Chrysophyllum panamense*

Sapote—See *Calocarpum borucanum* (Local). Plants with milky latex. Fruits grayish white.

Sapote—See *Pouteria triplarifolia* (Local). Plants with milky latex. Fruits brown.

Sapote—See *Tetrathylacium costaricense* (Costa Rica). Plants without milky latex.

Sapotillo—See *Licania operculipetala* (Local).

Saragundi—See *Cassia reticulata* (Local).

Saragundin—See *Cassia reticulata* (Local).

Sardinillo—See *Tecoma stans* (Nicaragua and Honduras).

Sardino—See *Xylopia sericophylla* (Local).

Sastra—See *Rheedia edulis* (Panama to Honduras).

SAURAUIA YASICAE Loes.—Actinidiaceae.

Shrubs or small bushy trees, to about 30 ft. in height, with somewhat scurfy branches, the alternate leaves averaging 5–6″ in length, with long slender petioles, the obovate-oblong, serrulate blades having acute or acuminate apices and long-decurrent bases. The small, white, fragrant flowers are produced in December in axillary panicles which about equal the leaves in length.—Pastures near the Esquinas Station, sea level, *Allen 5699*—Esquinas Forest, *Skutch 5386*.

Sayuc—See *Vochysia hondurensis* (Guatemala).

SCHEELEA ROSTRATA (Oerst.) Burret—Palmaceae—*Palma real* or *Manaca* (Local and Panama). (Pl. 6)

Slender or robust, single trunked palms, 30–75 ft. in height, with very

large, arching, pinnate fronds. The species is frequent in forests throughout the area, ascending the hills between Palmar and Boruca to about 2,500 ft., but sometimes forms groves of considerable extent in open pastures, particularly near Puerto Cortés. The large pendulous inflorescences are protected by a broad, corrugated boat-shaped spathe, and may bear (1) all staminate flowers, (2) all pistillate flowers, or (3) both, in which case the larger pistillate flowers are at the base of the individual strands, the terminal 3/4 of which bear the crowded smaller staminate form. The yellow, ellipsoidal fruits have a prominent beak at the apex, and are indistinguishable from those of the *Scheelea* palms of Panama, so that it seems probable that *S. zonensis* may be referable here. The fronds of this species are used to a considerable extent locally for thatch.—Pastures near Palmar Norte, 50 ft., *Allen 6661* —Pastures near Puerto Cortés, 25 ft., *Moore 6540* & *Allen 6754.*

Schizolobium parahybum (Vell.) Blake—Leguminosae; Caesalpinoideae—*Gallinazo* or *Gavilan* (Local)—*Indio* (Panama)—*Gavilan* (Nicaragua)—*Quon* (Nicaragua)—*Tambor, Plumajillo,* or *Zorra* (Honduras) —*Chapulaltapa* (Salvador)—*Copte, Cucte, Guanacaste, Plumajillo, Plumillo, Zorra,* or *Zorro* (Guatemala).

Tall, usually flat-topped forest trees, 75–120 ft. in height and up to about 30″ in trunk diameter above the prominently developed buttresses, with very large, alternate, bipinnate, fernlike leaves which are deciduous during the flowering season. The bright-yellow flowers usually first appear in late December, though blooming periods as soon as early November and as late as mid-January have been noted. The large, showy terminal panicles render the trees very conspicuous and attractive from a distance. The thin, brown, oblanceolate or spatulate, 1-seeded, indehiscent pods mature in late March or early April. The trees tend to form local colonies of from about 12 to 100 or more individuals, mostly in the 300–1,500 ft. belt on the forested hillsides, very conspicuous concentrations being found above Palmar Norte and on the hills fronting on Golfo Dulce between the delta of the Río Esquinas and Golfito, and above the Golfito townsite. The flowering season is not exactly correlated throughout the area, so that in some years stands near Palmar may bloom a week earlier than those elsewhere, probably because of the drier climate. A smaller second flowering frequently occurs from late January to early March. Young specimens are often seen on cutover land, where they develop very rapidly, with very large fronds up to 4 ft. or more in length, considerably resembling tree ferns in superficial

appearance. The wood varies in color and texture, apparently with the age of the tree, from nearly white and light in weight to pinkish or pale brown and fairly hard and heavy, but all types tend to saw woolly, and are not durable. The wood has no known uses, although it has been suggested as a source of paper pulp, on account of the rapid growth rate. It is surprising, considering the outstanding flowering and foliage characters, that the species is not more frequently planted for ornament. —Hills near Golfito, 250 ft., *Allen 5748.*

Sebo—See *Dialyanthera otoba* (Costa Rica).
Shila blanca—See *Pachira aquatica* (Salvador).
Sibicte—See *Lonchocarpus guatemalensis* (Guatemala).

SICKINGIA MAXONII Standl.—Rubiaceae—*Guaitil* or *Guaytil* (Local)—*Inkwood* (Limón)—*Guaitil colorado* or *Jagua de montaña* (Panama).

Small or medium-sized forest trees, 25–50 ft. in height, with very large, opposite, nearly glabrous leaves, the broadly obovate blades up to about 18″ long and 10″ wide, with abruptly acuminate apices and narrowed, subtruncate, shortly petiolate bases. The pale greenish-yellow flowers have conspicuous purple stamens and are produced in late June and early July in short, dense terminal panicles, and are followed by the brown, nearly globose, woody capsular fruits which are about 3″ in diameter and mature in December and January. The wood is pink or reddish when freshly cut, but fades upon exposure. It is used locally for bridge stringers, fence posts, and tool handles. Locally common in wet, lowland forests.—Esquinas Forest, 50 ft., *Allen 6257 & 6314.*

Sicunken—See *Hamelia patens* (Guatemala).
Siete cueros—See *Lonchocarpus sericeus* var. *glabrescens* (Panama).
Sigua—A general term, applied to various trees of the Lauraceae, notably to *Nectandra salicifolia* and *Phoebe costaricana,* which see.
Sigua amarillo—See *Nectandra glabrescens* (Costa Rica).
Sigua blanca—See *Nectandra salicifolia* (Local). Leaves with a single longitudinal nerve. Fruits with a deep, acorn-like cup.
Sigua blanco—See *Nectandra latifolia* (Panama). Leaves with a single longitudinal nerve. Fruits with a very shallow cup.
Sigua blanco—See *Phoebe costaricana* (Panama). Leaves with 3 longitudinal nerves.
Sigua canela or canelo—See *Ocotea veraguensis* (Costa Rica and Panama).

Sigua negro—See *Nectandra latifolia* (Panama).

Silacil—See *Ardisia revoluta* (Guatemala).

Silk cotton tree—See *Ceiba pentandra* (General).

SIMABA CEDRON Planch.—Simaroubaceae—*Cedron* (Local).

Small trees, 10–20 ft. in height, with very large, alternate pinnate leaves which are often as much as 4 ft. in length, the oblong-lanceolate, abruptly acute or acuminate leaflets about 6–7″ long and 1½–2″ wide. The green, slightly fragrant flowers are a little more than 1″ in length, and are produced in very large terminal panicles, and are followed by oval drupaceous fruits. All parts of the plant are intensely bitter, and the seeds have a local reputation as a remedy for snake bite. Cultivated as a medicinal plant in Palmar Norte, and naturalized as a very rare element along trails and in hedge rows.—Santo Domingo de Osa, *Brenes 12271*—Vicinity of Palmar Norte, *Allen 6738.*

SIMAROUBACEAE—KEY

1. Leaf petiole broadly and conspicuously winged........ *Quassia amara*
1. Leaf petiole not winged.
   2. Pinnate leaves very large, usually more than 3 ft. long........ *Simaba cedron*
   2. Pinnate leaves usually less than 1 ft. in length.
      3. Leaflets broadly rounded at the apex, conspicuously paler in color on the lower surface........ *Simarouba glauca*
      3. Leaflets acute or acuminate, not paler on the lower surface...... *Picramnia latifolia*

SIMAROUBA GLAUCA DC.—Simaroubaceae—*Aceituno* or *Olivo* (Costa Rica)—*Aceituno* or *Negrito* (Nicaragua to Guatemala)—*Chapascuapul, Jucumico, Jocote de mico, Pasac,* or *Zapatero* (Guatemala). (Pl. 17)

Medium-sized or large trees, to 120 ft. in height and up to about 3 ft. in trunk diameter, with alternate, pinnate leaves, the 9–13 alternate leathery, glabrous leaflets deep lustrous green above and much paler in color beneath, with obtuse or subapiculate apices and obliquely cuneate, nearly sessile bases. The small flowers are produced in rather large, diffuse terminal panicles, and are followed in March by the black, edible, olive-like fruits which are rich in oil, and are sometimes exploited commercially in Honduras and Salvador, the oil being used for cooking or for making soap. Both the bark and wood have a very bitter taste. The wood is white, light in weight and soft, but is sometimes used for fuel since it will burn when freshly cut.—Occasional, as very large specimens, in lowland rain forests near Palmar Norte, *Allen 5975.*

Sincahuite—See *Laguncularia racemosa* (Salvador).

SIPARUNA PATELLIFORMIS Perkins—Monimiaceae—*Limoncillo* (Costa Rica).

Shrubs or small, bushy trees, to about 18 ft. in height, with opposite, very short-petiolate leaves, the obovate-oblong blades tomentulose on both surfaces, with abruptly acute apices and subcordate bases, varying from about 3½ to 6″ in length. The very small yellow flowers are produced in nearly sessile axillary cymes in November and December, and are followed by the small, nearly globose drupaceous fruits. Frequent in brushy places throughout the area.—Palmar Norte, *Allen 5729*—Santo Domingo de Osa, *Tonduz 9956.*

Sirasil—See *Ardisia revoluta* (Guatemala).
Siril—See *Miconia argentea* (Guatemala).
Siril de shara—See *Miconia Schlimii* (Guatemala).
Sirin—A general term, applied in northern Central America to various species of *Miconia,* which see.
Sirin blanco—See *Miconia Schlimii* (Honduras).
Sirin cacal—See *Miconia argentea* (Guatemala).
Sirin macho—See *Miconia argentea* (Salvador).
Sirinon—See *Miconia argentea* (Honduras to Guatemala).
Sisin—See *Terminalia amazonia* (Guatemala).
Sisipince—See *Hamelia patens* (Salvador and Guatemala).

SLOANEA—KEY

1. Leaves usually less than 3″ long. Fruits densely covered with purple, needle-like spines........ *Sloanea picapica*
1. Leaves usually more than 4″ long. Fruits not spiny........ *Sloanea laurifolia*

SLOANEA LAURIFOLIA (Benth.) Benth.—Tiliaceae. (Pl. 32)

Tall forest trees, to about 140 ft. in height, the trunk strongly buttressed to about 12 ft. from the base, and about 4 ft. in diameter above the buttresses in mature specimens. The alternate, rather leathery leaves are very variable in size and shape, being sometimes lanceolate, elliptic-lanceolate, oblanceolate, or broadly elliptic-oblong and shortly acute to obtuse at the apex, the blades 4–9″ long and 1¾–3¾″ wide. The flowers have not been seen. The fruits are a woody, unarmed 4- to 5-valved capsule, and may vary from 1¼–2″ in length. The wood is reported locally to burn when freshly cut, and during heavy rain. Our specimens have been examined and named by Dr. C. E. Smith of the Gray Herbarium of Harvard University, who reports the previously known range of the species to have been from the Amazon to the northern part of South

America. Occasional in forests throughout the area.—Hills near the Golfito Airport, 100 ft., *Allen 6349 & 6560.*

SLOANEA PICAPICA Standl.—Tiliaceae—*Zopilote* (Local).

A large forest tree, to 125 ft. in height and about 4½ ft. in trunk diameter above the strongly developed buttresses, with alternate, lanceolate, acuminate, glabrous leaves 2–3″ in length. The flowers have not been seen. The woody 4- to 5-valved fruits are about 1″ in length and are densely covered with dark-purple, needle-like spines, and are produced in short axillary racemes in late March. The ellipsoidal seeds have a thin orange covering and are sweet, and pleasantly flavored, being eaten by roving bands of spider monkeys who break off short branches and drop them to the ground during feeding, the present specimens having been so obtained. The wood is rather hard, dark brown and lustrous, and is reported to be durable under water and in the ground.—Forested hills back of Farm 2, near Palmar, 100 ft., *Allen 6024.*

Snake seed—See *Posoqueria latifolia* (British Honduras).

Soap berry—See *Sapindus saponaria* (Southern U. S.).

Sobon—See *Malvaviscus arboreus* (Guatemala).

SOCRATEA DURISSIMA (Oerst.) Wendl.—Palmaceae—*Palmito* (Local)—*Stilt palm* (Local and general)—*Chonta* or *Palmilera* (Boruca)—*Maquenque* or *Palmito* (Costa Rica)—*Jira* (Panama). (Pl. 9)

Slender, single-trunked palms, 40–60 ft. in height, with very prominently developed prickly stilt roots which reach 6–8 ft. in height in mature specimens. Trunk unarmed, bearing 6–8 pinnate fronds, and with a cylindric, somewhat glaucous crownshaft about 4½ ft. long. Individual leaflets narrowly wedge-shaped, broadest near the apex, usually split longitudinally in 3 or 4 segments, the leaflets borne at an ascending angle with the rachis. Inflorescences usually 1 or 2, produced from the trunk below the crownshaft. Spathes apparently 6. Flowers white, with more than 50 pale yellow stamens. Seeds ellipsoidal, with an apical embryo. The palm cabbage from this species is edible, but somewhat bitter and considered inferior to that obtained from *Welfia Georgii.* Very common and widely distributed throughout the area.—Esquinas Forest, 200 ft., *Allen 6611 & 6715*—Tinoco Station, 50 ft., *Moore 6532* & *Allen 6747.*

Solacra—See *Mouriria parvifolia* (Panama).

SOLANACEAE—KEY

1. Leaves broadly cordate or asymmetrically lobed at the base. Flowers green, with purple stamens, in pendulous scorpioid cymes...................... *Cyphomandra costaricensis*

1. Leaves not broadly cordate or unequally lobed at the base. Flowers white, with yellow stamens, in erect, densely whitish corymbs.. *Solanum verbascifolium* var. *adulterinum*

SOLANUM VERBASCIFOLIUM L. var. ADULTERINUM (Ham.) G. Don—Solanaceae—*Zorillo* (Costa Rica)—*Friega plato* or *Hoja blanca* (Honduras)—*Friega plato* or *Tapalayote* (Salvador)—*Hediondilla* or *Tabaco bobo* (Guatemala).

Shrubs or small trees, often 12–18 ft. in height, with densely stellate-tomentose branches and alternate, elliptic-oblong, abruptly acuminate leaves up to about 10″ in length, the upper surface of the blade minutely tomentulose, the lower surface and slender petioles paler in color and densely stellate-tomentose. The small white flowers are produced in November and December in dense, rather attractive, terminal, long-pedunculate, densely whitish stellate-tomentose corymbs and are soon followed by the small globose fruits. Common along trails and on cutover land throughout the area.—Palmar Norte, 100 ft., *Allen 5440*.

Solpiem—See *Bursera simaruba* (Guatemala).
Sombra de armado—See *Casearia sylvestris* (Honduras).
Sombra de conejo—See *Casearia sylvestris* (Honduras).

SOMMERA GRANDIS (Bartl.) Standl.—Rubiaceae.

Small, bushy trees, about 30 ft. in height, with very large, opposite, petiolate leaves 9 to about 18″ long and up to 6½″ wide, the elliptic-oblong, shortly acuminate blades sparsely hispidulous on the upper surface and densely hispid, particularly on the veins, on the lower side. The small white flowers are produced in November and December in short axillary cymes, and are soon followed by the small, nearly globose fruits which bear the persistent calyx lobes at the apex. Locally frequent along fast-flowing streams.—Río Esquinas, near Kilometer 42, *Allen 5431*.

Soroncontil—See *Cassia reticulata* (Nicaragua).
Sotacaballo—See *Pithecolobium longifolium* (Costa Rica).
Souca—See *Trichilia acutanthera* (Chiriquí).
Spanish cedar—See *Cedrela mexicana* (General).
Spanish plum—See *Spondias purpurea* (Honduras).

SPONDIAS—KEY

1. Leaflets usually less than 3″ long, sour to the taste. Fruits in sessile clusters along the branches.. *Spondias purpurea*
1. Leaflets usually more than 3″ long, not sour. Fruits in terminal panicles.. *Spondias mombin*

SPONDIAS MOMBIN L.—Anacardiaceae—*Jobo* (Local)—*Hog plum* or *Wild plum* (Bocas del Toro and Limón)—*Ciruela de jobo, Jobo, Jocote, Jocote de jobo, Jocote jobo,* or *Jocote montero* (Nicaragua)—*Ciruela de monte* or *Jobo* (Honduras)—*Jobo, Jocote jobo, Jobo jocote, Poc,* or *Kinim* (Guatemala). (Pl. 17)

Medium-sized or sometimes very tall trees, 40–80 ft. in height, with rough, corrugated bark, the alternate, pinnate leaves often deciduous during the dry season. The 5–9 pairs of elliptic-lanceolate, acuminate leaflets are nearly glabrous and have strongly oblique, inequilateral, short-petiolate bases and vary from about 2½ to 5″ in length. The small, fragrant white flowers are produced in large terminal panicles in November and December and again in February and March, rendering the trees fairly conspicuous from a distance. The plumlike yellow, acidulous, 1-seeded fruits are edible but not much esteemed. They mature in August and December and January. The soft white wood has no local uses, but might be suitable for boxes. A very common species, found throughout the area.—Santo Domingo de Osa, *Tonduz 9924*—Palmar Norte, *Allen 5606 & 6240*—Golfito, *Allen 5859.*

SPONDIAS PURPUREA L.—Anacardiaceae—*Jobito* (Local)—*Jocote* or *Ciruelo* (Costa Rica)—*Wild plum* (Bocas del Toro and Limón)—*Jobo* (Panama)—*Ciruela* or *Jocote comun* (Nicaragua)—*Ciruela, Ciruelo, Jocote,* or *Spanish plum* (Honduras)—*Anum, Canum, Jocote, Rum, Run, Unum,* or *Xugut* (Guatemala). (Pl. 17)

Small trees, 20–40 ft. in height, commonly planted for living fence posts. The alternate, pinnate leaves are mostly clustered at the ends of the branches, and are usually deciduous during the dry season. The nearly glabrous leaflets occur as 5–15 alternate or subopposite pairs, and are variable in shape, but usually more or less obliquely obovate-oblong, with rounded or retuse apices and cuneate, inequilateral, subsessile bases. The tiny red flowers are produced in February and March in short, slender racemes from the defoliate lateral nodes, and are followed by the red (infrequently yellow) plumlike, large-seeded, acidulous, edible fruits. The leaves have a rather pleasant acid taste, and are sometimes cooked and eaten in northern Central America. The sap has a strong and characteristic aromatic odor. Very common in fence rows throughout the area.—Santo Domingo de Osa, *Tonduz 10076*—Palmar Norte, 50 ft., *Allen 6010.*

STEMMADENIA—KEY

1. Calyx about equaling the corolla tube in length............ *Stemmadenia Donnell-Smithii*
1. Calyx about ½ the length of the corolla tube or shorter.
   2. Calyx about ½ the length of the corolla tube..................... *Stemmadenia grandiflora*

2. Calyx less than ¼ the length of the corolla tube.
   3. Flowers bright yellow .... *Stemmadenia Alfari*
   3. Flowers yellowish, strongly suffused with pink .... *Stemmadenia nervosa*

STEMMADENIA ALFARI (Donn. Sm.) Woodson—Apocynaceae—*Guijarro* (Costa Rica).

Shrubs or small trees, to about 25 ft. in height, with opposite, short-petiolate, elliptic-oblong, long-acuminate, glabrous leaves 2½–4½″ in length and short, reduced terminal racemes of bright-yellow, salverform flowers, followed by paired, divaricate leathery follicles. All parts of the plant have an abundant milky latex. A characteristic species of wet, highland forests.—Cañas Gordas, 3,500 ft., *Pittier 11094.*

STEMMADENIA DONNELL-SMITHII (Rose) Woodson—Apocynaceae—*Huevos de caballo* (Local)—*Cojon de burro* or *Cojon de caballo* (Costa Rica)—*Cojon de burro* or *Cojon de mico* (Honduras)—*Cojon* or *Cojon de puerco* (Salvador)—*Cojon, Copal, Cojon de caballo, Cojon de puerco, Gutigamba,* or *Tonche* (Guatemala)—*Cojoton* (British Honduras). (Pl. 26)

Medium-sized forest trees, 40–60 ft. in height, with opposite, elliptic-oblong or lanceolate, short-petiolate, nearly glabrous leaves which vary from about 3 to 7″ in length, the blades having acute or falcate-acuminate apices and cuneate or decurrent bases. All parts of the plant have an abundant milky latex, which is reported to be used in Salvador for fastening cigarette wrappers. The rather large, salverform yellow flowers have a very large calyx, which about equals the floral tube in length. They are produced in late March and April in short axillary racemes, and are followed in late August and September by the very large, dark-green, fleshy paired follicles, whose appearance is amply suggested by the common names.—Very common in pastures and forest near Palmar Norte at low elevations, *Allen 5227 & 5611.*

STEMMADENIA GRANDIFLORA (Jacq.) Woodson—Apocynaceae.

Shrubs or small trees, to about 15 ft. in height, with opposite, elliptic-oblong, short-petiolate, glabrous leaves which vary from about 3½ to 7″ in length, the blades having shortly falcate-acuminate apices and rounded or cuneate bases. The pairs of leaves are apparently always of unequal size, at least in our specimens. All parts of the plant have an abundant milky latex. The rather large, bright-yellow salverform flowers have a conspicuous white calyx which about equals the floral tube in length. The plants bloom during the dry season from December until March, the flowers being produced in short axillary racemes.—Occasional along trails back of the residence in the Esquinas Forest, *Allen 5965.*

STEMMADENIA NERVOSA Standl. & L. Wms.—Apocynaceae.

Small forest trees, 20–35 ft. in height, with opposite, short-petiolate, glabrous leaves, the linear-lanceolate, falcately-acuminate blades with cuneate or decurrent bases, varying from about 4 to 6″ in length and 1¼—1¾″ in width. The slender salverform flowers are produced in sessile terminal fascicles in late January and February, the floral tube pink, with a very short calyx, the rather fleshy corolla lobes yellow, suffused with pink. All parts of the tree have milky latex, as in the other species of the genus.—Locally common on well-drained ridges in the Esquinas Forest, *Allen 5834.*

STERCULIACEAE—Key

1. Leaves digitately compound.
   2. Small trees, seldom exceeding 15 ft. in height. Flowers dark red, produced directly from the trunk. Fruits yellow, cacao-like .... *Herrania purpurea*
   2. Tall trees, usually more than 75 ft. in height. Flowers pink, in slender axillary panicles. Fruits not yellow or cacao-like .... *Sterculia mexicana*
1. Leaves not digitately compound.
   2. Leaves deeply palmately lobed .... *Sterculia apetala*
   2. Leaves not palmately lobed.
      3. Leaves broadly rounded or cordate at the base.
         4. Leaf blades conspicuously whitish on the lower surface ...... *Theobroma bicolor*
         4. Leaf blades not whitish on the lower surface.
            5. Leaf margins coarsely serrate .... *Guazuma ulmifolia*
            5. Leaf margins entire .... *Basiloxylon excelsum*
      3. Leaves not broadly rounded or cordate at the base.
         4. Leaves conspicuously paler on the lower surface.
            5. Flowers and fruits produced from the branches. Flowers orange .... *Theobroma angustifolium*
            5. Flowers and fruits produced from the trunk. Flowers red .... *Theobroma simiarum*
         4. Leaves not conspicuously paler on the lower surface .... *Theobroma cacao*

STERCULIA—Key

1. Leaves digitately compound .... *Sterculia mexicana*
1. Leaves deeply lobed, but not compound .... *Sterculia apetala*

STERCULIA APETALA (Jacq.) Karst.—Sterculiaceae—*Panama* (Local and Panama)—*Panama* (Nicaragua)—*Castaño* (Honduras). (Pl. 10)

Tall, straight-trunked, handsome trees, 75–100 ft. in height and 3–4 ft. in diameter above the very prominently developed buttresses, with large, alternate, long-petiolate leaves, the broad, 3- to 5-lobed blades often 1 ft. in diameter, the segments typically drooping, in aspect rather like a half-opened umbrella. The flowers are without petals, the broadly bell-shaped, 5-lobed calyx reddish brown on the outer surface and greenish yellow,

striped red, or maroon within. They are produced in subterminal panicles in November and December, and again in March in our area, the color effect being brown when seen from a distance. The fruits mature in December, and in April and May, and are clusters of 5 podlike carpels which split along the undersurface, freeing the numerous black or brown, peanut-like seeds, which are roasted and eaten and are the source of a fine edible oil in some of the Central American countries. The inner surface of the pods is covered with an attractive velvety-brown lining, composed of thousands of minute irritant hairs, which cause intense itching if rubbed into the skin. The nearly colorless wood is light in weight and soft, and is not used. This species which is so common in the lowlands of adjacent Chiriquí Province in Panama is thus far known in the Golfo Dulce area from 2 or 3 specimens in the pastures near Palmar Norte.—*Allen 6637.*

STERCULIA MEXICANA R. Br.—Sterculiaceae—*Castaño* (Guatemala). (Pl. 10)

Tall forest trees, 75–100 ft. in height and 3–4 feet in diameter above the very strongly developed buttresses, which often ascend the trunk for 15–18 ft. The bark is characterized by many vertical, more or less parallel lines of quite conspicuous corky lenticels. The large, alternate, long-petiolate leaves are digitately compound, with 7–9 linear or oblong-lanceolate, acute or acuminate, glabrous, rather resinous leaflets which may vary from 5 to about 12″ in length. The flushes of new leaves are red and flaccid, with handsomely contrasting green veining, and are very resinous to the touch. The flowers are without petals, the rose-pink, somewhat fragrant, deeply 5-lobed calyxes being produced in February in narrow, densely flowered axillary panicles, and are followed by large woody carpels similar to those of the preceding species. The wood is not used, but would appear to be suitable for concrete forms, boxes, and plywood corestock. The species is locally very common on forested hillsides near Palmar Norte, above about 1,000 ft., and forms one of the dominant elements in many places. The species is not otherwise known south of Guatemala, and almost completely replaces *Sterculia apetala* so common in the lowlands of Panama.—Hills above Palmar Norte, 1,500 ft., *Allen 5863.*

Stilt palm—See *Iriartea gigantea* or *Socratea durissima.*

Stinking toe—See *Cassia grandis* (Canal Zone, Bocas del Toro, and Limón).

Subin—See *Acacia farnesiana* (Guatemala). Thorns not hollow.

Subin—See *Acacia spadicigera* (Guatemala). Thorns hollow.
Subin blanco—See *Acacia spadicigera* (Guatemala).
Sucte—See *Carpotroche platyptera* (Honduras and Guatemala).
Supa—See *Guilielma utilis* (Nicaragua, Sumo).
Surra—See *Calycophyllum candidissimum* (Costa Rica).
Surtuba—See *Geonoma binervia* (Costa Rica).
Swamp kaway—See *Platymiscium pinnatum* (Bocas del Toro).
Swampwood—See *Pithecolobium longifolium* (Nicaragua).

SWARTZIA—KEY

1. Leaves simple .... *Swartzia simplex*
1. Leaves pinnate.
   2. Leaf rachis winged .... *Swartzia picramnoides*
   2. Leaf rachis not winged .... *Swartzia panamensis*

SWARTZIA PANAMENSIS Benth.—Leguminosae; Caesalpinoideae—*Comenegro* (Local)—*Carboncillo* or *Guayacan* (Costa Rica)—*Malvecino* (Panama)—*Chichipate* or *Paterno* (Honduras)—*Chichipate* (Salvador). (Pl. 12)

Medium-sized or large forest trees, 60–90 ft. in height, with alternate, pinnate leaves, the 5 ovate or elliptic-oblong, nearly glabrous leaflets with long-acuminate tips and broadly wedge-shaped or rounded, short-petiolulate bases, the blades varying from 4 to about 7″ in length. The rather large and attractive yellow flowers are produced in slender pendulous racemes up to about 16″ in length from the defoliate nodes of the woody branches, the single expanded fimbriate petal giving the flowers a rather orchidaceous appearance. The very large, leathery dehiscent pods may be as much as 10″ long and 3″ wide and contain 2 or 3 huge, fleshy, somewhat flattened seeds which average about 2½″ in diameter. The trees apparently flower and fruit at irregular intervals throughout the year, and produce a flush of new, pale-green foliage in early December that is very conspicuous from a distance. The hard, heavy, dark-brown wood is reported to be highly durable and is used in Honduras for fence posts, railroad ties, wheels, implements, and general heavy construction. A common species found in forests throughout the area.—Vicinity of Tinoco Station, 100 ft., *Allen 5474*—Pastures near Palmar Norte, 100 ft., *Allen 6642*.

SWARTZIA PICRAMNOIDES Standl. & L. Wms.—Leguminosae; Caesalpinoideae.

Small forest trees, 12–15 ft. in height, with alternate, pinnate leaves composed of 5 ovate to elliptic-lanceolate, abruptly obtuse-acuminate,

nearly glabrous leaflets which vary from about 1″ (the basal pair) to 2½″ (the terminal leaflet) in length. The petiole is terete and sericeous on the lower surface and narrowly winged above, the acute, terminal lobes of the wings covering the nearly sessile bases of the leaflets. The orange fruits are about 1¼″ long, and mature in late January and early February. The flowers have not been seen.—Frequent on well-drained ridges in the Esquinas Forest, *Allen 5815.*

SWARTZIA SIMPLEX (Swartz) Spreng.—Leguminosae; Caesalpinoideae —*Naranjillo* (Panama to Honduras).

Medium-sized forest trees, 25–50 ft. in height, with alternate, superficially simple (1-foliolate) leathery leaves, the elliptic-oblong blades 3–5″ long, with bluntly acute tips and rounded, short-petiolulate bases, the petiolules having narrowly auriculate margins. The relatively large, attractive, orchid-like flowers have a single expanded yellow petal, and are produced in slender few-flowered terminal or axillary racemes, followed by the bright-orange, subterete fruits. The trees apparently have 2 flowering seasons, at the beginning and again at the end of the dry season. There is a recent tendency to consider *Swartzia simplex* as being confined to the West Indies, and to refer material such as ours to *Swartzia ochnacea* DC., but the supposed differences would seem too slight and inconstant to warrant such a separation. The species is occasional in forests throughout the area, and locally very common along rocky beaches on Golfo Dulce.—Rocky beaches near the delta of the Río Esquinas, *Allen 5632*—Golfito, *Brenes 12311*—Hills above Palmar Norte, 1,500 ft., *Allen 5959.*

SWEETIA PANAMENSIS Benth.—Leguminosae; Papilionatae—*Guayacan* (Boruca)—*Carboncillo* (Costa Rica)—*Malvecino* (Panama)—*Chichipate* (Salvador, Honduras, and Guatemala)—*Quina silvestre* (Guatemala).

Medium-sized or tall forest trees, 50–90 ft. in height, with alternate, pinnate leaves, the 9–13 alternate, shortly petiolulate, elliptic-oblong, nearly glabrous leaflets with acute, subemarginate tips and rounded bases, the blades varying from 2–3″ in length. The small, creamy-white flowers are produced in September (?) in diffuse, subterminal, or axillary panicles which exceed the leaves in length, and are soon followed by the thin, flat, elliptic-oblong, 1- to 3-seeded indehiscent pods. The bark has a bitter taste, and is used in Salvador and Guatemala as a remedy for fevers. The yellowish or olive-brown wood is hard, heavy, tough, and strong, and is usually considered to be very durable. It is reported to be suitable for railroad ties, bridge timbers, wheels, implements, and

general heavy construction. The species is apparently common near Buenos Aires and in the General Valley, and bark samples have been brought from Boruca by Doris Stone.

Sweetwood—See *Inga marginata* (Bocas del Toro).

SYMPHONIA GLOBULIFERA L. f.—Guttiferae—*Cerillo* (Local and general)—*Marillo* (Local)—*Botoncillo* (Costa Rica)—*Bogum* or *Sambogum* (Bocas del Toro and Limón)—*Cero* or *Barillo* (Chiriquí)—*Barillo, Leche amarillo, Pimientillo,* or *Varillo* (Honduras and Guatemala). (Pl. 33)

Medium-sized, or sometimes tall trees, 60–100 ft. in height and 1–4 ft. in diameter, the lower trunk often with rather short, stout stiltlike roots. The lanceolate or oblong, acuminate, glabrous leaves are opposite on the stems, with short petioles and cuneate bases, the blades varying from 2–4″ in length. The waxy, blood-red, nearly globose flowers are produced in dense heads at the ends of the short, lateral branches from early July to the end of February, depending on location, individual trees remaining in bloom for nearly 2 months. Although the flowers are very handsome and showy when seen close at hand, only exceptional, advantageously placed trees are very conspicuous when seen from a distance. The fruit is a small, brown, ovoid, fleshy, 1-seeded edible berry about ¾″ in diameter. All parts of the tree have an abundant bright-yellow sticky latex, which is used for making candles and torches, caulking boats, and medicinally for the treatment of ulcers. It is reported to yield an ammonia-soluble khaki dye used in British Guiana to give a rich-brown color to leather. The grayish-, greenish-, or yellowish-brown heartwood is considered to be a second-class general utility timber, suitable for temporary railroad ties, bridge boards, boxes, crates, cooperage, plywood corestock, paper pulp, and general construction where protected from the weather. The largest trees are found in the mixed evergreen rain forest, from sea level to about 3,000 ft. elevation, somewhat smaller specimens forming nearly pure, often very extensive stands in areas of fresh water swamp, notable examples being in the Sierpe region back of Jalaca Farm, and in the coastal swamps near Tela, in Honduras.—Esquinas Forest, 200 ft., *Allen 5322*—Hills near Palmar, 200 ft., *Allen 6682*.

SYNECHANTHUS ANGUSTIFOLIUS Wendl.—Palmaceae.

Slender, unarmed palms, with solitary or rarely multiple trunks to about 8–12 feet in height, the canes with about 5 pinnate fronds. The elongate, erect or arching, broomlike spadices are produced either from

among or from below the fronds, and bear globose or ellipsoidal fruits which are orange or red at maturity.—Locally common in the Esquinas Forest, 200 ft., *Moore 6537* & *Allen 6677 & 6751*—Forested hills above Palmar Norte, 2,000 ft., *Allen 6760.*

Tabaco—See *Grias Fendleri* (Local). Flowers produced from the bark of the trunk and larger branches.

Tabaco—See *Triplaris melaenodendron* (Costa Rica). Flowers produced in conspicuous terminal panicles.

Tabaco bobo—See *Solanum verbascifolium* var. *adulterinum* (Guatemala).

Tabaco de monte—See *Triplaris melaenodendron* (Costa Rica).

Tabacon—See *Grias Fendleri* (Local).

Tabacon—See *Triplaris melaenodendron* (Costa Rica).

Tabaquillo—See *Urera caracasana* (Costa Rica).

TABEBUIA—KEY

1. Flowers pink or white. Leaves, petioles, and calyx glabrous...... *Tabebuia pentaphylla*
1. Flowers yellow. New leaves, petioles, and calyx with a brownish pubescence.............. .... *Tabebuia chrysantha*

TABEBUIA CHRYSANTHA (Jacq.) Nicholson—Bignoniaceae—*Cortes* or *Cortes amarillo* (Local)—*Corteza, Corteza amarilla,* or *Guayacan* (Costa Rica)—*Masicaran* or *Quebracho* (Honduras). (Pl. 11)

Medium-sized or tall trees, 60–90 ft. in height, with opposite (occasionally alternate) long-petiolate, digitately compound leaves, the 5 elliptic-oblong or obovate-oblong leaflets having elongate, slender petiolules, the blades with abruptly acute or acuminate tips and rounded bases, varying from 3 to about 9″ in length, the upper surfaces nearly glabrous, with the veins of the lower surface as well as the petiolules densely ferrugineous-hispid. The trees are usually deciduous during the dry season, the large bright-yellow flowers being produced in short, dense clusters from the ends of the leafless branches, rendering the trees very handsome and conspicuous for the 3 or 4 days of their individual duration. The flowering season is very irregular, a few specimens blooming as early as November, but with the majority appearing in late March. The flowers are soon followed by the long, slender, pendulous, dehiscent pods which are filled with the winged, papery-white seeds. The olive-brown, hard, heavy wood is extremely tough, strong, and durable, being suitable for railroad ties, house and fence posts, bridge timbers, and general heavy construction. Axe blades are often broken in the felling of the trees, and the wood

might be compared to lignum vitae in strength and texture. Branches up to 4 or 5″ in diameter will often strike root, and many may be seen as living fence posts in Palmar Norte. The species is fairly common in pastures and forests of the Terraba Valley, up to about 1,000 ft. in elevation, and is generally *reported* to be found throughout the area, although it would be well to remember that there is a strong local tendency to confuse this tree with the equally showy, yellow-flowered *Schizolobium parahybum.* Our specimens do not exactly match material of *Tabebuia chrysantha* in the United States National Museum. Mr. N. Y. Sandwith, of the Royal Botanic Gardens at Kew, who has been kind enough to examine our collections, is of the opinion that the species may be referable elsewhere. Very close comparisons can be made with *Tabebuia rufescens* J. R. Johnston of Colombia, but since there seems to be a truly remarkable degree of variation between juvenile and mature material, and we have not seen the Pittier collection, it is considered best to continue to use the name *chrysantha* until a more adequate series of specimens has been assembled for study.—Curres de Diquis, *Pittier 11963* —Palmar Norte, 50–100 ft., *Allen 5493, 6339, 6711, & 6731.*

TABEBUIA PENTAPHYLLA (L.) Hemsl.—Bignoniaceae—*Roble* or *Roble de la sabana* (Costa Rica and Panama)—*Macuelizo* (Nicaragua and Honduras).

Trees 50–100 ft. in height, with opposite, digitately compound, glabrous leaves. The showy terminal clusters of large pink, or rarely white flowers are produced in February and March while the trees are bare of leaves. The close-grained, hard, heavy wood somewhat resembles oak *(Quercus)* in superficial appearance, and is used to a considerable extent in Central America for general construction and carpentry. Rather rare, but widely distributed in our area, having been seen in the Coto Valley and near Palmar Norte.—Hills above Palmar Norte, 1,500 ft., *Allen 6730.*

TABERNAEMONTANA LONGIPES Donn. Sm.—Apocynaceae.

Shrubs or small trees, to about 15 ft. in height, with large, thin, opposite leaves 8–11″ in length, the elliptic-oblong, glabrous blades with abruptly acuminate tips and cuneate bases. The small, reddish-tan, salverform flowers are produced in late February and early March in remarkable long-pedunculate pendulous cymes which about equal the largest leaves in length. All parts of the plant have an abundant milky latex. Occasional in forested hills above Palmar Norte at about 1,500 ft.—Santo Domingo de Osa, *Tonduz 9980*—Palmar Norte, 1,500 ft., *Allen 5957.*

Tache—See *Tecoma stans* (Salvador).

TACHIGALIA VERSICOLOR Standl. & L. Wms.—Leguminosae; Caesalpinoideae—*Caña fistula, Plomo, Reseco,* or *Pellejo de toro* (Local). (Pl. 21)

Tall forest trees, 120–140 ft. in height and 3–4 ft. in trunk diameter above the very strongly developed buttresses, which frequently ascend to 18 or 20 ft. The rather conspicuous reddish-brown bark has a characteristic, almost granular surface. The large, alternate, pinnate leaves have 6 or 7 pairs of oblong-lanceolate, nearly glabrous leaflets with abruptly obtuse-acuminate tips and subtruncate to subcordate bases, varying from 2″ (the basal pair) to 6½″ (the terminal pair) in length. The small flowers are produced in large terminal paniculate racemes in late July and early August, and have pinkish-lavender buds and calyxes, contrasting with the yellow petals, the color effect being pinkish lavender when seen from a distance. It seems possible that the trees may have two flowering and fruiting seasons, since mature seeds, which are about 4″ long, almost paper thin, and narrowly elliptical in outline, have been collected in the Esquinas Forest in March. The hard, pale-brown wood is reported to be used locally for general construction. Another species of this genus has been tested for crossties in the relatively dry climate of South Africa and was found to have an average life of 9 years. The trees are solitary, or in small colonies on the forested hillsides from near Jalaca Farm to about Kilometer 42, but have not been seen elsewhere.—Esquinas Forest, 200 ft., *Allen 5594 & 6016.*

Tagualaiste—See *Tecoma stans* (Salvador).
Taixcaax—See *Casearia myriantha* (Guatemala).

TALISIA NERVOSA Radlk.—Sapindaceae—*Huesillo* (Local).

Slender forest trees, 20–40 ft. in height and 4–6″ in trunk diameter, with very large, alternate, even-pinnate leaves, which average about 3 ft. in length, the terete rachis having 10–14 subopposite and alternate leaflets, the linear-lanceolate or oblanceolate blades acute or abruptly short-acuminate at the apex and decurrent at the base, varying from about 5 to 10″ in length. The small, white, fragrant flowers are produced in February in very large terminal or axillary panicles, and are followed in June by the conspicuous pale-orange, ellipsoidal, 1-seeded indehiscent fruits. The white, fine-grained wood is extremely hard, a machete blade being considerably damaged in the process of felling one of the slender specimens. Although the species has no present local uses, the wood

might be suitable for axe handles, and generally as a substitute for United States hickory. Frequent in lowland and hillside forests throughout the area.—Woodlands near Jalaca, *Allen 5310*—Forested hills above Palmar Norte, 1,800 ft., *Allen 5917.*

Tallow wood—See *Ximenia americana* (Honduras).
Tamanchich—See *Malvaviscus arboreus* (Guatemala).
Tamarindo—See *Dialium guianense* (Panama to Guatemala). Leaves once-pinnate.
Tamarindo—See *Pithecolobium austrinum* (Local). Leaves bipinnate.
Tamarindo montero—See *Dialium guianense* (Nicaragua).
Tamarindo prieto—See *Dialium guianense* (Honduras and Guatemala).
Tambor—See *Genipa Caruto* (Salvador). Leaves simple, opposite.
Tambor—See *Ochroma lagopus* (Honduras). Leaves simple, alternate.
Tambor—See *Schizolobium parahybum* (Honduras). Leaves pinnate.
Tapabotija—See *Apeiba aspera* (Local and Nicaragua).
Tapacopal—See *Oreopanax capitatus* (Guatemala).
Tapaculo—See *Genipa Caruto* (Nicaragua). Leaves opposite.
Tapaculo—See *Guazuma ulmifolia* (Honduras and Guatemala). Leaves alternate.
Tapal—See *Byrsonima crassifolia* (Guatemala).
Tapalayote—See *Solanum verbascifolium* var. *adulterinum* (Salvador).
Tapasquit—See *Luehea Seemannii* (Honduras and Guatemala).
Tapon—See *Myriocarpa longipes* (Honduras).
Targua—See *Croton* sp. (Costa Rica).
Targua blanco—See *Croton xalapensis* (Costa Rica).
Targua colorado—See *Croton panamensis* (Costa Rica).
Tasto—See *Tecoma stans* (Salvador).
Tatascan—See *Zexmenia frutescens* (Honduras).
Te azteco—See *Hedyosmum mexicanum* (Guatemala).
Tecomajuche—See *Cochlospermum vitifolium* (Guatemala).

TECOMA STANS (L.) HBK—Bignoniaceae—*Candelillo, Vainilla,* or *Carboncillo* (Costa Rica)—*Copete* (Panama)—*Chilca, Flor amarilla,* or *Sardinillo* (Nicaragua)—*San Andres* or *Sardinillo* (Honduras)—*Marchucha, San Andres, Tasto, Tache,* or *Tagualaiste* (Salvador).

Shrubs or small trees, 10–35 ft. in height, with opposite, 2-ranked, odd-pinnate leaves, the 3–7 lanceolate, acuminate leaflets with serrate margins and inequilateral bases. The rather large, attractive, bright-yellow, faintly fragrant, tubular-campanulate flowers are produced in short terminal racemes, from November until about April. Frequently

planted for ornament, and rather doubtfully native in our area.—Forests of Boruca (?), *Tonduz 4731.*

Tecomasuche—See *Cochlospermum vitifolium* (Guatemala).
Tecomasuchil—See *Cochlospermum vitifolium* (Guatemala).
Tecomatillo—See *Cochlospermum vitifolium* (Guatemala).
Te de monte—See *Hedyosmum mexicanum* (Guatemala).
Te maya—See *Hedyosmum mexicanum* (Guatemala).
Tempate—See *Jatropha curcas* (Costa Rica to Guatemala).
Tepenance—See *Ximenia americana* (Guatemala).

TERMINALIA—KEY

1. Fruits fleshy. Mature leaves usually more than 7″ long.................. *Terminalia Catappa*
1. Fruits papery. Mature leaves usually less than 6″ long.
   2. Fruits less than 1″ wide.. *Terminalia amazonia*
   2. Fruits more than 1″ wide.
      3. Terminal, leafy portions of the branches conspicuously thickened. Bark not guava-like.. *Terminalia bucidioides*
      3. Terminal, leafy portions of the branches not thickened. Bark pale tan, guava-like.. *Terminalia lucida*

TERMINALIA AMAZONIA (Gmel.) Exell—Combretaceae—*Amarillo, Amarillon, Guayabo del monte,* or *Papayo* (Local)—*Amarillo real* or *Carboncillo* (Panama)—*Almendro, Bolador, Guayabillo, Guayabo, Membrillo, Naranjo,* or *Volador* (Honduras)—*Nargusta* (U. S. trade name)—*Canxan, Canxun, Guayabo, Guayabillo, Guayabo volador, Naranjo, Pucte, Quebracho, Sisin,* or *Volador* (Guatemala).

Tall forest trees, 80–120 ft. in height and up to about 4 ft. in diameter above the very strongly developed buttresses, with dark-brown, uniformly flaking bark and rather small, alternate, short-petiolate leaves which are mostly clustered at the ends of the branchlets. The lower portions of the trunks very frequently have slender whip suckers, with their characteristic leaves, by which the species may be readily recognized. The oblanceolate leaf blades have acute tips and decurrent bases, and vary from about 1½ to 3″ in length. The trees are briefly deciduous during February, the very small, greenish-tan, fragrant flowers being produced in fascicles of slender spikes from the ends of the leafless branches, and are followed within a few days by the papery, winged fruits which are about ¾″ long. The yellowish, moderately hard, heavy heartwood is usually irregularly streaked with red, and is variously reported as being reasonably durable to very durable, being considered suitable for boat framing, planking, and decking, house framing, flooring, furniture,

bridge decking, fence posts, railroad ties, general construction, plywood, and charcoal. The bark is reported to be used for tanning. Logs which have been allowed to lie too long in the forest are sometimes subject to attack by pinhole borers. The species is fairly frequent throughout the area on well-drained land. Usually listed as either *Terminalia Hayesii* Pittier or *Terminalia obovata* (Poir.) Eich.—Forested hills near Farm 18, 100 ft., *Allen 5874.*

TERMINALIA BUCIDIOIDES Standl. & L. Wms.—Combretaceae. (Pl. 33)

Trees 50–75 ft. in height, the unbuttressed trunk with brown, uniformly flaking bark, rather like that of *Terminalia amazonia.* The narrowly oblanceolate, subacute leaves have strongly decurrent bases and are about 5–6″ long and 1¼–2″ wide, and are clustered at the ends of the conspicuously thickened branchlets, as in the allied genus *Bucida.* The small flowers are produced in slender, subterminal racemes and are followed by an abundance of dark-brown, papery, winged seeds about 1½″ in width.—Fairly frequent in pastures and forest near Golfito at low elevations, *Allen 6282.*

TERMINALIA CATAPPA L.—Combretaceae—*Tropical almond* or *Almendro* (Local and general)—*Alcornoque* (Nicoya).

An introduced species, native of the Asiatic tropics, frequently planted for ornament and naturalized on the sea beaches of the Río Coto delta. Mature specimens may reach 30–40 ft. in height, with conspicuous horizontal whorls of branches. The obovate, obtuse or abruptly short-acuminate leaves are usually about 6–12″ long, the trees usually rendered conspicuous from a distance by numbers of bright-red, senescent blades. The small flowers are produced in slender subterminal racemes, and are followed by fleshy fruits which contain an almond-like edible seed.

Terminalia chiriquensis—See *Terminalia lucida.*
Terminalia Hayesii—See *Terminalia amazonia.*

TERMINALIA LUCIDA Hoffm.—Combretaceae—*Guayabo del monte, Guayabon,* or *Quiura* (Local)—*Guayabo de monte* (U. S. trade name)—*Guayabon* (Nicaragua)—*Guayabo* or *Volador* (Guatemala).

Tall, handsome forest trees, 80–120 ft. in height and up to about 3 ft. in trunk diameter above the prominently developed buttresses, with smooth, pale-tan or sometimes nearly white bark that often peels off in irregular sheets, in general aspect very reminiscent of *Psidium Guajava.* The rather small, alternate leaves are nearly identical with those of *Terminalia amazonia,* some trees being completely deciduous during the heavy rains in October and November. The small flowers are pro-

duced in late December and early January in slender axillary spikes while the trees are in full leaf, and are soon followed by the papery winged seeds 1½–1¾″ in length. The yellowish or brownish heartwood is moderately hard and heavy, and is reported to season readily, finish well, and to be fairly durable, but is often somewhat cross-grained and difficult to work. It is considered suitable for siding, flooring, furniture, dragline mats, and general construction. The species is listed in *Flora of Costa Rica* as *Terminalia chiriquensis* Pittier. Frequent in forests and pastures throughout the area.—Vicinity of Palmar Norte, 100 ft., *Allen 5876.*

Terminalia obovata—See *Terminalia amazonia.*

Terre—See *Croton xalapensis* (Costa Rica).

TESSARIA MUCRONATA DC.—Compositae.

Shrubs or small, very straight-trunked trees, to about 15 ft. in height with alternate, grayish-green leaves, the elliptic-lanceolate blades 2½–3½″ in length, with acute tips and long-decurrent bases. The small, attractive lavender or pinkish-lavender flowers are produced in November in long-pedunculate terminal cymes. The plants form thickets on gravel bars along fast-flowing streams and rather resemble young willows when seen from a distance. There are conspicuous stands along the Río Esquinas near Kilometer 40 that can be seen from a passing motor car.—Palmar Norte, *Allen 5356*—Río Esquinas near Kilometer 42, *Allen 6660.*

Teta negra—See *Theobroma simiarum* (Costa Rica).

TETRAGASTRIS PANAMENSIS (Engler) Kuntze—Burseraceae—*Carbon* or *Copal* (British Honduras). (Pl. 18)

Medium-sized or large forest trees, 40–90 ft. in height, with alternate, pinnate leaves, the 3–7 elliptic-oblong, abruptly obtuse-acuminate, glabrous leaflets with short petiolules and rounded bases, the blades varying from 4–6½″ in length. The small, greenish-yellow flowers have united petals, and are produced in June in slender axillary panicles, and are followed in February and March by the broadly ovoid, more or less top-shaped, 3- to 4-seeded fruits. All parts of the tree have a pleasant and distinctive aromatic odor. The reddish-brown, fine-textured wood is hard, strong, and fairly durable. It often has a rather irregular grain, but is reported to finish smoothly and take a good polish. It is used for furniture and other cabinet work in British Guiana. The species is locally very common in the hills above Palmar Norte, at 1,200–1,800 ft.,

averaging 3–5 trees per acre in some places and forming one of the dominant elements in the forest.—*Allen 5932.*

TETRATHYLACIUM COSTARICENSE Standl.—Flacourtiaceae—*Sapote* (Costa Rica). (Pl. 26)

Small or medium-sized, often spreading trees, 20–50 ft. in height, with very large, alternate, short-petiolate, oblong, leathery leaves 10–12″ in length, with abruptly acuminate apices and cordate bases, the parallel margins of the blades rather inconspicuously serrate. The small, dark-red flowers are produced at irregular intervals during the year in dense axillary panicles, and are followed by small, ovoid fruits. Common on the margins of mangrove swamps near the Golfito dairy, and along small streams throughout the area.—Llanuras de Curris, Rio Diquis (Terraba), *Pittier 11968*—Río Esquinas near Kilometer 42, *Allen 5428*—Golfito, on road to dairy, *Allen 5615*—Hills near Golfito, *Allen 6689.*

THEACEAE—One genus, *Pelliciera.*

THEOBROMA—KEY

1. Leaves conspicuously paler in color on the lower surface.
   2. Leaves with 5–7 nerves radiating from the center of the broadly rounded or subcordate base........*Theobroma bicolor*
   2. Leaves with 3 nerves radiating from the center of the narrowed base, the lateral pair short and inconspicuous.
      3. Flowers and fruits produced from the branches. Flowers orange........*Theobroma angustifolium*
      3. Flowers and fruits produced from the trunk. Flowers red...*Theobroma simiarum*
1. Leaves not conspicuously paler in color on the lower surface........*Theobroma Cacao*

THEOBROMA ANGUSTIFOLIUM DC—Sterculiaceae—*Cacao de mico* (Local)—*Caco silvestre* (Costa Rica)—*Cacao cimarron* (Chiriquí)—*Cacao de Costa Rica* (Guatemala). (Pl. 26)

Medium-sized or tall forest trees, 45–80 ft. in height, with alternate, elliptic-oblong or oblanceolate, shortly acute leaves, which may vary from 4–8½″ in length, the blades with strongly contrasting whitish undersurfaces, more or less serrate margins and narrowly rounded bases. The young branches, petioles, and veins of the lower leaf surface are covered with a rather scurfy pale-tan tomentum. The relatively large, bright-orange flowers are produced in great profusion in several successive flowerings from November until February from the axils of the slender, younger branches, and are followed in August and September by the large, woody, brown-tomentose, cacao-like pendulous pods which are from 4 to about 7″ in length. The species is locally frequent in

lowland forests throughout the area.—Terraba, 750 ft., *Tonduz 4074*—Boruca, *Pittier s.n.*—Swampy woodlands near the Tinoco Station, 50 ft., *Allen 6259*—Lowland forest near Palmar Norte, 100 ft., *Allen 6341.*

THEOBROMA BICOLOR Humb. & Bonpl.—Sterculiaceae—*Pataste, Pataiste,* or *Cacao pataste* (Costa Rica).

Small or medium-sized trees, with large, alternate, oblong or broadly cordate leaves, the blades nearly white on the lower surface and with 5–7 longitudinal nerves radiating from the base of the blade. The reddish flowers are produced in the leaf axils, and are followed by ellipsoidal fruits about 6″ long which have a hard, gourdlike shell. Occasionally cultivated, and perhaps persistent near old house sites.—Puerto Jiménez, *Brenes 12333.*

THEOBROMA CACAO L.—Sterculiaceae—*Cacao* (Local and general)—*Cacau, Caco, Kicou, Kicob, Cuculat,* or *Pacxoc* (Guatemala).

Small, spreading trees, 18–35 ft. in height, with alternate, short-petiolate, elliptic-oblong leaves 6–12″ in length, the blades with abruptly acuminate apices and rounded bases. The small pinkish flowers are produced in few-flowered fascicles directly from the trunk and older branches, and are followed by the familiar large yellow or red, glabrous pods. Frequently cultivated, and sometimes persistent near old house sites.

Theobroma purpureum—See *Herrania purpurea.*

THEOBROMA SIMIARUM Donn. Sm.—Sterculiaceae—*Cacao de mico* or *Teta negra* (Costa Rica).

Medium-sized to tall forest trees with spreading branches, having alternate, obovate-oblong leaves which are paler in color on the lower surface. The red flowers are produced directly from the trunk, as in *T. Cacao,* and are followed by slender, terete, pubescent pods about 1 ft. long.—Terraba, 800 ft., *Pittier & Tonduz 3925*—Boruca, 1,500 ft., *Tonduz 6852*—Area between the Estero Azul and the Río Sierpe, sea level, *Allen & Chittenden 6600.*

TILIACEAE—KEY

1. Leaf axils with very large, circular, peltate stipules. Flowers about 3″ in diameter.......... *Dicraspidia Donnell-Smithii*
1. Leaf axils without circular, peltate stipules. Flowers much less than 3″ in diameter.
   2. Leaf blades very asymmetrical at the base, the longest lobe frequently overlapping the stem. Fruit a red, many-seeded edible berry.......... *Muntingia calabura*
   2. Leaf blades often somewhat asymmetrical at the base, but never with the largest lobe overlapping the stem. Fruit not red or edible.

3. Leaves with a single longitudinal nerve.
 4. Leaves usually more than 4″ long. Fruits not spiny.................. *Sloanea laurifolia*
 4. Leaves usually less than 3″ long. Fruits densely covered with needle-like spines.................. *Sloanea picapica*
3. Leaves with 3–5 longitudinal nerves. at least at the base of the blade.
 4. Base of the leaf blade with a pair of auriculate lobes which usually have glandular margins.................. *Heliocarpus appendiculatus*
 4. Base of the leaf blade without auriculate lobes.
  5. Fruits densely covered with spines or bristles.
   6. Spines long, hairy, and flexible.................. *Apeiba Tibourbou*
   6. Spines short, hard, and conical.................. *Apeiba aspera*
  5. Fruits not armed with spines or bristles.
   6. Fruits completely enveloped in a conspicuous 5-lobed calyx.................. *Pentaplaris Doroteae*
   6. Fruits not enveloped in the calyx.
    7. Leaves conspicuously reddish brown on the lower surface.................. *Luehea Seemannii*
    7. Leaves not reddish brown on the lower surface.
     8. Fruits flat, 2-celled, usually broader than long.................. *Belotia*
     8. Fruits never flat; 3- to 5-celled, about as broad as long or longer than broad.
      9. Fruits composed of 3 oblong, broadly winged seeds fused together along the central axis.................. *Goethalsia meiantha*
      9. Fruit a 3- to 5-celled capsule, each cell with 2–3 seeds.................. *Mortoniodendron*

Timbersweet—See *Nectandra salicifolia* (British Honduras).
Timber sweetwood—See *Persea pallida* (Bocas del Toro).
Tinajito—See *Miconia laevigata* (Guatemala).
Tizate—See *Zexmenia frutescens* (Salvador).
Tiñadientes—See *Genipa Caruto* (Salvador).
Tocote de monte—See *Ximenia americana* (Guatemala).

TOCOYENA PITTIERI (Standl.) Standl.—Rubiaceae.

Medium-sized forest trees, 50–75 ft. in height, with large, opposite, slender-petiolate leaves, the obovate-oblong blades 6–12″ in length and 4–7″ in width, with a blunt terminal apicule and a decurrent base. The attractive butter-yellow flowers are produced in late March and early April in broad, dense terminal cymes, and are followed from August to December by the large globose fruits. The species is reminiscent of a larger, yellow-flowered *Posoqueria* and would be well worth growing for ornament. Occasional on the margins of forest near the Golfito air field and in the hills near Palmar Norte.—Golfito, 100 ft., *Allen 5234.*

Tolte—See *Miconia argentea* (Guatemala).

Tonche—See *Stemmadenia Donnell-Smithii* (Guatemala).

Tontol—See *Protium neglectum* var. *sessiliflorum* (Guatemala).

Torchwood—See *Bursera simaruba* (Honduras).

Torete or toreto—See *Rollinia Jimenezii* (Panama).

Torito—See *Plumeria rubra* forma *acutifolia* (Costa Rica).

Torolillo—See *Alibertia edulis* (Salvador). Ends of flowering branches never armed with spines.

Torolillo—See *Randia armata* (Guatemala). Ends of flowering branches armed with short spines.

Tortugo—See *Crataeva tapia* (Honduras and Guatemala).

Totumbo or Totumo—See *Crescentia cujete* (Panama).

TOVOMITOPSIS—KEY

1. Leaves very large, often more than 10″ long.......................... *Tovomitopsis grandifolia*
1. Leaves usually less than 7″ long.
   2. Fresh flowers about ¼″ in diameter.................................. *Tovomitopsis costaricana*
   2. Fresh flowers about ⅜″ in diameter.. *Tovomitopsis glauca*

TOVOMITOPSIS COSTARICANA Tr. & Pl.—Guttiferae.

Shrubs or small trees, to about 15 ft. in height, nearly identical with the following species, differing principally in the smaller flowers.—Esquinas Forest, *Skutch 5286*.

TOVOMITOPSIS GLAUCA Oersted ex Planch & Triana—Guttiferae.

Shrubs or small trees, to about 15 ft. in height, with opposite, elliptic-lanceolate glabrous leaves, the blades varying from 3 to about 7″ in length with abruptly acuminate tips and decurrent bases. The small white flowers are about ⅜″ in diameter and are produced in March in terminal panicles, and are followed by the small red, fleshy obovoid fruits. Occasional in lowland forests throughout the area.—Esquinas Forest, *Skutch 5403*—Jalaca, 50 ft., *Allen 5214*.

TOVOMITOPSIS GRANDIFOLIA Standl. & L. Wms.—Guttiferae.

Small forest trees, 15 to about 25 ft. in height, with large, slender-petiolate, opposite leaves, the leathery elliptic-oblong blades 9 to about 15″ in length with subacute apices and decurrent bases. The pale-yellow flowers are produced from late December until early March in broad, dense terminal panicles, and are followed by the relatively large red, fleshy, ellipsoid fruits. The fine-grained wood is dark red, and of an attractive appearance when freshly cut.—Occasional on well-drained ridges in the Esquinas Forest, *Allen 5771 & 5968*.

Tree fern—See *Hemitelia choricarpa*.

Trema micrantha (L.) Blume—Ulmaceae—*Capulin blanco* (Local and general)—*Capulin, Juco,* or *Vara blanca* (Costa Rica)—*Capulin macho* (Chiriquí)—*Capulin negro* (Honduras)—*Capulin macho, Capulin montes, Capulincillo,* or *Churrusco* (Salvador)—*Capulin* or *Kib* (Guatemala)—*White capulin, Wild bay cedar,* or *Bastard bay cedar* (British Honduras).

Common, medium-sized weed trees, 20 to about 65 ft. in height, with alternate, slender-petiolate leaves 3–6″ in length, the lanceolate, long-acuminate blades rounded at the base and paler in color on the lower surface. The minute greenish flowers are produced in short axillary cymes in late March and early April, and are followed by the very small red drupaceous fruits. The pale-brown or pinkish wood is light and soft, and is not used. The bark contains a tough fiber that is sometimes used for making cordage, and the supple twigs can be woven into baskets. The trees are of extremely rapid growth, and are found in thickets on cutover land, on gravel bars, and on the margins of forest throughout the area.—Boruca, 1,400 ft., *Tonduz 4418*—Cleared land along the railroad near Jalaca, *Allen 5215.*

TRICHILIA—Key

1. Leaves simple (1-foliolate) .... *Trichilia Skutchii*
1. Leaves pinnate, with 3 or more leaflets.
   2. Fruits tuberculate .... *Trichilia tuberculata*
   2. Fruits not tuberculate.
      3. Lower leaf surface densely pubescent .... *Trichilia acutanthera*
      3. Lower leaf surface glabrous .... *Trichilia montana*

Trichilia acutanthera C. DC.—Meliaceae—*Cedrillo* (Local)—*Conejo colorado, Huesito,* or *Souca* (Chiriquí)—*Mata piojo* (Nicaragua)—*Cedrillo, Cedro espino, Coligallo,* or *Mata piojo* (Honduras)—*Cola de pavo* or *Jocotillo* (Salvador)—*Carbon, Carboncillo, Cedrillo, Cedro colorado, Cola de pavo, Coligallo, Mapaguite, Mapahuite, Mapahuito, Mata piojo, Napahuite,* or *Trompillo* (Guatemala).

Small or medium-sized forest trees, 12 to about 60 ft. in height, with alternate, pinnate leaves, the 5–21 short-petiolate, oblong-lanceolate to oblanceolate, obtusely acute or shortly acuminate leaflets 2½–6″ in length. The small white or greenish-white, fragrant flowers are produced in February and March in slender axillary panicles, and are soon followed by the ovoid, 3-celled red-seeded capsular fruits. The seeds are reported to be used in Nicaragua and Honduras to kill head lice. The soft white wood is apparently not used. Our specimens do not resemble one another very closely, yet all will match specimens identified as *T. acutanthera* in the larger herbaria in the eastern United States. In such a large and

exceedingly technical group it is impossible to decide without prolonged study of a monographic nature whether these differences are merely varietal, or specific in character.—Forests of Curris, Diquis (Terraba) Valley, *Pittier 12036*—Palmar Norte, sea level to about 1,500 ft., *Allen 5222, 5862, & 5939.*

TRICHILIA MONTANA HBK—Meliaceae—*Carbon de rio* (British Honduras).

Small or medium-sized trees, to about 65 ft. in height, with alternate, pinnate leaves, the 3–5 short-petiolulate, elliptic-oblong or oblanceolate, glabrous, rather leathery leaflets 3–5″ in length, with acute or acuminate apices and more or less rounded bases. The small pale-yellow flowers are produced in short axillary panicles in October and are followed in late February by the small, 3-celled, densely pilose fruits.—Santo Domingo de Osa, *Tonduz 10031*—Hills near Palmar Norte, *Allen 5946 & 6597.*

TRICHILIA SKUTCHII Morton & Allen, sp. nov.—Meliaceae.

Small trees, about 12 ft. in height, with alternate, 1-foliolate leaves which have 10–12 pairs of lateral nerves, the broadly elliptic-oblong, abruptly acuminate, glabrous blades 3½–9″ long and 1½–3½″ wide, nearly sessile on the ¼–⅜″ petioles. The small white flowers are produced in December in few-flowered, short-pedunculate, axillary clusters which are usually less than ½″ in length, the peduncles, branches of the inflorescence, shallowly lobed floral calyxes, and unopened buds with a very dense brownish puberulence. All the other species of this genus in our area have manifestly pinnate leaves with 3 or more leaflets.—Esquinas Forest, between Golfo Dulce and Río Terraba, Prov. Puntarenas, alt. 30 ft., *Skutch 5373* TYPE (Herbarium Escuela Agrícola Panamericana).

TRICHILIA TUBERCULATA (Tr. & Pl.) C. DC.—Meliaceae—*Camfin* (Local)—*Alfajillo colorado* (Chiriquí)—*Camfin* or *Fosforito* (Chiriquí and Bocas del Toro)—*Alfaje* (Panama, Chepo).

Medium-sized trees, 30 to 40 ft. in height, with alternate, pinnate leaves, the 5–7 short-petiolulate, elliptic-lanceolate, acuminate, nearly glabrous leaflets varying from 2–5″ in length. The small white or pale-yellow flowers are produced in short axillary panicles and are followed in late August and early September by the ellipsoidal, tuberculate fruits. The hard, heavy brownish wood is reported to burn when freshly cut.—Jalaca, 100 ft., *Allen 5305.*

Triplaris americana—See *Triplaris melaenodendron.*

Triplaris Macombii—See *Triplaris melaenodendron.*

TRIPLARIS MELAENODENDRON (Bertol.) Standl. & Steyerm.—Polygonaceae—*Hormigo, Tabaco, Tabaco de monte,* or *Tabacon* (Costa Rica)—*Guayabo hormiguero, Palo santo,* or *Vara santa* (Panama)—*Flor de arco, Flor de garrobo, Guayabito,* or *Santa Rosa* (Nicaragua)—*Mulato* (Honduras)—*Canilla de mula* or *Gallito* (Salvador)—*Hormigo, Mulato,* or *Palo mulato* (Guatemala).

Slender trees, 40–65 ft. in height and 10–18″ in trunk diameter, with pale, smooth, mottled bark that flakes off in fairly large, irregular sheets. The large, alternate, elliptic-oblong leaves may vary from about 8 to 16″ in length, with subacute apices and rounded or subcordate, sometimes inequilateral, short-petiolate bases. The upper surfaces of the leaves are usually glabrous, while the lower surfaces may be either glabrous or strigillose, with more or less strigose veins and petioles. The dioecious flowers are produced in February and March in large terminal panicles, the creamy-white staminate form small and rather inconspicuous, while in the pistillate form the lobes of the 3-winged accrescent calyx become pink or red, and render the trees very conspicuous and attractive from a distance. The smaller branches are usually hollow and inhabited by colonies of stinging ants. The hard, heavy, white, fine-grained wood is sometimes used for carpentry and general construction. Collections from Costa Rica have been previously listed under the more familiar name of *T. americana* which, according to Standley, was applied by Linnaeus to some at present unknown species from South America. A quite striking form with strigillose lower leaf surfaces and unusually long calyx lobes is fairly frequent in the forested hills between Palmar Norte and Buenos Aires at about 2,000 ft., the straight, white trunks being very conspicuous from a distance. A single staminate tree with glabrous leaves has been seen near Palmar Norte.—Trail to Buenos Aires, 1,800–2,500 ft., *Allen 5923* (pistillate form)—Pastures near Palmar Norte, 100 ft., *Allen 6012* (staminate form).

Trompillo—See *Alibertia edulis* (Costa Rica). Leaves simple. Shrubs or small trees with white flowers.

Trompillo—See *Lafoensia punicifolia* (Salvador). Leaves simple. Medium-sized or large trees with greenish-yellow or red flowers.

Trompillo—See *Trichilia acutanthera* (Guatemala). Leaves pinnate.

Trompito—See *Alibertia edulis* (Panama).

Trompo—See *Alibertia edulis* (Panama).

Tronador—See *Hura crepitans* (Panama). Trunks spiny.

Tronador—See *Oreopanax capitatus* (Guatemala). Trunks not spiny.

TROPHIS—KEY

1. Leaves rough to the touch. Fruits smooth..*Trophis racemosa*
1. Leaves smooth to the touch. Fruits tuberculate..............................*Trophis chorizantha*

TROPHIS CHORIZANTHA Standl.—Moraceae—*Palo morillo* (Guatemala).

Shrubs or small trees, to about 45 ft. in height, with alternate, lanceolate, short-petiolate, glabrous leaves 2½–4½″ in length, with long-acuminate apices, cuneate bases, and entire margins. The minute white dioecious flowers are produced in March in slender axillary spikes.—Woodlands near the Tinoco Station, sea level, *Allen 5492.*

TROPHIS RACEMOSA (L.) Urban—Moraceae—*Ojoche macho* (Chiriquí)—*Lechosa, Breadnut,* or *Ramoon* (Bocas del Toro)—*Gallote, Morillo,* or *Ramon* (Panama)—*Hoja tinta, Ramon,* or *San Ramon* (Honduras)—*Chilujushte, Chulujushte, Pilijushte, Raspa lengua, Ojushte,* or *Ujushte* (Salvador)—*Catalox, Ramon colorado,* or *Yaxox* (Guatemala).

Small or medium-sized trees, to about 50 ft. in height, with alternate, lanceolate or elliptic-oblong, slender petiolate leaves 4–7″ in length, with strigillose upper surfaces which are rough to the touch, the blades with long acuminate apices and serrate margins. The dioecious flowers are produced in February, the minute green staminate form in slender axillary catkins, while the pistillate flowers are nearly sessile, followed by the subglobose, red, fleshy fruits which are reported to be edible. The yellowish-brown, hard, heavy wood is not used locally. The branches are cut as forage for cattle during the dry season in northern Central America.—Fairly frequent along small streams near Palmar Norte, *Allen 5882.*

Tropical almond—See *Terminalia Catappa* (Local and general).

Tsuyuy—See *Cochlospermum vitifolium* (Guatemala).

Tuburus—See *Enterolobium cyclocarpum* (Nicaragua, Sumo).

Tucuico—See *Ardisia revoluta* (Costa Rica).

Tuete—See *Vernonia patens* (Local).

Tuete blanco—See *Vernonia patens* (Local).

Tulipancillo—See *Malvaviscus arboreus* (Guatemala).

Tuna—See *Castilla fallax* (Nicaragua).

Turro—See *Myrcia Oerstediana* (Costa Rica).

Tuskra—See *Corozo oleifera* (Boruca).

Uacoot—See *Pachira aquatica* (Guatemala).

Uapake—See *Dialium guianense* (Guatemala).

Uca—See *Calycophyllum candidissimum* (Guatemala).

Ujushte—See *Brosimum alicastrum* (Guatemala). Leaves not rough to the touch.

Ujushte—See *Trophis racemosa* (Salvador). Leaves rough to the touch.

Ujushte blanco—See *Brosimum alicastrum* (Guatemala).

Ule—See *Castilla fallax* (Chiriquí).

ULMACEAE—One genus, *Trema*.

Um—See *Persea americana* (Guatemala).

Un—See *Persea americana* (Guatemala).

Unonopsis Pittieri Safford—Annonaceae—*Yaya blanca* (Chiriquí).

Small or medium-sized forest trees, 10 to about 65 ft. in height, with alternate, very short-petiolate, linear or elliptic-oblong leaves 6–12″ in length, the thin, nearly glabrous blades with acute or shortly acuminate apices and cuneate bases. The small white flowers are produced in February in few-flowered, short-stalked umbels directly from the bark of the trunk and larger branches, and are soon followed by the globose fruits which are black and nearly ¾″ in diameter at maturity. Occasional in climax forests throughout the area.—Esquinas Forest, 250 ft., *Allen 5839*—Forested hills above Palmar Norte, trail to Buenos Aires, 1,800–2,500 ft., *Allen 5938*—Climax forest near Coto Junction, 50 ft., *Allen 6658*.

Unum—See *Spondias purpurea* (Guatemala).

URERA—Key

1. Leaves broadly ovate, rounded or more or less cordate at the base... *Urera caracasana*
1. Leaves narrowly elliptic-lanceolate, wedge-shaped at the base............ *Urera alceifolia*

Urera alceifolia Gaud.—Urticaceae—*Mata cartago* (Costa Rica)—*Chichicaste* or *Chichicaste de montaña* (Guatemala).

Shrubs or small trees, to about 10–20 ft. in height, with alternate, narrowly lanceolate or elliptic-lanceolate leaves which have serrate margins, the blades wedge-shaped at the base. The small, bright-orange fruits are produced in short, diffuse cymes from the defoliate nodes.—Pastures and margins of forest near Palmar Norte, 100 ft., *Allen 6325*.

Urera caracasana (Jacq.) Griseb.—Urticaceae—*Crespon, Ortiga, Ortiga blanca,* or *Tabaquillo* (Costa Rica)—*Chichicaste, Chichicaste de hormiga, Chichicaston,* or *La* (Guatemala).

Small trees, 20–25 ft. in height, with alternate, long-petiolate leaves which have broadly elliptic-lanceolate or ovate leaves, the blades rounded

or subcordate at the base, and with serrate margins. The small orange fruits are produced in short, compact pedunculate cymes from the defoliate nodes.—Forested hills near Palmar Norte, 1,000 ft., *Allen 6330.*

URTICACEAE—KEY

1. Inflorescences threadlike and pendulous, usually more than 1 ft. in length........................ *Myriocarpa longipes*
1. Inflorescences short, compact branching cymes, usually less than 2″ in length...*Urera*

Utaxcaax—See *Casearia myriantha* (Guatemala).
Uva—See *Ardisia revoluta* (Honduras and Salvador).
Uvero—See *Hamelia patens* (Panama).
Uvito—See *Ardisia revoluta* (Panama).
Vaco—See *Brosimum utile* (Local).
Vainilla—See *Tecoma stans* (Costa Rica).

VANTANEA BARBOURII Standl.—Humiriaceae—*Ira chiricana* or *Nispero* (Costa Rica). (Fig. 20)

Tall forest trees, 90–120 ft. in height and up to 5 ft. in diameter above the basal buttresses, with alternate, short-petiolate, leathery leaves, the oblong-elliptic blades 3–5″ in length, with rounded or emarginate apices and cuneate bases. The trees are completely deciduous for a brief period at about the end of the heavy fall rains in November or December. The small white flowers are produced in June, and again from about November until January in dense terminal or subterminal corymbs, followed by ellipsoidal fruits about 1″ in length. The seeds have a very distinctive appearance, and are frequently seen on the forest floor. The strong, reddish-brown wood is fairly hard and heavy, and is reported to be very durable, having been used for bridge timbers on the Pan-American highway. Locally very common on forested hillsides up to about 2,500 ft. in elevation.—Golfo Dulce, without definite locality, *H. J. Marks, s.n. (Yale School of Forestry 16822)*—Esquinas Forest, 250 ft., *Allen 6546*—Hills near Palmar Sur, 200 ft., *Allen 6681.*

Vara blanca—See *Hedyosmum mexicanum* (Costa Rica). Leaves opposite.
Vara blanca—See *Trema micrantha* (Costa Rica). Leaves alternate.
Vara colorada—See *Zexmenia frutescens* (Guatemala).
Vara santa—See *Triplaris melaenodendron* (Panama).
Varillo—See *Calophyllum braziliense* var. *Rekoi* (Salvador). Latex white.

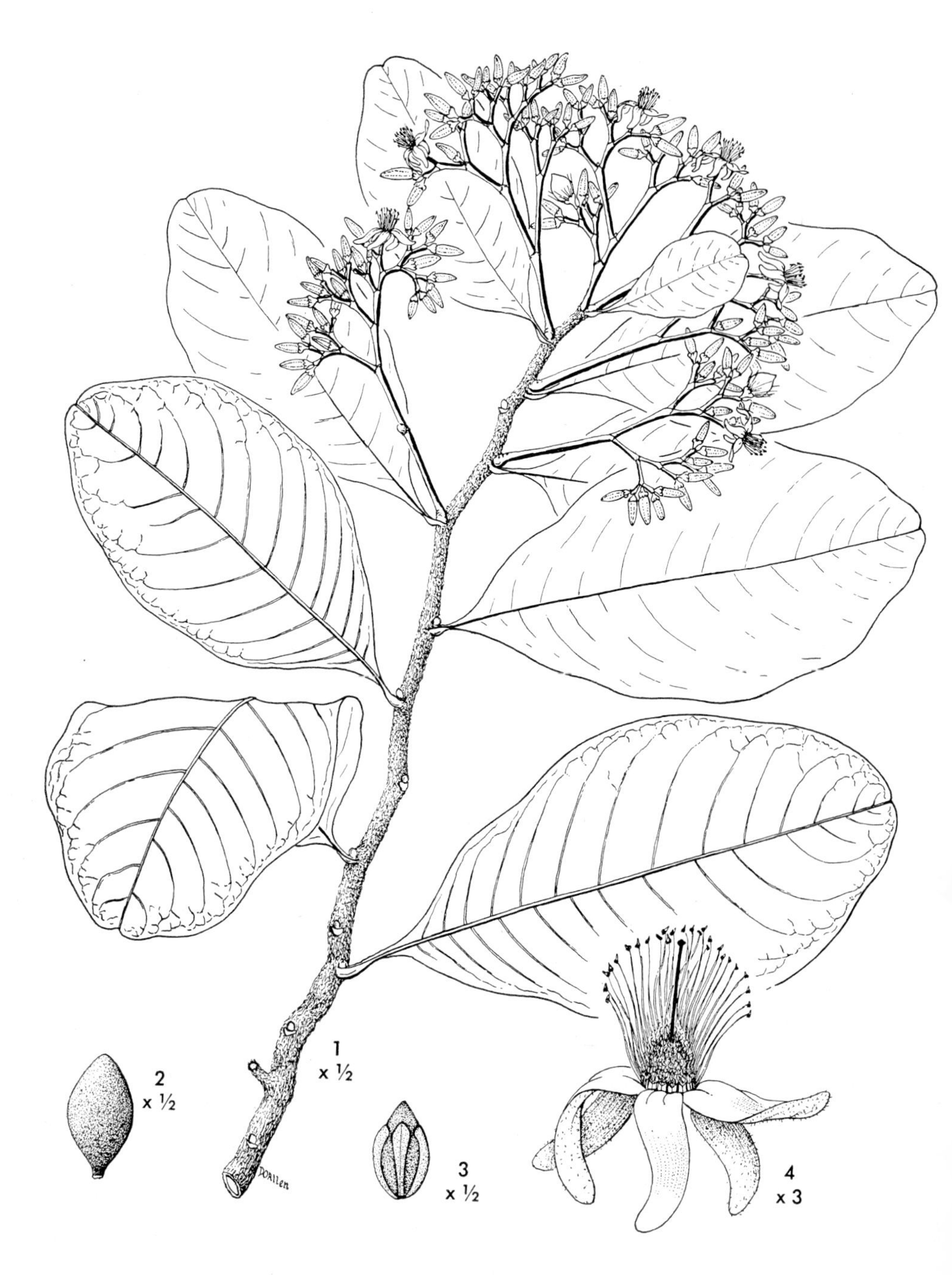

Fig. 20. Vantanea Barbourii. 1 — Flowering branch. 2 — Individual fruit. 3 — Individual seed. 4 — Individual flower.

Varillo—See *Symphonia globulifera* (Honduras and Guatemala). Latex yellow.

VERBENACEAE—KEY

1. Leaves digitately compound........*Vitex Cooperi*
1. Leaves simple.
  2. Leaf margins serrate.
    3. Inflorescences short, nearly globose heads of closely overlapping bracts, produced in pairs from the leaf axils........*Lippia oxyphyllaria*
    3. Inflorescences elongate terminal panicles........*Cornutia grandifolia*
  2. Leaf margins not serrate.
    3. Leaves rounded at the apex. Trees confined to coastal swamps...*Avicennia nitida*
    3. Leaves acute or acuminate at the apex.
      4. Flowers produced in unbranched spikes or racemes........*Citharexylum viride*
      4. Flowers produced in branching cymes or panicles........*Aegiphila*

VERNONIA PATENS HBK—Compositae—*Tuete* or *Tuete blanco* (Local)—*Palo blanco* or *Sanalego* (Chiriquí)—*Boton de pega-pega, Lengua de buey, Lengua de vaca,* or *Salvia* (Panama)—*Cucunango* (Honduras).

Shrubs or small trees, to about 18 ft. in height and 6″ in trunk diameter, with rather rough, alternate, oblong-lanceolate, acuminate, grayish-green leaves 4–6″ in length, and dense terminal clusters of whitish, intensely fragrant flowers which are produced during the dry season in February and March. The crushed leaves are used (thrust up each nostril) to stop nosebleed. Extremely common, often forming fairly extensive thickets on cutover land.—Hills above Palmar Norte, 1,800 ft., *Allen 5909*.

VIOLACEAE—KEY

1. Flowers produced in slender terminal spikes or racemes........*Rinorea pubipes*
1. Flowers produced in very short, branching axillary panicles........*Gloeospermum diversipetalum*

VIROLA—KEY

1. Lower leaf surface brown or tan in color, contrasting strongly with the green upper surface.
  2. Leaves with 12–15 pairs of lateral nerves, the apex of the blade tapering gradually to a long point........*Virola sebifera*
  2. Leaves with 20–30 pairs of lateral nerves, the apex of the blade abruptly acute or acuminate........*Virola Koschnyi*
1. Lower leaf surface not brown or tan.
  2. Leaves with 14–21 pairs of lateral nerves, the blades very long and narrow, tapering gradually to an acute tip.
    3. Seeds about 1″ long........*Virola guatemalensis*
    3. Seeds about ¾″ long........*Virola surinamensis*

2. Leaves with about 30–35 pairs of lateral nerves, the blades abruptly short-acuminate at the tip..*Virola nobilis*

VIROLA GUATEMALENSIS (Hemsl.) Warb.—Myristicaceae—*Fruta dorado* (Local)—*Palo de sebo, Cacao volador,* or *Cacao cimarron* (Guatemala).

Forest trees, 65–100 ft. in height, with alternate, narrowly lanceolate, acuminate leaves 6–12″ long and 1½–2½″ wide, the lower surface paler in color and with 14–21 pairs of lateral nerves. The small yellow flowers are produced in brownish axillary panicles, and are followed by clusters of ellipsoidal, nutmeg-like fruits about 1″ in length, the thick pericarp typically splitting open only on one side.—Esquinas Forest, 50–200 ft., *Allen 5763 & 6727.*

VIROLA KOSCHNYI Warb.—Myristicaceae—*Fruta dorada* (Local)—*Bogamani* (Chiriquí)—*Banak* or *Sangre drago* (Nicaragua)—*Sangre* (Honduras)—*Cedrillo, Drago,* or *Sangre* (Guatemala)—*Banak* (British Honduras).

Tall forest trees, mature specimens 90–120 ft. in height and up to 4 ft. in trunk diameter with nearly horizontal branches and flat-topped crowns, younger specimens having a much more pointed or conical profile. The large, alternate oblong or oblanceolate leaves are 6 to about 12″ in length and densely stellate-tomentose on the lower surface, the blades with acute to acuminate apices and cordate bases. The small yellow flowers are produced in short axillary panicles in December together with the new leaves and are followed in June by clusters of ellipsoidal, yellow, 1-seeded fruits, the mottled glossy-brown nutmeg-like seeds enveloped in a scarlet laciniate aril. The sap of the trunk is red, which explains the common name of "Sangre" used in Honduras and Guatemala. Many specimens are briefly deciduous in November or December, showing the horizontal conifer-like branching habit. The flushes of bright-green new leaves render the trees quite conspicuous at this season. The soft, pale-brown wood is light in weight and not durable, but is used for boxes, concrete forms, interior construction, and plywood corestock. A very common species of lowland and hillside forests, ascending to about 2,000 ft.—Esquinas Forest, 50–200 ft., *Allen 5554.*

Virola merendonis—See *Virola Koschnyi.*

VIROLA NOBILIS A. C. Smith—Myristicaceae—*Fruta dorada* (Local).

Tall forest trees, 60–100 ft. in height, with alternate, pale-green, 2-ranked, short-petiolate leaves 6–10″ in length, the oblong or oblanceolate blades with shortly acuminate apices and broadly cuneate bases, the lower surface with a sparse tomentum of minute stellate hairs. The small

yellow flowers are produced in January in slender lax axillary panicles. The fruits have not been seen, but undoubtedly more or less resemble those of the preceding species.—Esquinas Forest, 50 ft., *Allen 5774.*

Virola panamensis—See *Virola sebifera.*

VIROLA SEBIFERA Aubl.—Myristicaceae—*Fruta dorada* (Local)—*Bogamani* (Chiriquí)—*Fruta dorada* (Panama)—*Banak* or *Sangre drago* (Nicaragua).

Tall forest trees, 65–90 ft. in height, with alternate, short-petiolate leaves, the linear-lanceolate or elliptic-lanceolate, long-acuminate blades 4–9″ in length, the lower surface variously pale and glabrate or brown-tomentulose, probably depending on age. The bark contains a sticky red sap. The small brownish flowers are produced in January and February in short axillary panicles, and are followed in June by the 1-seeded, nutmeg-like fruits which are much relished by forest animals. The seeds contain up to 60% fat and can be used for making candles and soap, and are strung on splinters and burned for illumination by the Indians of Darien and the San Blas coast in Panama. The light, soft, pale-brown wood is easy to work but tends to sapstain badly and is subject to pinhole borer injury if the cut logs are allowed to lie in the forest. The wood is considered suitable for boxes, crates, concrete forms, plywood, and cheap interior construction. The species is common throughout the area.—Esquinas Forest, 250 ft., *Skutch 5401* & *Allen 5832.*

VIROLA SURINAMENSIS (Rol.) Warb.—Myristicaceae—*Fruta dorada* (Local). (Fig. 21)

Slender, narrow-crowned trees about 75 ft. in height, with alternate, oblong-lanceolate, rather abruptly acuminate leaves which are mostly 6½–7½″ long and 2–2½″ wide, the blades with about 14–21 pairs of lateral nerves. The flowers have not been seen, but are probably produced in short axillary panicles, as in the genus, and are followed by clusters of ellipsoidal or nearly globose, nutmeg-like fruits about ¾″ in length.—Vicinity of Palmar Sur, 50 ft., *Allen 6723.*

VISMIA—KEY

1. Flowers pale orange, the floral calyx about ¼″ long, the sepals spreading in fruit. Lower leaf surface pale greenish tan .... *Vismia guianensis*
1. Flowers white, the floral calyx more than ¼″ long, the sepals erect in fruit. Lower leaf surface reddish brown .... *Vismia ferruginea*

VISMIA FERRUGINEA HBK—Guttiferae—*Sangre de perro* (Panama).

Small trees, 15 to about 45 ft. in height, with opposite, short-petiolate, ovate, acute or acuminate leaves, the blades green above and rather

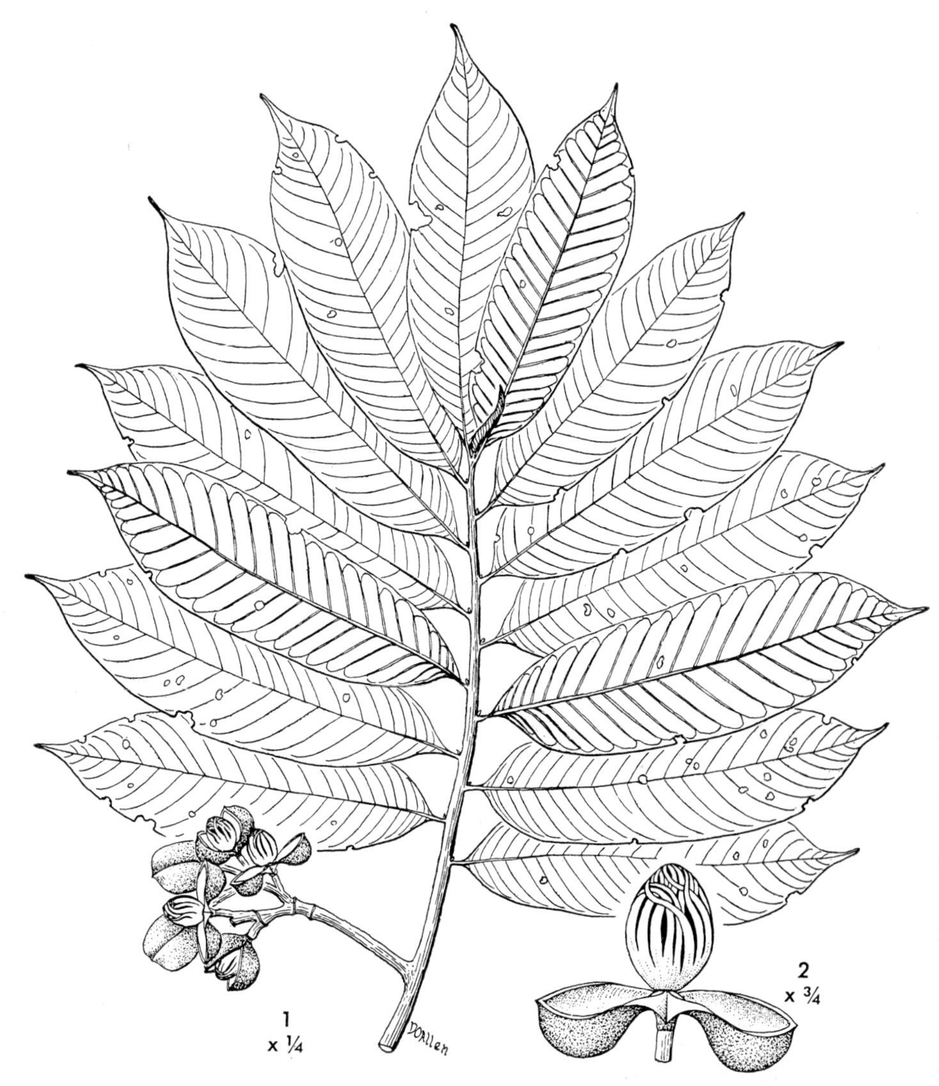

Fig. 21. VIROLA SURINAMENSIS. 1 — Fruiting branch. 2 — Individual fruit.

coarsely brown-tomentose on the lower surface. The relatively large, white flowers are produced in short terminal cymes in March and April. All parts of the plant have a rather sticky orange latex that dries blood red on contact with the air.—Altos de Mano de Tigre, valley of the Diquis (Terraba), 2,100 ft., *Pittier 12045*—Buenos Aires, *Tonduz 3973 & 6695.*

VISMIA GUIANENSIS (Aubl.) Pers.—Guttiferae—*Achiotillo, Achotillo,* or *Sangrillo* (Local)—*Sangrillo* (Panama).

Small trees, 15 to about 40 ft. in height, with opposite, slender-petiolate leaves 5–7″ in length, the ovate to elliptic-oblong blades with acuminate apices and broadly rounded bases, the lower surface with a fine, pale-brown or whitish tomentum. The small flowers are produced in February and March in short terminal cymes, the relatively short brown sepals contrasting handsomely with the pale-orange petals when seen close at hand, but appearing brown in color when seen from a short distance. All parts of the plant have a pale-orange latex that dries lacquer-red on contact with the air. The species is fairly common in the Terraba Valley on the margins of forest and on cutover land, sometimes forming almost pure stands.—Boruca, *Tonduz 3939*—Buenos Aires, *Tonduz 3973, 3997, & 6695*—Margins of forest near Palmar Norte, *Allen 5258 & 6026.*

VITEX COOPERI Standl.—Verbenaceae—*Cuajada* (Local)—*Yellow manwood* (Bocas del Toro)—*Rajate bien* (U. S. trade name)—*White manwood* (Honduras)—*Bimbayan* (Nicaragua). (Pl. 11)

Medium-sized or large trees, 50–80 ft. in height and 1–3 ft. in trunk diameter, with opposite, long-petiolate, digitately 3-foliolate leaves, the elliptic-oblong or obovate-oblong, acute or abruptly acuminate leaflets very variable in size, from about 1–7″ in length. The foliage is apparently always a distinctive shade of yellowish- or brownish-green, contrasting noticeably with the darker blue-greens of the surrounding forest. The small, pale-blue, fragrant flowers may be produced from June to August in profuse, long-pedunculate axillary cymes, and are soon followed by the small, black, ellipsoidal, drupaceous fruits. The hard, reasonably heavy yellowish wood is reported to be very durable in contact with the ground, and is suitable for first-class fence posts, house posts, ships decking, flooring, house framing, professional instruments, and general durable construction. Sometimes used for railroad ties but has the serious defect of splitting badly when spiked. Sawed boards can be used for bridge decking. A fairly frequent species, found in lowland and hillside forests throughout the area.—Forested hills near the Golfito air field, *Allen 5294*—Pastures near Palmar Norte, *Allen 5605*—Esquinas rest pastures, *Allen 6250*—Esquinas Forest, 50 ft., *Allen 6251.*

VOCHYSIA—KEY

1. Leaves in whorls of 3 .... *Vochysia hondurensis*
1. Leaves in pairs.
    2. Leaves densely ferrugineous-tomentose on the lower surface .... *Vochysia ferruginea*
    2. Leaves essentially glabrous on the lower surface .... *Vochysia Allenii*

VOCHYSIA ALLENII Standl. & L. Wms.—Vochysiaceae—*Mayo* (Local). (Fig. 22)

Tall, flat-topped forest trees, about 80 ft. in height, with opposite, slender-petiolate leaves 3½–5″ in length, the lanceolate, acuminate blades glabrous on both surfaces. The relatively large, dark-yellow flowers are produced in slender terminal and subterminal panicles in June, rendering the trees conspicuous and attractive. The heartwood obtained from this specimen was dark pink in color.—Hills near Piedras Blancas, 200 ft., *Allen 6256.*

VOCHYSIA FERRUGINEA Mart.—Vochysiaceae—*Mayo, Mayo colorado* (?), or *Palo de Mayo rosado* (?) (Local)—*Flor de Mayo, Mecri, Pegle, Palo malin,* or *Yemeri macho* (Panama)—*Barba chele* or *Yemeri* (Nicaragua).

Medium-sized or tall forest trees, 40 to about 80 ft. in height, with opposite, short-petiolate, mostly linear-lanceolate, acuminate leaves 3½–6″ in length, the blades glabrous on the upper surface and densely brown-tomentose below. The orange flowers are produced in slender terminal and subterminal panicles, and have an odor reminiscent of freshly cut pumpkin. Individual specimens remain in flower for about 2 months during the blooming season, which may extend from late March until June, the trees being very conspicuous and attractive on the hillsides. The slender, 3-angulate capsular fruits are about 1″ in length and mature in August and September. Mixed lots of *Vochysia* logs may have either white or pink wood, corresponding to the common names of "Mayo blanco" or "Mayo colorado." The darker type is reported to be easier to work and to be more durable than the white, and is the only kind exploited commercially. Present evidence would indicate that the colors probably do not represent specific differences, but are rather the result of individual growing conditions. "Mayo colorado" is considered suitable for general nondurable carpentry, construction, and plywood, while "Mayo blanco" is mostly limited to local use for boxes, crates, concrete forms, and siding. *Vochysia ferruginea* is apparently common throughout the area, but since there are 2 other species of the genus that flower at the same season it is impossible to estimate the percentage of *ferruginea* in the stands.—Buenos Aires, 1,500 ft. (Cited in *Flora of Costa Rica,* but without a definite specimen.)—Forests near Palmar Norte, 100 ft., *Allen 5257 & 6241.*

VOCHYSIA HONDURENSIS Sprague—Vochysiaceae—*Mayo* or *Mayo blanco* (Local) (?)—*Magnolia* or *Palo de chancho* (Costa Rica)—*Yamery* (Bocas del Toro)—*Yemeri* (Panama)—*San Juan* (?) (Honduras)—*Barba chele* or *Palo chancho* (Nicaragua)—*Palo bayo, Robanchab, San*

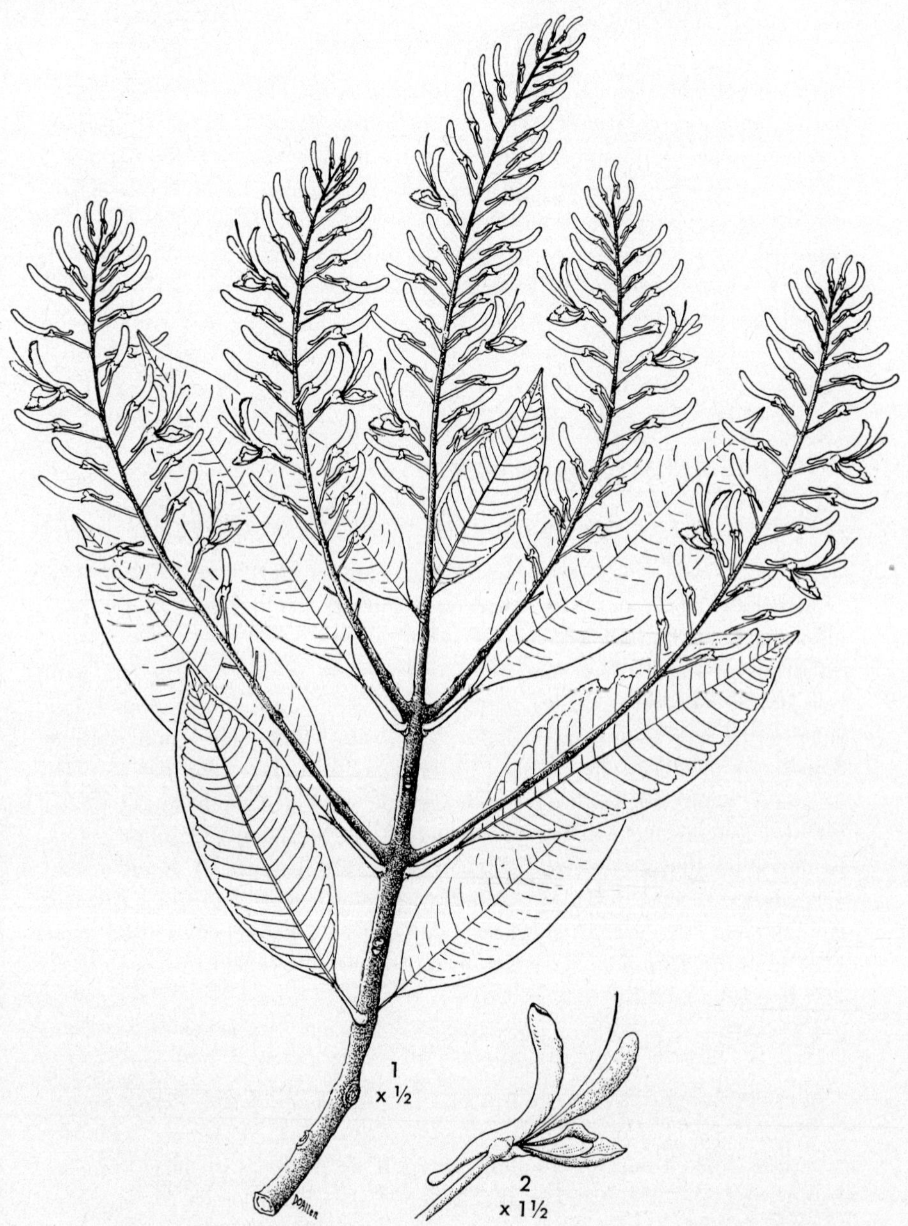

Fig. 22. Vochysia Allenii. 1 — Flowering branch. 2 — Individual flower.

Juan, Sanpedrano, or *Sayuc* (Guatemala)—*Quaruba* (U. S. trade name).

Medium-sized or tall forest trees, 60 to about 90 ft. in height, each node with a whorl of three, slender-petiolate leaves 3½–5½″ in length, the

leathery, elliptic-oblong, glabrous blades with obtuse or emarginate apices and cuneate bases. The bright-yellow flowers have a pleasant fragrance, not reminiscent of freshly cut pumpkin (as in *V. ferruginea*), and are produced in slender terminal panicles in April and May, rendering the trees conspicuous and attractive from a distance. The wood is reported to vary in color from white to pink, the latter type being easier to work and more durable. It is used for untreated railroad ties in British Honduras, which are reported to last 3 or 4 years. It is also suitable for siding and general construction, and has been exported to the United States to some extent for use as plywood. Apparently frequent throughout the area.—Forests near Palmar Norte, *Allen 6239.*

VOCHYSIACEAE—One genus, *Vochysia.*

Volador—See *Terminalia amazonia* or *lucida* (Honduras and Guatemala).

WARSCEWICZIA COCCINEA (Vahl) Klotzsch—Rubiaceae—*Pastora de montaña* (Local). (Pl. 26)

Shrubs or small trees, sometimes reaching 30 or 40 ft. in height, with very large, opposite, nearly sessile leaves, the obovate or oblanceolate acute or shortly acuminate blades 8–20″ in length. The small orange flowers are produced during most of the year in elongate terminal panicles of many small, dense cymes, each cyme provided with an elongate, blood-red petal-like calyx lobe, the effect of the complete inflorescence being rather like a *Poinsettia* when seen from a distance. A common and very showy species, found on well-drained clay ridges and along railroad embankments throughout the area, with a very conspicuous colony near El Alto.—Río Esquinas, near Kilometer 42, *Allen 5249*—Esquinas Forest, 200 ft., *Allen 6714.*

Waterwood—See *Cassipourea podantha* (British Honduras).

WELFIA GEORGII Wendl. ex Burret—Palmaceae—*Palmito, Palma conga,* or *Palma real* (Local)—*Palma conga* or *Palma real* (Panama). (Pl. 10)

Single-trunked, unarmed palms, 50–65 ft. in height, with large, pinnate fronds, the newest member in the center of the crown a rich, conspicuous reddish brown, contrasting very handsomely with the dark-green older foliage. Inflorescences are usually 1 or 2 relatively short pendulous spathes and spadices which are produced below the bases of the fronds. The white flowers are borne in deep pits in the thick rachillae, which are octagonal in cross section, and are followed by the almond-like

fruits, which are dark purple at maturity. The crown yields a sweet, edible palm cabbage, but each utilized means a specimen destroyed. Common in climax forest, particularly in the Esquinas District.—Esquinas Forest, 250 ft., *Allen 6346 & 6770* & *H. E. Moore 6557.*

White capulin—See *Trema micrantha* (British Honduras).

White cedar—See *Dialyanthera otoba* (Bocas del Toro). Trunk unarmed.

White cedar—See *Hura crepitans* (Bocas del Toro and Limón). Trunk heavily armed with spines.

White mangrove—See *Avicennia nitida* (Chiriquí and Bocas del Toro). Leaf petiole without glands.

White mangrove—See *Laguncularia racemosa* (British Honduras). Leaf petiole with 2 glands.

White manwood—See *Vitex Cooperi* (Honduras).

Whitewood—See *Psychotria chiapensis* (British Honduras).

Wild bay cedar—See *Trema micrantha* (British Honduras).

Wild cacao—See *Herrania purpurea* (Bocas del Toro and Limón).

Wild calabash—See *Crescentia cujete* (British Honduras). Leaves in clusters at each node.

Wild calabash—See *Enallagma latifolia* (Bocas del Toro to British Honduras). One leaf at each node.

Wild calabash—See *Parmentiera macrophylla* (Bocas del Toro). Leaves digitately compound.

Wild cashew—See *Anacardium excelsum* (Bocas del Toro).

Wild cherry—See *Byrsonima crassifolia* (Chiriquí).

Wild cocoplum—See *Hirtella* sp. (British Honduras).

Wild coffee—See *Posoqueria latifolia* (Bocas del Toro).

Wild cotton—See *Cochlospermum vitifolium* (British Honduras).

Wild cucumber—See *Jacaratia dolichaula* (Panama).

Wild fig—See *Ficus.* sp. (General). No glands at base of leaf.

Wild fig—See *Sapium jamaicense* (Bocas del Toro). Glands at base of leaf.

Wild guava—See *Alibertia edulis* (British Honduras).

Wild jasmine—See *Aegiphila martinicensis* (Panama).

Wild pigeon plum See *Hirtella triandra* (British Honduras).

Wild plum—See *Spondias mombin* (Bocas del Toro and Limón). Leaflets more than 3″ long. Fruits yellow.

Wild plum—See *Spondias purpurea* (Bocas del Toro and Limón). Leaflets less than 3″ long. Fruits red.

Wild sage—See *Casearia sylvestris* (British Honduras).
Wild senna—See *Cassia reticulata* (Bocas del Toro).
Xacago-que—See *Bursera simaruba* (Guatemala).
Xaxmujin—See *Lonchocarpus guatemalensis* (Guatemala).
Xayau—See *Bixa Orellana* (Guatemala).
Xcanan—See *Hamelia patens* (Guatemala).
Xigal—See *Crescentia cujete* (Guatemala).

XIMENIA AMERICANA L.—Olacaceae—*Pepenance* (Costa Rica)—*Cagalera, Cagalero, Chocomico,* or *Tallow wood* (Honduras)—*Pepenance* (Salvador)—*Abalche, Chocomico, Limoncillo, Manzanilla, Membrillo de monte, Putzil, Saaxnic, Tepenance,* or *Tocote de monte* (Guatemala).

Shrubs or small bushy trees, to about 30 ft. in height, with smooth reddish bark and spiny branches. The alternate, elliptic-oblong, glabrous leaves vary from 1¼–2½″ in length, with rounded apices and short-petiolate, rounded bases. The small white flowers are fragrant, and have densely barbate petals. They are produced in February and March in short, subterminal few-flowered cymes, and are followed by the small yellow or orange plumlike fruits, which are reported to be edible. The hard yellow wood is fragrant and is sometimes used in India as a substitute for sandalwood.—Boruca, *Tonduz 6856* & *Pittier 11983*—Santo Domingo de Osa, *Tonduz 9925.*

Xugut—See *Spondias purpurea* (Guatemala).
Xuyuy—See *Guazuma ulmifolia* (Guatemala).

XYLOPIA—KEY

1. Lower leaf surface silky. Individual carpels of the fruiting clusters more than 4 times as long as broad .... *Xylopia sericophylla*
1. Lower leaf surface glabrous. Individual carpels of the fruits nearly globose .... *Xylopia frutescens*

XYLOPIA FRUTESCENS Aubl.—Annonaceae—*Majagua* or *Malagueta* (Costa Rica)—*Malagueto hembra* (Panama).

Shrubs or slender trees, 10 to about 45 ft. in height, with elongate, usually more or less pendulous branches and alternate, lanceolate, long-acuminate, rather leathery, essentially glabrous leaves and short axillary clusters of fragrant white flowers which are usually produced in May, and are followed in November and December by quite conspicuous reddish stipitate fruits. What appears to be this species is common on the margins of grassland near Potrero Grande and Boruca.

XYLOPIA SERICOPHYLLA Standl. & L. Wms.—Annonaceae—*Sardino* or *Malagueto* (Local). (Pl. 34)

Tall, slender forest trees, 65 to about 80 ft. in height, with alternate, short-petiolate leaves 2–4″ in length, the linear-lanceolate, acuminate blades pale tan in color on the lower surface. The pinkish-yellow flowers are produced in short, few-flowered axillary cymes in January and February, and are followed by dense, short-stalked umbels of terete fruits. Young plants of what appear to be this species develop as tall, slender poles which are used locally for "chusos" for cutting bananas. Fairly frequent in forests of the Esquinas area, forming nearly pure stands in some places.—Esquinas Forest, 250 ft., *Allen 5837.*

XYLOSMA EXCELSUM Standl. & L. Wms.—Flacourtiaceae—*Paiputo* or *Huitite* (Local). (Pl. 34)

Medium-sized trees, 40 to about 75 ft. in height, the trunks usually armed with dense fascicles of needle-like spines. The alternate, short-petiolate, glabrous leaves are 3½–6″ in length, the linear or oblong, acuminate blades with coarsely serrate margins. The small white flowers are produced in late August and early September in short axillary racemes, and are soon followed by the small, dark-green, globose fruits.—Fairly frequent in hedgerows and along small streams near Palmar Norte, *Allen 5650 & 6712.*

Yahal—See *Curatella americana* (Costa Rica).

Yaja—See *Curatella americana* (British Honduras).

Yamery—See *Vochysia hondurensis* (Bocas del Toro).

Yaya blanca—See *Unonopsis Pittieri* (Chiriquí).

Yayo—See *Licaria Cervantesii* (Local). Leaf margins not serrate.

Yayo—See *Luehea Seemannii* (Honduras and Guatemala). Leaf margins serrate.

Yaxox—See *Trophis racemosa* (Guatemala).

Yellow manwood—See *Vitex Cooperi* (Bocas del Toro).

Yema de huevo—See *Chimarrhis latifolia* (Local).

Yemeri—See *Vochysia ferruginea* (Nicaragua). Leaves in pairs at each node.

Yemeri—See *Vochysia hondurensis* (Panama). Leaves in whorls of 3 at each node.

Yemeri macho—See *Vochysia ferruginea* (Panama).

Yigualti—See *Genipa Caruto* (Nicaragua).

Yolillo—See *Raphia taedigera* (Local).

Yos—See *Sapium jamaicense* (Costa Rica).

Yoxcha—See *Cedrela mexicana* (Guatemala).

Yupur—See *Jatropha curcas* (Guatemala).

ZANTHOXYLUM PROCERUM Donn. Sm.—Rutaceae—*Lagartillo* (Local)—*Palo de lagarto* (Costa Rica)—*Cedrillo* (Honduras)—*Ceibillo, Choonte, Kiixche, Lagarto, Limoncillo,* or *Naranjillo* (Guatemala)—*Black prickly yellow* (British Honduras).

Small or medium-sized trees, 15 to about 65 ft. in height, with grayish-white, mottled bark, the trunk armed at irregular intervals with broad-based pyramidal or conical spines. The large, alternate pinnate leaves have 4–17 opposite, short-petiolate leaflets, the elliptic-oblong, shortly acuminate blades with entire margins and varying from about 3–7″ in length. The small white or yellowish-white flowers are produced in late August and early September in short axillary or terminal panicles, and are soon followed by the small green fruits. The rather hard and heavy yellow wood is suitable for fine cabinet work.—Santo Domingo de Osa, *Pittier & Tonduz 9920*—Boruca, 2,100 ft., *Pittier 12080*—Pastures near Palmar Norte, 50 ft., *Allen 5662.*

Zapatero—See *Hieronyma tectissima* (Local). Leaves simple.
Zapatero—See *Simarouba glauca* (Guatemala). Leaves pinnate.
Zapote—See *Calocarpum borucanum* (Local). Fruits grayish white.
Zapote—See *Pouteria triplarifolia* (Local). Fruits brown.
Zapote bobo—See *Pachira aquatica* (Guatemala).
Zapote del monte—See *Pouteria neglecta* (Local).
Zapote de mico—See *Diospyros ebenaster* (Salvador).
Zapote negro—See *Diospyros ebenaster* (Guatemala).
Zapotillo—See *Pouteria* sp. var. (Local).
Zapoton—See *Pachira aquatica* (Honduras and Guatemala).
Zaragoza—See *Conocarpus erecta* (Panama).

ZEXMENIA FRUTESCENS (Mill.) Blake—Compositae—*Tatascan* (Honduras)—*Tizate* (Salvador)—*Flor amarilla, Cambrillo, Palo de escoba,* or *Vara colorada* (Guatemala).

Shrubs or small trees, to about 25 ft. in height, with opposite, ovate, acuminate leaves 2½–6″ in length, the blades with short petioles and serrate margins. The showy yellow flowers are produced in umbelliform cymes from December until February. The slender trunks are used for broom handles in Honduras. The species is very common and conspicuous on cutover land near Boruca, often forming pure stands of considerable extent.—Boruca, 1,500 ft., *Tonduz 4522 & 4757*—Buenos Aires, *Tonduz 4921.*

Zopilote—See *Cornutia grandifolia* (Honduras). Shrubs or small trees. Flowers blue.

Zopilote—See *Cyphomandra costaricensis* (Local). Shrubs or small trees. Flowers greenish.

Zopilote—See *Sloanea picapica* (Local, Sierpe). Very large trees.

Zorillo—See *Hamelia patens* (Costa Rica). Leaves in whorls at each node.

Zorillo—See *Solanum verbascifolium* var. *adulterinum* (Costa Rica). Leaves alternate.

Zorillo colorado—See *Hamelia magnifolia* (Costa Rica).

Zorillo real—See *Hamelia patens* (Costa Rica).

Zorra—See *Schizolobium parahybum* (Honduras and Guatemala).

Zorro—See *Astronium graveolens* (Panama). Leaves once pinnate.

Zorro—See *Schizolobium parahybum* (Guatemala). Leaves bipinnate.

Zschokkea panamensis—See *Lacmellea panamensis.*

Zuncilla—See *Licania arborea* (Guatemala).

GLOSSARY

ACCRESCENT — Increasing in size.
ACUMINATE — Tapering to a slender point.
ACUTE — Distinctly pointed but not drawn out.
ADVENTITIOUS — Produced from an abnormal or unusual place.
ALATE — Furnished with a marginal extension or wing.
ANTHER — The pollen-bearing part of the stamen.
APICULATE — Terminating in a short, sharp point.
APICULE — A short, sharp point.
ARIL — A fleshy covering.
AURICLE — An ear-shaped lobe or projection.
AURICULATE — Provided with an ear-shaped or lobed projection.
BARBATE — Having relatively long, weak, tufted hairs.
BIFID — Divided longitudinally into two parts.
BINOMIAL — A name composed of two parts, as in *Cedrela mexicana.*
BIPINNATE — Having a central nonwoody axis which in turn bears lateral pinnae.
BIVALVATE — Two-valved.
BRACT — A small, leaflike structure.
BRACTEOLATE — Bearing a small bract.
BRACTEOLE — A small bract.
CALYX — The outer, or lowest series of floral segments.
CANOPY — The uppermost layer of the forest.
CAPITATE — Growing in a compact, headlike cluster.
CAPSULE — A dry, compound fruit having more than 1 cell.
CARPEL — A unit of a compound pistil or fruit.
CAUDATE — Having a tail-like appendage.
CAUDATE-ACUMINATE — Having a slender point which terminates in a tail-like appendage.
CHRISM — Consecrated oil used in religious ceremonies.
CLIMAX — The maximum possible development of a plant community on a given site.
CONSOCIATION — A nearly pure stand of a single species.
CORDATE — Heart-shaped in outline.
COROLLA — The inner series of floral segments, either divided or united.
CORYMB — A broad, flat-topped, indeterminate flower cluster.
CRENATE — Having short, rounded teeth.
CROWNSHAFT — In palms, the sheathing basal portions of a frond which envelope the upper trunk.
CUNEATE — Wedge-shaped.
CUPULAR — Cup-shaped.
CUPULE — A small cup.
CYME — A broad, more or less flat-topped determinate inflorescence in which the main and secondary axes always terminate in a single flower.
DECIDUOUS — Falling off at maturity, as leaves in the early dry season.
DECURRENT — Running down the stem or petiole.
DEFOLIATE — Leafless.
DEHISCENT — Opening spontaneously at maturity.
DENTATE — Provided with teeth.
DENTICULATE — Minutely toothed.
DIGITATELY COMPOUND — Having several leaflets which are borne in a more or less radiate cluster on the apex of the petiole.

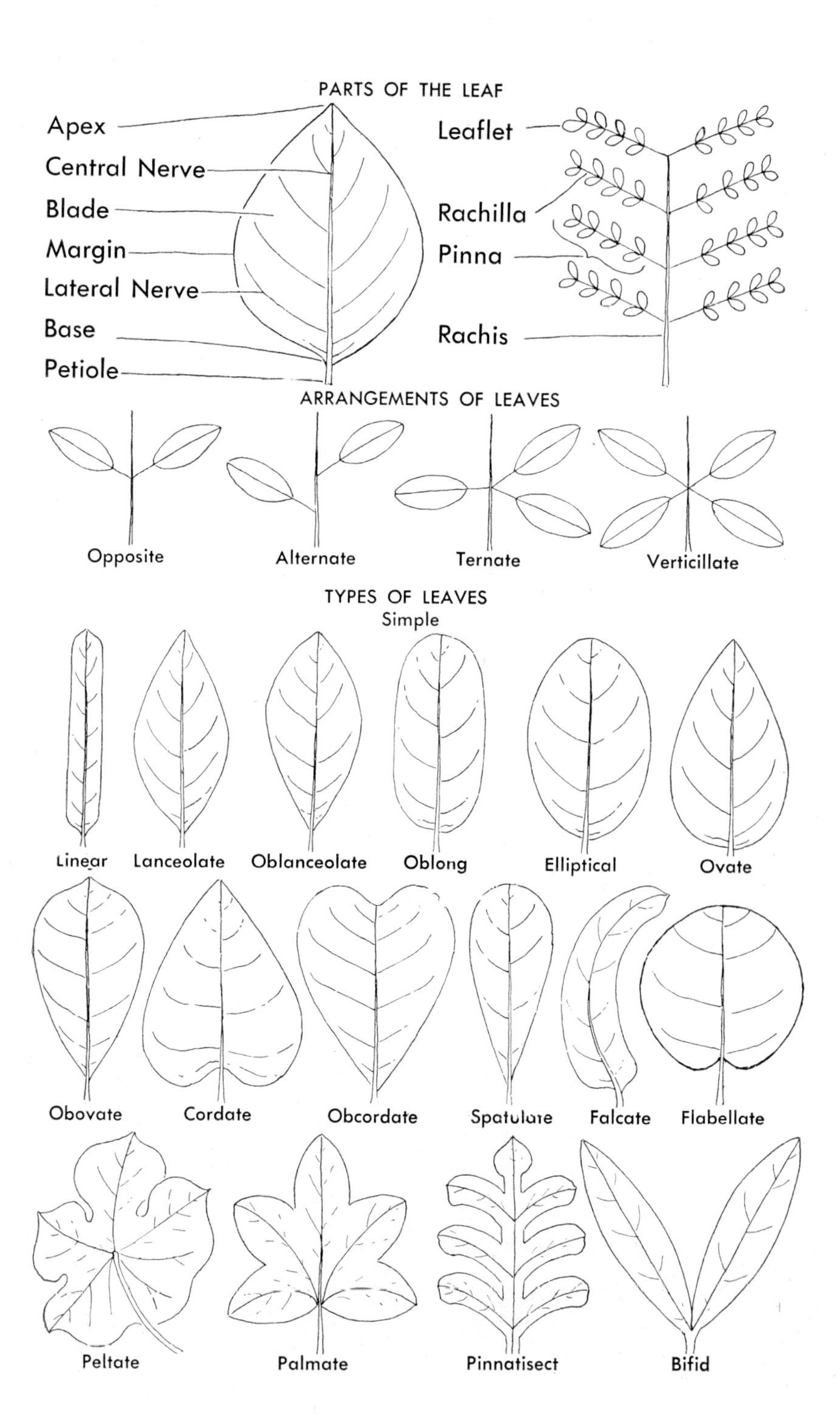
PARTS OF THE LEAF
Apex
Central Nerve
Blade
Margin
Lateral Nerve
Base
Petiole
Leaflet
Rachilla
Pinna
Rachis
ARRANGEMENTS OF LEAVES
Opposite
Alternate
Ternate
Verticillate
TYPES OF LEAVES
Simple
Linear
Lanceolate
Oblanceolate
Oblong
Elliptical
Ovate
Obovate
Cordate
Obcordate
Spatulate
Falcate
Flabellate
Peltate
Palmate
Pinnatisect
Bifid

DIMORPHIC — Occuring in two different forms.
DIOECIOUS — Having male and female flowers on separate plants.
DIVARICATE — Widely spreading.
DRUPACEOUS — Resembling a drupe.
DRUPE — A fruit having a fleshy or leathery pericarp surrounding a single seed, as in a plum.
ELLIPSOID — A solid, such as a fruit which is elliptical in profile.
EMARGINATE — Having a notched apex.
ENTIRE — Having an even margin without teeth or divisions.
EVEN-PINNATE — Having an equal number of leaflets on each side of the rachis. Terminal leaflets 2.
EPIPHYTIC — A nonparasitic plant which grows on the trunk or branches of another without contact with the ground.
EXFOLIATING — Scaling or peeling off in thin layers.
EXPECTORANT — Any substance which clears the throat and chest by stimulating expectoration.
EXSERTED — Projecting beyond or out of.
FACIATION — A local, specialized association of plants.
FALCATE — Curving, sickle-like.
FALCATE-ACUMINATE — Having a slender, curving termination.
FASCICLE — A small bundle or cluster.
FASCICULATE — Occurring in a small cluster.
FERRUGINEOUS — Rust-colored.
FLABELLATE — Fan-shaped.
FLACCID — Limp and weak.
FOLIACEOUS — Leaflike.
FOLIOLATE — Having leaflets.
FOLLICLE — A dry, 1-celled fruit which opens only on 1 side.
FORMATION — A group of plants inhabiting a particular type of environment.
FROND — A (usually) compound leaf, particularly of a fern or palm.
GLABRATE — Nearly glabrous.
GLABRESCENT — Becoming, or slightly glabrous.
GLABROUS — Smooth; destitute of visible or microscopic hairs.
HASTULA — In palms, a more or less wedge-shaped structure at the base of a fan-shaped frond.
HIRTELLOUS — Minutely hairy.
HISPIDULOUS — Covered with minute bristles.
IMBRICATE — Overlapping, as fish scales or tiles on a roof.
INDEHISCENT — Not opening, as of fruits which remain closed at maturity.
INFLORESCENCE — A flower cluster of any type.
INFRA-FOLIACEOUS — Borne below the foliage, particularly the inflorescences of palms.
INVOLUCEL — A secondary, and usually smaller involucre.
INVOLUCRE — A ring or whorl of small leaves or bracts below or supporting a flower or flower cluster.
LACINIATE — Deeply cut into narrow, pointed lobes.
LANCEOLATE — Lance-shaped.
LEGUME — A simple, beanlike pod which (usually) opens spontaneously along both margins at maturity.
LENTICEL — Corky, more or less lens-shaped, spots on bark.
NODE — A joint which has, has had, or may have a leaf or whorl of leaves.
OBCORDATE — Inversely heart-shaped. (Broadest toward the apex.)
OBLANCEOLATE — Inversely lance-shaped. (Broadest toward the apex.)
OBOVATE — Inversely egg-shaped. (Broadest near the apex.)

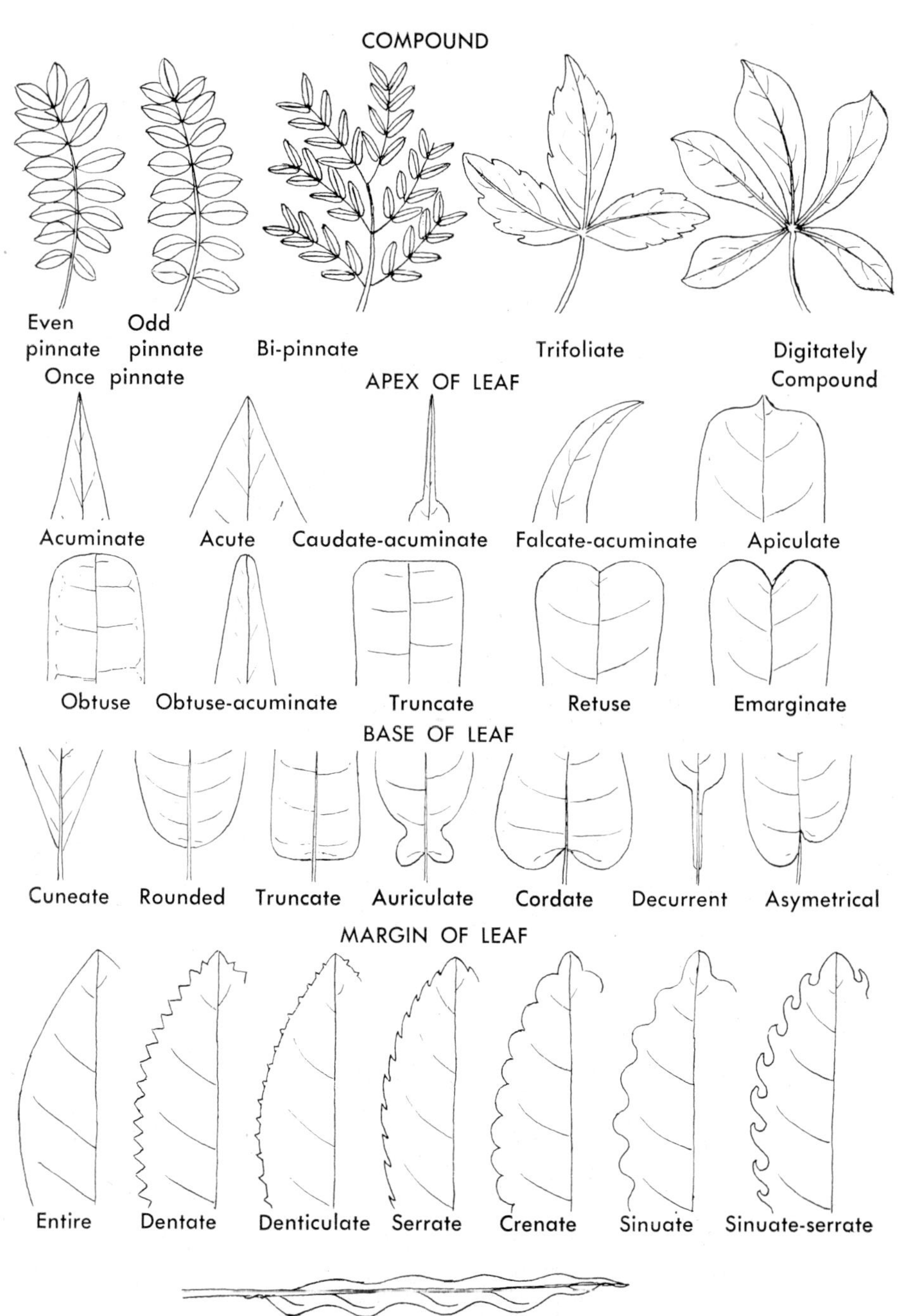
COMPOUND
Even pinnate
Odd pinnate
Once pinnate
Bi-pinnate
Trifoliate
Digitately Compound
APEX OF LEAF
Acuminate
Acute
Caudate-acuminate
Falcate-acuminate
Apiculate
Obtuse
Obtuse-acuminate
Truncate
Retuse
Emarginate
BASE OF LEAF
Cuneate
Rounded
Truncate
Auriculate
Cordate
Decurrent
Asymetrical
MARGIN OF LEAF
Entire
Dentate
Denticulate
Serrate
Crenate
Sinuate
Sinuate-serrate
UNDULANT
Seen from the Side

OBOVOID — An egg-shaped solid which is broadest near the apex.
OBTUSE — Rounded.
OBTUSE-ACUMINATE — Having a slender point, with a blunt or rounded termination.
OCREA — A tubular stipule which envelops the stem.
ODD-PINNATE — Having an uneven number of leaflets. (Terminal leaflet 1.)
OVATE — Egg-shaped in outline and broadest near the base.
OVOID — An egg-shaped solid.
PALMATE — Having lobes or divisions which radiate from the apex of the petiole.
PANICLE — A more or less elongate central axis which bears lateral racemose branches.
PANICULATE — Having the form of a panicle.
PEDICEL — The support of a single flower in a cluster.
PEDUNCLE — The stalk of a solitary flower or of a flower cluster.
PEDUNCULATE — Provided with a stalk.
PELTATE — Having the stalk or petiole attached inside the leaf margin.
PERICARP — The outer, often fleshy covering of certain types of fruits.
PETALOID — Petal-like.
PETIOLATE — Provided with a stalk.
PETIOLE — A leaf stalk.
PETIOLULE — The stalk of a leaflet.
PILOSE — Covered with soft hairs.
PINNAE — Lateral divisions of a bipinnate leaf.
PINNATE — Feather-like. Having a central nonwoody axis with lateral leaflets.
PINNATISECT — Pinnately cut or divided, but not completely to the central axis.
PISTIL — The female organ of a flower.
PISTILLATE — Exclusively female (having a pistil but no stamens).
PNEUMATOPHORES — Erect, finger-like breathing organs attached to roots.
POLYNOMIAL — A compound name of more than two elements.
PUBERULENCE — A minute, downy covering.
PUBERULENT — Having a minute, downy covering.
PUBESCENT — Covered with short, soft hairs.
PYRIFORM — Pear-shaped.
RACEME — A central axis which bears an indefinite number of stalked flowers.
RACEMOSE — Raceme-like.
RACHILLAE — A small, secondary rachis or axis.
RACHIS — The axis of a compound leaf, frond, or inflorescence.
RECEPTACLE — Usually a more or less fleshy structure which supports the flowers, parts of a flower, or fruits.
REPENT — A prostrate, creeping plant, usually rooting at the nodes.
RETICULATE — Netlike in pattern.
RETUSE — Having a shallow notch at the apex.
RUFOUS — Reddish.
SALVERFORM — Narrowly tubular, with abruptly expanded terminal lobes.
SAMARA — An indehiscent fruit which is provided with a single wing, as in *Myroxylon*.
SCORPIOID — Coiled like the tail of a scorpion.
SCURFY — Covered with small scales.
SENESCENT — Growing old.
SERRATE — Provided with sawlike teeth.
SESSILE — Without a stalk.
SINUATE — Having a wavy margin.
SINUATE-SERRATE — Having a wavy margin, with rounded indentations and sawlike teeth.
SPADICES — Plural of spadix.

PARTS OF THE FLOWER

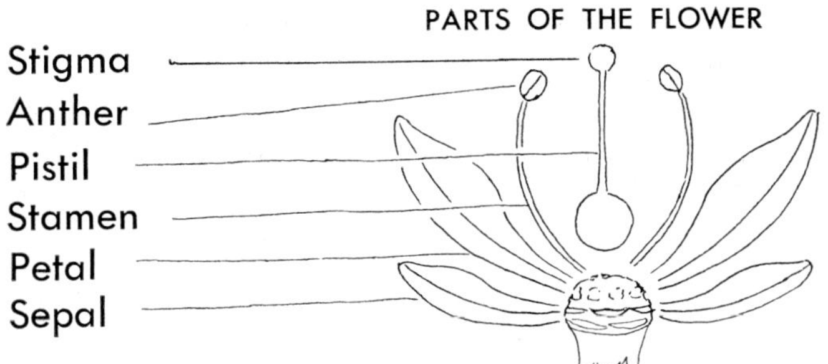

TYPES OF INFLORESCENCES

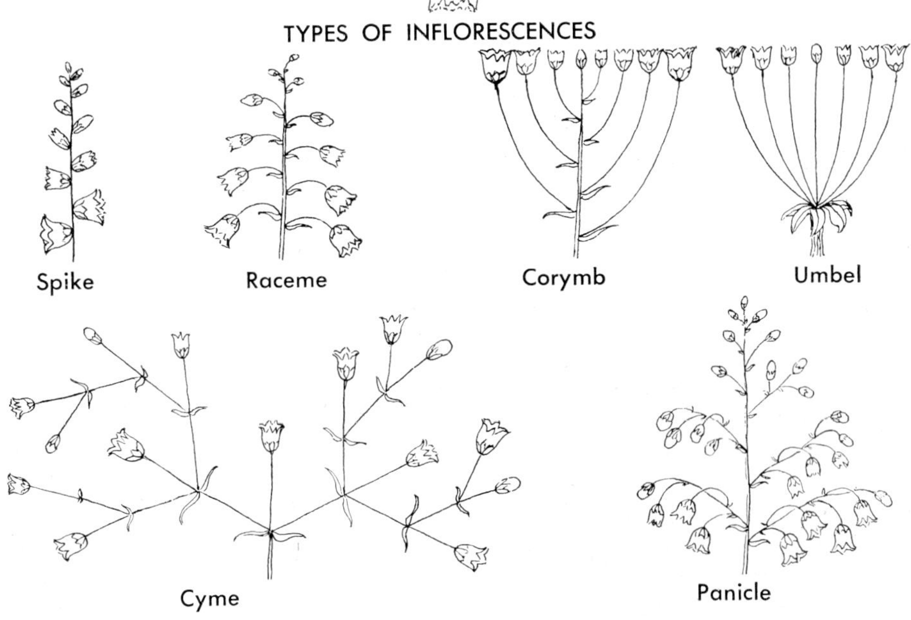

SPADIX — A central spike covered or more or less enveloped by a spathe.

SPATHACEOUS — Spathelike.

SPATHE — In palms, the more or less boat-shaped covering of the inflorescence.

SPATULATE — Rounded and broadest near the apex, tapering gradually to a slender, narrow base.

SPIKE — A central axis along which are distributed an indefinite number of stalkless flowers.

STAMEN — The male, or pollen-bearing, organ of a flower.

STAMINATE — Exclusively male (having stamens but no pistil).

STAMINODIA — A sterile or abortive stamen, often a more or less petal-like structure.

STELLATE — Star-shaped, or radiating like the points of a star.

STELLATE-TOMENTOSE — Having a densely woolly covering composed of minute star-shaped clusters of hairs.

STIGMA — The part of the pistil which receives the pollen.

STIPE — A slender stalk.

STIPITATE — Having a stalk.

STIPULE — A (usually) more or less leafy appendage at the base of the leaf petiole.
STOMACHIC — A drug or tonic which strengthens the stomach.
STRIGILLOSE — Covered with straight, stiff, sharp-pointed hairs which are closely appressed to the surface.
SUBCORDATE — Somewhat heart-shaped.
SUBTEND — To extend below.
SUBTERETE — Nearly terete.
SUPRA-AXILLARY — Produced above the axil.
TAXONOMIC — Pertaining to classification.
TAXONOMY — The science of classification.
TERETE — Rodlike and usually tapering at one or both ends.
TERNATE — In threes.
TERTIARY — Of the third order or rank.
TOMENTOSE — Having a densely woolly covering.
TOMENTULOSE — Somewhat or minutely woolly.
TOMENTUM — A dense, woolly covering.
TRANSLUCENT — Imperfectly transparent.
TRUNCATE — Terminating abruptly and squarely.
TUBERCULATE — Covered with rounded protuberances.
UMBEL — A cluster of an indefinite number of flowers in which pedicels of about equal length arise from a common point like the ribs of an umbrella.
UNDULANT — Wavy.
VALVATE — Valved. Opening by or pertaining to valves.
VELUTINOUS — Velvety.
VERTICILLATE — Whorled.
VISCID — Sticky.

PLATE 2

UPPER – Townsite of Palmar, showing the Río Terraba and the forested Cerros de Retinto in the background.

LOWER – Typical recent clearings for pasture on hills near Palmar Norte.

PLATE 3

Typical hillside rain forest, with mature specimens of *Brosimum utile*, near Palmar Norte.

PLATE 4

Dr. H. E. Moore with characteristic palms of the Esquinas Forest, showing *Asterogyne Martiana* (lower plants with lighter foliage) and *Geonoma congesta* (taller plants in center).

PLATE 5

UPPER HALF – Typical buttressing in mature specimens of *Pelliciera rhizophorae.*
LOWER LEFT – *Huberodendron Allenii,* showing characteristic plank buttresses.
LOWER RIGHT – Flying buttresses found in old specimens of *Ficus lapathifolia.*

PLATE 6

UPPER LEFT – *Scheelea rostrata.*
LOWER LEFT – *Oenocarpus panamanus.*
UPPER RIGHT – *Astrocaryum Standleyanum.*
LOWER RIGHT – *Cryosophila guagara.*

PLATE 7

LEFT – *Brosimum utile*, showing flow of milky latex. RIGHT – Old specimen of *Hippomane mancinella*.

PLATE 8

UPPER LEFT – Spathe and fruits of *Bactris balanoidea*.

LOWER LEFT – Staminate inflorescences and fruit clusters of *Corozo oleifera*.

UPPER RIGHT – Fruit cluster of *Astrocaryum alatum*.

LOWER RIGHT – Fruits of *Raphia taedigera*.

PLATE 9

UPPER LEFT – *Iriartea gigantea,* showing typical hornlike spathes.

LOWER LEFT – Groves of *Cocos nucifera* along sea beaches.

UPPER RIGHT – Short, compact stilt roots characteristic of *Iriartea gigantea.*

LOWER RIGHT – Tall, open stilt roots characteristic of *Socratea durissima.*

PLATE 10

UPPER LEFT – *Welfia Georgii*.

UPPER RIGHT – *Bactris militaris*.

LOWER LEFT – *Sterculia mexicana*.

LOWER RIGHT – *Sterculia apetala*.

PLATE 11

UPPER LEFT – *Cecropia peltata.* UPPER RIGHT – *Cecropia Sandersoniana.*
LOWER LEFT – *Vitex Cooperi.* LOWER RIGHT – *Tabebuia chrysantha.*

PLATE 12

Upper left – Terminal portion of inflorescence of *Swartzia panamensis*.

Upper right – *Cochlospermum vitifolium*.

Lower left – *Ochroma lagopus*.

Lower right – *Didymopanax Morototoni*.

PLATE 13

UPPER LEFT – *Pithecolobium austrinum*.
LOWER LEFT – *Pachira aquatica*.
UPPER RIGHT – *Phyllocarpus septentrionalis*.
LOWER RIGHT – *Cassia reticulata*.

PLATE 14

UPPER LEFT – *Eschweilera calyculata.*
LOWER LEFT – *Anacardium occidentale.*

UPPER RIGHT – Open fruit of *Carapa Slateri*, showing seeds.
LOWER RIGHT – *Duguetia panamensis.*

PLATE 15

UPPER LEFT – *Pterocarpus Hayesii* (flowering specimen).

UPPER RIGHT – *Pterocarpus officinalis* (fruiting specimen).

LOWER LEFT – *Dussia mexicana* (flowering specimen).

LOWER RIGHT – *Dussia macrophyllata* (fruiting specimen).

PLATE 16

UPPER LEFT – *Hibiscus tiliaceus.* UPPER RIGHT – *Hippomane mancinella.*

LOWER LEFT – *Mora oleifera.* LOWER RIGHT – *Enallagma latifolia.*

PLATE 17

Upper left – *Simarouba glauca.*
Lower left – *Spondias mombin.*
Upper right – *Spondias purpurea.*
Lower right – *Diphysa robinioides.*

PLATE 18

Upper left – *Hampea Allenii.* Upper right – *Tetragastris panamensis.*
Lower left – *Croton panamensis.* Lower right – *Dipterodendron costaricense.*

PLATE 19

UPPER LEFT – *Cassia spectabilis.*
LOWER LEFT – *Gliricidia sepium.*

UPPER RIGHT – *Cassia grandis.*
LOWER RIGHT – *Acacia farnesiana.*

PLATE 20

UPPER LEFT – *Conocarpus erecta.*
LOWER LEFT – *Laguncularia racemosa.*
UPPER RIGHT – *Rhizophora mangle.*
LOWER RIGHT – *Pelliciera rhizophorae.*

PLATE 21

UPPER LEFT – *Andira inermis.*
UPPER RIGHT – *Tachigalia versicolor.*
LOWER LEFT – *Cedrela mexicana.*
LOWER RIGHT – *Albizzia filicina.*

PLATE 22

UPPER LEFT – Mature specimen of *Bravaisia integerrima*, showing typical stilt roots.

UPPER RIGHT – Flowers of *Bravaisia integerrima.*

LOWER LEFT – Old specimen of *Dracaena americana* (see man at base for scale).

LOWER RIGHT – Fruiting branch of *Dracaena americana.*

PLATE 23

UPPER LEFT – *Cordia alliodora.*

UPPER RIGHT – *Belotia macrantha.*

LOWER LEFT – *Byrsonima crassifolia.*

LOWER RIGHT – *Guazuma ulmifolia.*

PLATE 24

UPPER LEFT – *Chrysophyllum panamense.* UPPER RIGHT – *Diospyros ebenaster.*

LOWER LEFT – *Goethalsia meiantha.* LOWER RIGHT – *Batocarpus costaricense.*

PLATE 25

UPPER LEFT – *Casearia arguta*.
LOWER LEFT – *Guatteria lucens*.

UPPER RIGHT – *Guatteria chiriquiensis*.
LOWER RIGHT – *Guatteria aeruginosa*.

PLATE 26

UPPER LEFT – *Stemmadenia Donnell-Smithii.*
LOWER LEFT – *Tetrathylacium costaricense.*

UPPER RIGHT – *Theobroma angustifolium.*
LOWER RIGHT – *Warscewiczia coccinea.*

PLATE 27

Upper half – Fruits of the wild form of *Persea americana*.

Lower left – *Ocotea Williamsii*. Lower right – *Licaria Cufodontisii*.

PLATE 28

UPPER LEFT – *Rollinia Jimenezii.* UPPER RIGHT – *Neea elegans.*

LOWER LEFT – *Muntingia calabura.* LOWER RIGHT – *Quararibea guatemalteca.*

PLATE 29

UPPER LEFT – *Miconia argentea.*

UPPER RIGHT – *Heliocarpus appendiculatus.*

LOWER LEFT – *Homalium eurypetalum.*

LOWER RIGHT – *Hieronyma tectissima.*

PLATE 30

UPPER LEFT – *Pouteria neglecta.*

UPPER RIGHT – *Ladenbergia Brenesii.*

LOWER LEFT – *Pouteria heterodoxa.*

LOWER RIGHT – *Pouteria triplarifolia.*

PLATE 31

UPPER LEFT – *Apeiba aspera.*

LOWER LEFT – *Casearia banquitana var. laevis.*

UPPER RIGHT – *Calycophyllum candidissimum.*

LOWER RIGHT – *Compsoneura Sprucei.*

PLATE 32

UPPER LEFT – Mature specimen of *Sloanea laurifolia* (note man at base for scale).

LOWER LEFT – *Macrocnemum glabrescens.*

UPPER RIGHT – Leaves and fruits of *Sloanea laurifolia.*

LOWER RIGHT – *Perebea trophophylla.*

PLATE 33

UPPER LEFT – *Mortoniodendron anisophyllum.* UPPER RIGHT – *Terminalia bucidioides.*

LOWER LEFT – *Meliosma Allenii.* LOWER RIGHT – *Symphonia globulifera.*

PLATE 34

UPPER LEFT – *Xylopia sericophylla.*

UPPER RIGHT – *Sapindus saponaria.*

LOWER LEFT – *Xylosma excelsum.*

LOWER RIGHT – *Dicraspidia Donnell-Smithii.*

APPENDIX

LATIN DESCRIPTIONS OF NEW SPECIES

ARDISIA DUNLAPIANA P. H. Allen, *sp. nov. Arbores graciles, 12–15 m. altitudine. Folia petiolata; lamina lanceolata vel elliptico-lanceolata, acuta vel breviter acuminata, basi anguste cuneata, plerumque glabra, 7.5–15 cm. longa, 2.5–5 cm. lata: petiolus maxima pro parte alatus. Inflorescentia terminalis, angusta, paniculata, 7.5–10 cm. longa. Flores parvi, puniceo-fusci, calycis lobis late rotundatis valde imbricatibus distincti.*

ARDISIA STANDLEYANA P. H. Allen, *sp. nov. Arbores 6–8 m. in altitudine. Folia lanceolata vel elliptico-oblonga, acuta vel breviter acuminata, basi anguste rotundata, subsessila, glabra, 17.5–25 cm. longa, 6.25–7.5 cm. lata. Inflorescentiae terminales, paniculatae, circa 12.5 cm. longae. Flores parvi, albi, cum calyce profunde lobata.*

CALLIANDRA GRANDIFOLIA P. H. Allen, *sp. nov. Arbores parvae, circa 6 m. in altitudine. Folia alterna, pinnata, cum foliolis duobus; foliola opposita, elliptico-lanceolata, obliqua, subabrupte acuminata, coriacea, glabra, 18.5–23.5 cm. longa, 6.25–11.25 cm. lata, cum nervis tribus longitudinalibus valde prominentibus. Inflorescentiae terminales vel axillares, subglobosae, circa 6.25 cm. in diametro. Flores coccinei.* CALLIANDRAE ARBOREAE Standl. *et* C. BRENESII Standl. *ut videture affinis, sed foliis multo majoribus etc. valde distincta.*

CECROPIA STANDLEYANA P. H. Allen, *sp. nov. Arbores valde variabiles, maturae pomiferae 25–30 m. in altitudine et usque ad 90 cm. in diametro supra basim. Arbores juveniles et arborum vetustiorum rami conspicue nigri et sine annulis prominentibus. Folia matura octolobata, 37.5–45 cm. in diametro; superficies superior atroviridis, nitens, glabra; superficies inferior canescens, minute puberula. Inflorescentiae pistillatae sunt fasciculi 4–6 laminarum teretium citrinarum, 7.5–10 cm. longarum, primo bractea rubra, spathacea mox decidua cincti compositi pedunculo subtereti hispidulo paulo longiores.*

COCCOLOBA STANDLEYANA P. H. Allen, *sp. nov. Arbores graciles, silvestres, 12–20 m. in altitudine. Folia late elliptica vel obovata, acuta vel breviter acuminata, basi cordata, plerumque glabra, 35–50 cm. longa, 20–30 cm. lata, cum venis petiolisque dense hispidulis. Affinis* COCCOLOBAE HIRSUTAE Standl. *et* C. BELIZENSI Standl. *ut videtur, sed ocreis multo minoribus etc. valde distincta.*

FARAMEA SESSIFOLIA P. H. Allen, *sp. nov. Arbores silvestres, parvae, usque ad circa 4.5 m. in altitudine. Folia oblonga vel oblongo-lanceolata, glabra, abrupte et saepe falcato-acuminata, subsessilia, 17.5–22.5 cm. longa, 3.75–7.5 cm. lata, cum nervo unico longitudinali. Inflorescentiae terminales, longe pedunculatae, graciles. Flores parvi, caerulei.*

NEEA ELEGANS P. H. Allen, *sp. nov. Arbores parvae, circa 4.5–6 m. in altitudine. Folia vel alterna vel opposita vel verticillata, anguste lanceolata, acuminata, 17.5–25 cm. longa, 5–7.5 cm. lata, cum nervorum lateralium 12–16 paribus in superficie inferiorum et petiolis et ramulis dense ferrugineo-tomentellis. Inflorescentiae terminales, cymosae, pedunculatae, comparate densae. Flores parvi. Fasciculi fructus penduli. Fructus anguste ellipsoidales.*

APPENDIX

NEEA POPENOEI P. H. Allen, *sp. nov. Arbores silvestres, patentes, usque ad circa 20 m. in altitudine. Folia alterna vel opposita, elliptica vel lineari-lanceolata, acuta vel acuminata, glabra, tenuia, 10–20 cm. longa, cum nervorum lateralium 6–10 paribus. Inflorescentiae terminales vel axillares, cymosae, longe pedunculatae. Flores parvi, pallide aurei.*

OCOTEA WILLIAMSII P. H. Allen, *sp. nov. Arbores silvestres, usque circa ad 25 m. in altitudine. Folia lanceolata, elliptico-lanceolata vel oblanceolata, acuta vel brevissime acuminata, basi fere cuneata, 6.8–13.1 cm. longa, 2.4–4.3 cm. lata, maxima pro parte glabra, sed flocculis fuscis in venorum axillis distinctis in superficie inferiore praedita. Inflorescentiae subterminales, diffusae, paniculatae, 11.25–15 cm. longae. Flores parvi, pallide aurei. Fructus nigri, ellipsoidales, circa 5 cm. longi, cupula brevi rubra que fructus diametro conspicue minor est subtenta.*

TRICHILIA SKUTCHII P. H. Allen, *sp. nov. Arbores silvestres, parvae, circa 4 m. in altitudine. Folia alterna, late elliptico-oblonga, abrupte acuminata, glabra, 8.75–22.5 cm. longa, 3.75–8.75 cm. lata, cum nervorum lateralium 10–12 paribus. Inflorescentiae axillares, breviter pedunculatae, fasciculi minores quam 1.25 cm. in longitudine saepe. Flores parvi, albi.*

REFERENCES

ALLEN, Paul H.

1943 Poisonous and injurious plants of Panama. *The American Journal of Tropical Medicine*, 23:1, 76 pp. illus.

1947 Preliminary index to the timber woods of Panama. Washington, D. C. (For. Ser., U.S.D.A.), 76 pp., mimeographed.

1953 Two new fan palms from Central America. *Ceiba* 3(3):173-178. Illus.

BEARD, J. S.

1945 The natural vegetation of Trinidad. *Oxford Forestry Memoirs*, No. 20.

1948 The natural vegetation of the Windward & Leeward Islands. *Oxford Forestry Memoirs*, No. 21.

CARR, Archie

1950 Outline for a classification of animal habitats in Honduras. *Bull. Amer. Mus. Nat. Hist.*, 94(10).

COOPER, G. Proctor

1928 Some interesting trees of Western Panama. *Tropical Woods*, 14:1-8.

1928 The forests of Western Panama. *Tropical Woods*, 16:1-35.

CRONQUIST, A.

1945 The North American species of Chrysophyllum. *Bull. Torrey Bot. Club*, 72(2):191-204.

CURRAN, Hugh M.

1929 The lands of Loba, Colombia: An annotated list of the trees. *Tropical Woods*, 19:11-38.

DAHLGREN, B. E.

1936 Index of American Palms. *Field Museum of Natural History, Bot. Series*, 14.

EVANS, H. H.

1942 List of timber woods used by the Panama Canal. (Personal communication.)

FANSHAWE, D. B.

1948 Forest products of British Guiana. Principal timbers. *Forestry Bull. No. 1. Forest Department.*

1950 Forest products of British Guiana. Minor forest products. *Forestry Bull. No. 2. Forest Department.*

FRANCIS, G. W. E.

1926 Native woods used for railway crossties in British Honduras. *Tropical Woods*, 7:30-32.

GLEASON, H. A.

1926 The individualistic concept of the plant associations. *Bull. Torrey Bot. Club*, 53:7-26.

HOWARD, Alexander L.

1948 A manual of the timbers of the world, their characteristics and their uses. London (Macmillan).

JOHNSTON, I. M.

1949 The botany of San Jose Island (Gulf of Panama). *Sargentia*, 8.

KLUGE, H. C.

1926 Trees of the Bayano river watershed, Panama. *Tropical Woods*, 5:4-13.

KRUKOFF, B. A.

1939 The American species of Erythrina. *Brittonia*, 3(2):205-337.

LITTLE, Elbert L.

1948 A collection of tree specimens from Western Ecuador. *The Caribbean Forester*, 9(3):215.

MERKER, C. A., W. R. Barbour, J. A. Scholten, and W. A. Dayton

1943 The Forests of Costa Rica. Washington, D. C. (For. Serv. U.S.D.A.)

REFERENCES

MONACHINO, Joseph

1944 A revision of Lacmellea, and the transfer of Zschokkea (Apocynaceae). *Lloydia,* 7:275-302.

PAX, F.

1912 Sapium. *Pflanzenreich,* 4:147; 5:199-258.

PITTIER, H.

1898 Primitiae Florae Costaricensis. Brussels (Jardin Botanique de L'Etat.)

1908 Ensayo sobre las plantas usuales de Costa Rica. Washington, D. C. (H. L. and J. B. McQueen, Inc.)

1908 The genus Sapium in Mexico and Central America. *Contr. U. S. Nat. Herb.,* 12(4).

1910 A preliminary treatment of the genus Castilla. *Contr. U. S. Nat. Herb.,* 13.

1910 New or noteworthy plants from Colombia and Central America. *Contr. U. S. Nat. Herb* (through 1918).

1916 Preliminary revision of the genus Inga. *Contr. U. S. Nat. Herb.,* 18(5).

1917 The middle American species of Lonchocarpus. *Contr. U. S. Nat. Herb.,* 20(2).

1921 Notes on the genus Swartzia in Panama and Guatemala. *Journ. Wash. Acad. Science,* 11(7).

1927 The Lecythidaceae of Costa Rica. *Contr. U. S. Nat. Herb.,* 12.

RECORD, S. J.

1927 Trees of Honduras. *Tropical Woods,* 10:10-47.

1927 Trees of Gualán, Guatemala. *Tropical Woods,* 11:10-18.

1944 Random observations on tropical American timbers (Humiriaceae in Central America). *Tropical Woods,* 77:8-9.

————, S. J., and R. W. Hess

1943 Timbers of the New World. New Haven (Yale University Press).

————, S. J., and H. Kuylen

1926 Trees of the lower Río Motagua Valley, Guatemala. *Tropical Woods,* 7:10-29.

————, S. J., and C. D. Mell

1924 Timbers of tropical America. New Haven (Yale University Press).

RICHARDS, Paul W.

1952 The tropical rain forest, an ecological study. London (Cambridge University Press).

SAFFORD, W. E.

1914 Classification of the genus Annona with descriptions of new and imperfectly known species. *Contr. U. S. Nat. Herb.,* 18(1).

SCHIMPER, A. F. W.

1903 Plant-geography upon a physiological basis. Oxford (Clarendon Press).

SPRUCE, Richard

1908 Notes of a botanist on the Amazon and in the Andes. London (Macmillan). 2 vols.

STANDLEY, Paul C.

1917 The Mexican and Central American species of Ficus. *Contr. U. S. Nat. Herb.,* 20(1).

1920 Trees and shrubs of Mexico. *Contr. U. S. Nat. Herb.* 23(1-5). (Completed 1926.)

1928 Flora of the Panama Canal Zone. *Contr. U. S. Nat. Herb.,* 27.

1929 Trees and shrubs collected by F. C. Englesing in northeastern Nicaragua. *Tropical Woods,* 17:18-38.

1930 Flora of Yucatan. *Field Museum of Natural History, Bot. Series,* 3(3).

1930 A second list of the trees of Honduras. *Tropical Woods,* 21:9-41.

1931 Flora of the Lancetilla Valley, Honduras. *Field Museum of Natural History, Bot. Series,* 10.

1933 The Flora of Barro Colorado Island, Panama. *Contr. Arnold Arboretum of Harvard Univ.*, 5.
1934 Additions to the trees of Honduras. *Tropical Woods*, 37:27-39.
1937 Flora of Costa Rica. *Field Museum of Natural History, Bot. Series*, 18(1-4). (Completed in 1938.)
———, Paul C., and S. J. Record
1936 The forests and flora of British Honduras. *Field Museum of Natural History, Bot. Series*, 12.
———, Paul C., and Julian A. Steyermark.
1946 Flora of Guatemala. *Fieldiana: Botany*. Parts III-VI (Incomplete.)
———, Paul C., and Louis O. Williams
1950 Plantae Centrali-Americanae. *Ceiba* (through 1954).
STONE, Doris Z.
1949 The Boruca of Costa Rica. *Peabody Museum of Harvard Univ.*, 24(2).
WARMING, E., and P. Graebner
1933 Lehrbuch der okologischen Pflanzengeographie. Berlin.
WHEELER, W. M.
1942 Studies of neotropical ant-plants and their ants. *Bull. Mus. Comp. Zool., Harvard*. 90(1).
WILLIAMS, Llewelyn
1936 Woods of northeastern Peru. *Field Museum of Natural History, Bot. Series*, 15.
WOLCOTT, George N.
1950 An index to the termite resistance of woods. *Univ. Puerto Rico, Agr. Exp. Sta. Bull.*, 85.
WOODSON, Robert E.
1951 An interim revision of the genus Aspidosperma Mart. & Zucc. *Ann. Mo. Bot. Gard.*, 38:119-206.
———, Robert E., and R. W. Schery
1937 Contributions toward a flora of Panama. *Ann. Mo. Bot. Gard.*, 1-6. (Completed in 1942.)
1943 Flora of Panama. *Ann. Mo. Bot. Gard.* (Incomplete.)
YUNCKER, T. G.
1940 Flora of the Aguan Valley, Honduras. *Field Museum of Natural History, Bot. Series*, 9(4).

INDEX

NOTE: See Alphabetical Index in text (pages 116-365) for all common and scientific names.

Abrasives, 114
Acrostichum aureum, 69
Airplanes, model, 113
Allergy-producing pollens, 115
Ants, stinging, 98
Archery bows, 111
Axe handles, 111

Banana prop poles, 114
Barbascos, 114
Bark: bitter, 97-98; distinctive, 93-94; poisonous, 115
Bark cloth, 111
Barrel hoops, 111
Barrels, 111
Baskets, 114
Beams, 106
Bitter bark, 97-98
Bitter wood, 97-98
Black gum, 97
Black sap, 97
Boat caulking, 114
Boat decking, 110
Boat framing, 110
Boat planking, 110
Boat work, interior, 110
Boards, bridge, 107
Bodies: motor car, 110; truck, 110
Bows, archery, 111
Boxes: cigar, 112; ornamental, 109; rough, 109
Bridge boards, 107; timbers, 107
Buttresses, 91-93

Cabbage, palm, 112
Cabinet woods, 108-9
Calathea insignis, 69; *lutea*, 69
Candles, 113
Canes, 110
Carpentry, general, 107-8
Casks, 111
Cauliflorous species, 95
Caulking, 114
Caustic sap, 115
Cements, 113
Charcoal, 114
Cigar boxes, 112
Climatic formations, 10
Cloth, bark, 111
Combs, 111
Common names, 101-2
Concentrated stands, 98-100
Concrete forms, 109
Construction: heavy, 105; interior, 107-8
Cooperage: slack, 111; tight, 111
Cordage, 111
Corestock, plywood, 109
Conspicuous or distinctive species, 86
Conspicuous flowers, 86-91
Conspicuous vegetative characters, 91-98
Cotton, silk (kapok), 112
Crates, 109
Crossties, railroad, 106
Cyperus canus, 69

Decking, boat, 110
Dock fenders, 106
Dodge, C. W., 3
Doors, 108
Dragline mats, 107
Drugs, 115
Durables, 86
Dyes, 114

Ecological formations, 10-85
Ecology, 7-9
Economic species, 105-14
Edaphic formations, 62
Edible flowers, 112
Edible fruits, 112
Edible leaves, 113
Ephemerals, 86
Evergreen lowland forest, 10

Fence posts, 107
Fence posts, living, 78-79
Fence rows and trailside thickets, 77-81
Fenders, dock, 106
Fishing rods, 110
Fish poisons (barbascos), 114
Flood plain forests (Evergreen lowland forest), 10
Flooring, 108
Flowering trees, conspicuous, 86-91

Flowers eaten by stock, 113
Flowers, edible, 112
Fodder for stock, 113
Forms, concrete, 109
Foundations, mill, 106
Framing, boat, 110
Fruits: edible, 112; poisonous, 115
Fruits relished by stock, 113
Furniture, 108-9

Gallery woodland, 69-70
General carpentry, 107-8
Girders, 106
Grasslands, 70
Gum, black, 97
Gun stocks, 110
Gynerium sagittatum, 70

Handles: axe, 111; tool, 111
Heliconia bihai, 69; *imbricata*, 69; *latispatha*, 69
Herbaceous swamp, 69
Highland forests (Upper montane rain forest), 60
Hillside forests (Lower montane rain forest), 24-60
Hoops, barrel, 111
House: posts, 106; roofing (thatch), 113; siding, 108; walls, 113

Incense, 114
Instruments, professional, 111
Insulation, 113
Interior construction and trim, 107-8

Kapok, 112

Latex: milky, 95-97; for caulking, 113; yellow or orange, 97
Leaves: edible, 113; poisonous, 115; for polishing, 114; for thatch, 113
Life rafts, 110
Living fence posts, 107
Lower montane rain forest, 24-60

Mangrove swamp, 66-69
Mangrove woodland, 66-69
Mastate, *see* Bark cloth
Match sticks, 113
Mats, dragline, 107
Mayna grandiflora, 259-60
Medicinal species, 115
Milk substitutes, 114
Milky latex, 95-7
Mill foundations, 106
Model airplanes, 113
Moore, H. E., 3
Motor car bodies, 110
Myrmecophyllous species, 98

Names: common, 101-2; scientific, 103-4
Neurolaena lobata, 82

Oars, 110
Oily seeds, 113
Orange latex, 97
Orange sap, 97
Ox yokes, 110

Pack saddles, 113
Palm cabbage, 112
Palmito, 112
Palms, 281-84
Palm: sugar, 112; wine, 112
Palm swamp, 66
Pancratium littorale, 69
Paneling, 108
Paper pulp, 111
Paspalum fasciculatum, 69
Pastures, 70-77
Pencils, 112
Perfumes, 114
Physiography, 4-6
Piling, 106
Pink sap, 97
Pistia stratiotes, 62
Pittier, Henri, 3
Planking, boat, 110
Plywood corestock, 109
Poisonous species, 115
Poles, 114
Polishing materials, 114
Pollen, allergy-producing, 115
Posts: fence, 107; house, 106; living, 107
Professional instruments, 111
Prop poles, 114
Pulp, paper, 111

Rafters, 106
Rafts, life, 110
Railroad ties, 106
Red sap, 97
Refrigerators, insulation for, 113
Rhabdadenia biflora, 69
River bank trees (Gallery woodland), 69-70

INDEX

Rocky seashores, 23-24
Rods, fishing, 110
Roots, stilt, 95
Rope, 111
Rough boxes, 109

Saddles, pack, 113
Sap, caustic, 115
Sap, colors of: black, 97; milky, 95-97; orange, 97; pink, 97; red, 97; yellow, 97
Sap, poisonous, 115
Sap, uses of, for clarifying sugar, 114
Savanna, 70
Sawdust, poisonous, 115
Scientific names, 103-4
Sea beaches, 22-23
Second growth, 82-85
Seeds: poisonous, 115; rich in oil, 113
Siding, 108
Silk cotton (kapok), 112
Skutch, Alexander, 3
Slack cooperage, 111
Soap substitutes, 114
Species which harbor stinging ants, 98
Species which occur in concentrated stands, 98-100
Spines, 94
Stains, 114
Stilt roots, 95
Stocks, gun, 110
Sugar clarifiers, 114
Sugar, palm, 112
Swamp forest, 62-65
Swamp: herbaceous, 69; mangrove, 66-69; palm, 66

Tanbark, 111
Thatch, 113
Thorns, 94
Ties, railroad, 106
Tight cooperage, 111
Timbers, bridge, 107
Tonduz, Adolfo, 3
Tool handles, 111
Torches, 113
Transitional formations, 70
Trim, interior, 108
Truck bodies, 110
Trunks; distinctive, 94
Twine, 111

Upper montane rain forest, 60
Utilization lists, 105-15

Veneer, 110

Walls, 113
Wheelstock, 110
Windows, 108
Wine, palm, 112
Wood, bitter, 97-98
Woodland, gallery, 69-70
Woods: cabinet, 108-9; furniture, 108-9; burnable when green, 113

Yellow latex, 97
Yellow sap, 97
Yokes, ox, 110

S0-ELG-633